Advances in Intelligent Systems and Computing

Volume 806

Series editor

Janusz Kacprzyk, Polish Academy of Sciences, Warsaw, Poland
e-mail: kacprzyk@ibspan.waw.pl

The series "Advances in Intelligent Systems and Computing" contains publications on theory, applications, and design methods of Intelligent Systems and Intelligent Computing. Virtually all disciplines such as engineering, natural sciences, computer and information science, ICT, economics, business, e-commerce, environment, healthcare, life science are covered. The list of topics spans all the areas of modern intelligent systems and computing such as: computational intelligence, soft computing including neural networks, fuzzy systems, evolutionary computing and the fusion of these paradigms, social intelligence, ambient intelligence, computational neuroscience, artificial life, virtual worlds and society, cognitive science and systems, Perception and Vision, DNA and immune based systems, self-organizing and adaptive systems, e-Learning and teaching, human-centered and human-centric computing, recommender systems, intelligent control, robotics and mechatronics including human-machine teaming, knowledge-based paradigms, learning paradigms, machine ethics, intelligent data analysis, knowledge management, intelligent agents, intelligent decision making and support, intelligent network security, trust management, interactive entertainment, Web intelligence and multimedia.

The publications within "Advances in Intelligent Systems and Computing" are primarily proceedings of important conferences, symposia and congresses. They cover significant recent developments in the field, both of a foundational and applicable character. An important characteristic feature of the series is the short publication time and world-wide distribution. This permits a rapid and broad dissemination of research results.

More information about this series at http://www.springer.com/series/11156

Paulo Novais · Jason J. Jung
Gabriel Villarrubia González
Antonio Fernández-Caballero
Elena Navarro · Pascual González
Davide Carneiro · António Pinto
Andrew T. Campbell · Dalila Durães
Editors

Ambient Intelligence – Software and Applications –, 9th International Symposium on Ambient Intelligence

 Springer

Editors
Paulo Novais
ALGORITMI Centre/Departamento de
Informatica
University of Minho
Braga, Portugal

Jason J. Jung
Department of Computer Engineering
Chung-Ang University
South Korea, Korea (Republic of)

Gabriel Villarrubia González
Departamento de Informática y Automática,
Facultad de Ciencias
Universidad de Salamanca
Salamanca, Spain

Antonio Fernández-Caballero
Departamento de Sistemas Informáticos
University of Castilla-La Mancha
Albacete, Spain

Elena Navarro
Departamento de Sistemas Informáticos
Universidad de Castilla-La Mancha
Albacete, Spain

Pascual González
Departamento de Sistemas Informáticos
Universidad de Castilla-La Mancha
Albacete, Spain

Davide Carneiro
Instituto Politécnico do Porto
Escola Superior de Tecnologia e Gestão
Felgueiras, Portugal

António Pinto
ESTG, Politécnico do Porto and CRACS
& INESC TEC
Porto, Portugal

Andrew T. Campbell
Department of Computer Science
Dartmouth College
Hanover, NH, USA

Dalila Durães
Department of Artificial Intelligence
Technical University of Madrid
Madrid, Spain

ISSN 2194-5357　　　　　ISSN 2194-5365　(electronic)
Advances in Intelligent Systems and Computing
ISBN 978-3-030-01745-3　　　　ISBN 978-3-030-01746-0　(eBook)
https://doi.org/10.1007/978-3-030-01746-0

Library of Congress Control Number: 2018957286

This Springer imprint is published by the registered company Springer Nature Switzerland AG
The registered company address is: Gewerbestrasse 11, 6330 Cham, Switzerland

Preface

This volume contains the proceedings of the 9th International Symposium on Ambient Intelligence (ISAmI 2018). The Symposium was held in Toledo, Spain, on June 20–22 at Campus Tecnológico de la Fábrica de Armas, University of Castilla-La Mancha.

ISAmI has been running annually and aiming to bring together researchers from various disciplines that constitute the scientific field of Ambient Intelligence to present and discuss the latest results, new ideas, projects, and lessons learned, namely in terms of software and applications, and aims to bring together researchers from various disciplines that are interested in all aspects of this area.

Ambient Intelligence is a recent paradigm emerging from Artificial Intelligence, where computers are used as proactive tools assisting people with their day-to-day activities, making everyone's life more comfortable.

After a careful review, 48 papers from 15 different countries were selected to be presented in ISAmI 2018 at the conference and published in the proceedings. Each paper has been reviewed by, at least, three different reviewers, from an international committee composed of 165 members from 23 countries.

June 2018

Paulo Novais
Jason J. Jung
Gabriel Villarrubia
Antonio Fernández-Caballero
Elena Navarro
Pascual González
Davide Carneiro
Antonio Pinto
Andrew T. Campbell
Dalila Duraes

Organization

Program Committee Chairs

Paulo Novais	Universidade do Minho, Portugal
Jason J. Jung	Chung-Ang University, Republic of Korea
Gabriel Villarrubia González	University of Salamanca, Spain

Program Committee

Emmanuel Adam	Univ Lille Nord de France, France
Sami Albouq	Oakland University, USA
Ana Almeida	ISEP/IPP, Portugal
Ana Alves	Centre for Informatics and Systems, University of Coimbra, Portugal
Victor Alves	University of Minho, Portugal
Ricardo Anacleto	ISEP, Portugal
Cesar Analide	University of Minho, Portugal
Cecilio Angulo	Universitat Politècnica de Catalunya, Spain
José Antonio Castellanos	University of Salamanca, Spain
Mário Antunes	Polytechnic Institute of Leiria, ESTG & INESC TEC, CRACS, University of Porto, Portugal
Javier Bajo	Universidad Politécnica de Madrid, Spain
Emilia Barakova	Eindhoven University of Technology, Netherlands
Manuel Barbosa	Portugal
Massimo Bartoletti	Dipartimento di Matematica e Informatica, Università degli Studi di Cagliari, Italy

Marta Fernandes

GECAD, Research Group on Intelligent Engineering and Computing for Advanced Innovation and Development, Polytechnic of Porto, Portugal

Antonio Fernández-Caballero
Universidad de Castilla-La Mancha, Spain

João Carlos Ferreira
ISCTE, Portugal

Lino Figueiredo
ISEP, Portugal

Adina Magda Florea
University Politehnica of Bucharest, AI-MAS Laboratory, Romania

Daniela Fogli
Università di Brescia, Italy

Miguel Frade
Instituto Politécnico de Leiria, Portugal

Juan Francisco De Paz
University of Salamanca, Spain

Jorge G. Barbosa
University of Porto, Portugal

Arturo García
University of Castilla-La Mancha, Spain

Jose García-Alonso
University of Extremadura, Spain

Ayse Goker
AmbieSense, UK

Hélder Gomes
Escola Superior de Tecnologia e Gestão de Águeda, Universidade de Aveiro, Portugal

Celestino Goncalves
Instituto Politecnico da Guarda, Portugal

Pascual González
University of Castilla-La Mancha, Spain

Alfonso González Briones
BISITE Research Group, Spain

Sérgio Gonçalves
University of Minho, Portugal

David Griol
Universidad Carlos III de Madrid, Spain, Spain

Junzhong Gu
East China Normal University, China

Esteban Guerrero
Umeå University, Sweden

Hans W. Guesgen
Massey University, New Zealand

Daniel Hernández de La Iglesia
Universidad de Salamanca, Spain

Francisco Javier Hinojo Lucena
University of Granada, Spain

Javier Jaen
Universitat Politècnica de València, Spain

Jean-Paul Jamont
LCIS, Université de Grenoble, France

Javier Jaén
UPV, Spain

Rui José
University of Minho, Portugal

Vicente Julian
Universitat Politècnica de València, Spain

Jason Jung
Chung-Ang University, South Korea

Leszek Kaliciak
AmbieSense, UK

Anastasios Karakostas
Aristotle University of Thessaloniki, Greece

Alexander Kocian
University of Pisa, Italy

Igor Kotenko
St. Petersburg Institute for Informatics and Automation of the Russian Academy of Sciences (SPIIRAS), Russia

Sergii Kushch
University of Salamanca, Spain

Bogdan Kwolek
AGH University of Science and Technology, Poland

Joyca Lacroix	Philips Research, Netherlands
Latif Ladid	UL & IPv6 Forum, IOT subC, Luxembourg
Rosa Lanzilotti	Dipartimento di Informatica, University of Bari, Italy
Guillaume Lopez	Aoyama Gakuin University, College of Science and Technology, Japan
Álvaro Lozano Murciego	University of Salamanca, Spain
Alberto López Barriuso	University of Salamanca, Spain
José Machado	University of Minho, Portugal
João Paulo Magalhaes	ESTGF, Porto Polytechnic Institute, Portugal
Diego Manuel JimÉnez Bravo	University of Salamanca, Spain
Antonio Marques	Instituto Politécnico do Porto, Escola Superior de Tecnologia da Saúde do Porto, Portugal
Goreti Marreiros	ISEP/IPP-GECAD, Portugal
Rafael Martinez Tomas	Universidad Nacional de Educación a Distancia, Spain
Constantino Martins	Knowledge Engineering and Decision Support Research (GECAD), Institute of Engineering, Polytechnic of Porto, Porto, Portugal
Rolando Martins	.
Rubén Martín García	Universidad de Salamanca, Spain
Arturo Martínez	University of Castilla-La Mancha, Spain
Rene Meier	Lucerne University of Applied Sciences, Switzerland
Antonio Meireles	ISEP, Portugal
Ahlem Melouah	University of Annaba, Algeria
Imran Memon	Zhejiang University, China
Jose M. Molina	Universidad Carlos III de Madrid, Spain
José Pascual Molina Massó	Universidad de Castilla-La Mancha, Spain
Elena Navarro	University of Castilla-La Mancha, Spain
María Navarro	BISITE, Spain
German Nemiorvskij	Albstadt-Sigmaringen University, Germany
Jose Neves	University of Minho, Portugal
Paulo Novais	University of Minho, Portugal
Andrei Olaru	University Politehnica of Bucharest, Romania
Miguel Oliver	Universidad Castilla-La Mancha, Spain
Jaderick Pabico	University of the Philippines Los Banos, Philippines
Juan José Pantrigo Fernández	Universidad Rey Juan Carlos, Spain
Jose Manuel Pastor Garcia	Universidad de Castilla-La Mancha, Spain
Miguel Angel Patricio	Universidad Carlos III de Madrid, Spain
João Paulo Magalhães	ESTGF, Porto Polytechnic Institute, Portugal
Juan Pavón	Universidad Complutense de Madrid, Spain
Hugo Peixoto	University of Minho, Portugal

Antonio Pereira   Escola Superior de Tecnologia e Gestão do
                           IPLeiria, Portugal
Ruben Pereira   ISCTE, Portugal
Antonio Piccinno   University of Bari, Italy
António Pinto   ESTG, P.Porto, Portugal
Tiago Pinto   University of Salamanca, Portugal
Filipe Portela   University of Minho, Portugal
Isabel Praça   GECAD/ISEP, Portugal
Javier Prieto   University of Salamanca, Spain
Francisco Prieto-Castrillo Massachusetts Institute of Technology, USA
Javier Pérez Marcos   BISITE Research Group, University
                           of Salamanca, Spain
Carlos Ramos   Instituto Superior de Engenharia do Porto,
                           Portugal
João Ramos   University of Minho, Portugal
Juan Ramos   University of Salamanca, Spain
Ricardo Rato   TAGUSPARK, OEIRAS, Av. Prof. Dr. Cavaco
                           Silva, 33, 2740-120 Porto Salvo, Portugal
Rogério Reis   University of Porto, Portugal
Alberto Rivas   BISITE Research Group, University
                           of Salamanca, Spain
Sara Rodríguez   University of Salamanca, Spain
Teresa Romão   Faculdade de Ciências e
                           Tecnologia/Universidade NOVA de Lisboa
                           (FCT/UNL), Portugal
Albert Ali Salah   Bogazici University, Turkey
Altino Sampaio   Instituto Politécnico do Porto, Escola Superior de
                           Tecnologia e Gestão de Felgueiras, Portugal
Altino Sampaio   Portugal
Nayat Sanchez-Pi   Universidade do Estado do Rio de Janeiro, Brazil
Manuel Filipe Santos   University of Minho, Portugal
Ricardo Santos   Portugal
Ichiro Satoh   National Institute of Informatics, Japan
Enzo Pasquale Scilingo   University of Pisa, Italy
Ali Shoker   HASLab, INESC TEC & University of Minho,
                           Portugal
Amin Shokri Gazafroudi   Universidad de Salamanca, Spain
Fernando Silva   Department of Informatics Engineering; School
                           of Technology and Management; Polytechnic
                           Institute of Leiria, Portugal
Fábio Silva   University of Minho, Portugal
Rui Silva   Lab UbiNET, Segurança Informática
                           e Cibercrime/IPBeja, Portugal
S. Shyam Sundar   Pennsylvania State University, USA,
                           and Sungkyunkwan University, Korea

Miguel A. Teruel	University of Castilla-La Mancha, Spain
Rafael Tomas	Universidad Nacional de Educación a Distancia, Spain
Radu-Daniel Vatavu	University Stefan cel Mare of Suceava, Romania
Jorge Vieira Da Silva	MTA—Mobility, Ticketing and Applications—Portugal
Lawrence Wai-Choong Wong	National University of Singapore, Singapore
Ansar-Ul-Haque Yasar	Universiteit Hasselt, IMOB, Belgium
Roberto Zangroniz	Universidad de Castilla-La Mancha, Spain
Roberto Zunino	University of Trento, Italy
José Ramón Álvarez-Sánchez	Universidad Nacional de Educación a Distancia, Spain

Organizing Committee Chairs

Antonio Fernández-Caballero	University of Castilla-La Mancha, Spain
Elena Navarro	University of Castilla-La Mancha, Spain
Pascual González	University of Castilla-La Mancha, Spain
Davide Carneiro	Intelligent Systems Laboratory, Universidade do Minho, Portugal

Local Organization Committee

José Carlos Castillo	Universidad Carlos III de Madrid, Spain
Beatriz García-Martínez	Universidad de Castilla-La Mancha, Spain
María Teresa López	Universidad de Castilla-La Mancha, Spain
Víctor López-Jaquero	Universidad de Castilla-La Mancha, Spain
Arturo Martínez-Rodrigo	Universidad de Castilla-La Mancha, Spain
José Pascual Molina	Universidad de Castilla-La Mancha, Spain
Francisco Montero	Universidad de Castilla-La Mancha, Spain
Encarnación Moyano	Universidad de Castilla-La Mancha, Spain
Miguel Oliver	Universidad de Castilla-La Mancha, Spain
José Manuel Pastor	Universidad de Castilla-La Mancha, Spain
Miguel Ángel Teruel	Universidad de Castilla-La Mancha, Spain
Francisco José Vigo Bustos	Universidad de Castilla-La Mancha, Spain

Workshops

AM: Agents and Mobile Devices

Organizing Committee

Andrew Campbell	Dartmouth College, USA
Gabriel Villarrubia	University of Salamanca, Spain
Javier Bajo	Polytechnic University of Madrid, Spain

Program Committee

Antonio Juan Sánchez	University of Salamanca, Spain
Juan Francisco De Paz	University of Salamanca, Spain
Cristian Pinzón	Technical University of Panama, Panama
Montserrat Mateos	Pontifical University of Salamanca, Spain
Luis Fernando Castillo	University of Caldas, Colombia
Miguel Ángel Sánchez	Indra, Spain
Roberto Berjón	Pontifical University of Salamanca, Spain
Encarnación Beato	Pontifical University of Salamanca, Spain
Fernando De la Prieta	University of Salamanca, Spain
Alvaro Lozano Murciego	University of Salamanca, Spain
Alvaro López Barriuso	University of Salamanca, Spain

AIfeH: Ambient Intelligence for e-Healthcare

Organizing Committee

Antonio Fernández-Caballero	Universidad de Castilla-La Mancha, Spain
Elena Navarro	Universidad de Castilla-La Mancha, Spain
Pascual González	Universidad de Castilla-La Mancha, Spain

Program Committee

Jose Carlos Castillo	Universidad Carlos III de Madrid, Spain
Javier Jaen	Universitat Politécnica de Valencia, Spain
S. Shyam Sundar	Sungkyunkwan University, Korea
Albert Salah	Bogazici University, Turkey
Jose Manuel Pastor	Universidad de Castilla-La Mancha, Spain
Jesus Favela	CICESE, Mexico
Sylvie RattÈ	Ecole de technologie supérieure, Canada
Enzo Pasquale Scilingo	University of Pisa, Italy

José Ramón Álvarez-Sánchez   Universidad Nacional de Educación a Distancia,
                               Spain
José Ramón Álvarez-Sánchez   Universidad Carlos III de Madrid, Spain
Rafael Martínez-Tomás   Universidad Nacional de Educación a Distancia,
                               Spain
Ilias Maglogiannis   University of Piraeus, Greece
Julie Doyle   Dundalk Institute of Technology, Ireland
Bogdan Kwolek   AGH University of Science and Technology,
                               Poland

BTS: Blockchain Technologies and Security

Organizing Committee

Antonio Pinto   Politécnico do Porto and CRACS & INESC TEC,
                               Portugal
Ricardo Costa   Politécnico do Porto, Portugal

Program Committee

Afef Mdhaffar   University of Sousse, Tunisia
Ali Shoker, Haslab   INESC TEC, Portugal
Altino Sampaio   Politécnico do Porto, Portugal
André Zúquete   Universidade de Aveiro, Portugal
David G. Rosado   Universidad de Castilla-La Mancha, Spain
Hélder Gomes   Universidade de Aveiro, Portugal
Imran Memon   Zhejiang University, China
João Paulo Magalhães   Politécnico do Porto, Portugal
Luís Coelho Antunes   Universidade do Porto, Portugal
Manuel Eduardo Correia   Universidade do Porto, Portugal
Manuel Barbosa   Universidade do Porto, Portugal
Massimo Bartoletti   University of Cagliari, Italy
Miguel Frade   Instituto Politécnico de Leiria, Portugal
Mário Antunes   Instituto Politécnico de Leiria, Portugal
Pedro Pinto   Instituto Politécnico de Viana do Castelo,
                               Portugal
Ricardo Santos   Politécnico do Porto, Portugal
Rogério Reis   Universidade do Porto, Portugal
Rolando Martins   Universidade do Porto, Portugal
Roberto Zunino   University of Trento, Italy
Rui Silva   Instituto Politécnico de Beja, Portugal
Sami Albouq   Oakland University, USA

Stefano Bistarelli   University of Perugia, Italy
Satya Lokam   Microsoft Research, India
Sergii Kushch   University of Salamanca, Spain

Ni-AFCiA: Non-intrusive Acquisition and Fusion of Context Information in Ambient Intelligence

Organizing Committee

Davide Carneiro   Instituto Politécnico do Porto, Portugal
Ângelo Costa   Universidade do Minho, Portugal
José Carlos Castillo   Universidad Carlos III de Madrid, Spain
Gabriel Villarrubia González University of Salamanca, Spain

Program Committee

Javier Pérez Marcos   University of Salamanca, Spain
Paulo Novais   Universidade do Minho, Portugal
Cesar Analide   University of Minho, Portugal
Fábio Silva   Universidade do Minho, Portugal
Marco Gomes   Universidade do Minho, Portugal
Manuel Rodrigues   Universidade do Minho, Portugal
Sérgio Manuel Gonçalves   Universidade do Minho, Portugal
Álvaro Castro González   Universidad de Castilla-La Mancha, Spain
Fernando Martín Monar   Universidad de Castilla-La Mancha, Spain
Fernando Alonso Martín   Universidad de Castilla-La Mancha, Spain
Juan Alberola   Universitat Politècnica de València, Spain
Elena del Val   Universitat Politècnica de València, Spain
Jaume Jordan   Universitat Politècnica de València, Spain
Jaime Rincon   Universitat Politècnica de València, Spain

TLAmIS: Teaching and Learning with Ambient Intelligence Systems

Organizing Committee

Dalila Durães   Technical University of Madrid, Spain
Juan Manuel Trujillo Torres University of Granada, Spain
Julio Ruiz Palmero   University of Málaga, Spain

Program Committee

Ângelo Costa	University of Minho, Portugal
António Costa	University of Minho, Portugal
David Carneiro	University of Minho, Portugal
Fábio Silva	University of Minho, Portugal
Javier Bajo	University of Politécnica de Madrid, Spain
Paulo Novais	University of Minho, Portugal
Ricardo Costa	Polytechnic of Porto, Portugal
Tiago Oliveira	University of Minho, Portugal
Pilar Cáceres	University of Granada, Spain
Francisco Hinojo	University of Granada, Spain
Rebeca Soler Costa	University of Zaragoza, Spain
José Sánchez Rodríguez	University of Málaga, Spain
Eduardo Chaves Barboza	Universidad Nacional de Costa Rica, Costa Rica
Melchor Gómez García	Universidad Autónoma de Madrid, Spain

ISAMI 2018 Sponsors

University of Minho
School of Engineering

Acknowledgments

A special thanks to the editors of the workshops on:

- Agents and Mobile Devices (AM)
- Ambient Intelligence for e-Healthcare (AIfeH)
- Blockchain Technologies and Security (BTS)
- Non-intrusive Acquisition and Fusion of Context Information in Ambient Intelligence (Ni-AFCiA)
- Teaching and Learning with Ambient Intelligence Systems (TLAmIS)

This event was partially supported by the project, "MOVIURBAN: Máquina social para la gestión sostenible de ciudades inteligentes: movilidad urbana, datos abiertos, sensores móviles; REF: SA070U 16". The project is co-financed with Junta Castilla y León, Consejería de Educación, and FEDER funds.

We want to thank all the sponsors of ISAmI 2018: IEEE Sección España, CNRS, AFIA, AEPIA, APPIA, AI*IA, and Junta de Castilla y León.

ISAmI 2018 would not have been possible without an active program committee. We would like to thank all the members for their time and useful comments and recommendations.

We would like also to thank all the contributing authors and the local organizing committee for their hard and highly valuable work.

Your work was essential to the success of ISAmI 2018.

Contents

Guimarães: Innovative and Engaged City

Ricardo Costa, Ricardo Machado[⊠], and Sérgio Gonçalves

Câmara Municipal de Guimarães, Lg. Cónego José Maria Gomes, 4804-534 Guimarães, Portugal
{Ricardo.costa,ricardo.machado,sergio.goncalves}@cm-guimaraes.pt

Abstract. This paper addresses the theme of smart cities and all their influence on citizens. The concept of intelligent city is diverse, as it is influenced by the different technologies and solutions that provide citizens well-being. The solutions presented intend to go beyond the smart city definition, focusing on issues of citizen engagement, commitment and participation, walking towards the happy city definition. Artificial Intelligence, Cognitive Computing and IoT as basis for integrating innovative solutions.

Keywords: Smart cities · Information technology · Participation · Engagement Governance · Citizen · Citizenship

1 Introduction

More than half the population (54%) of the planet Earth lives today in urban areas. It is estimated that by 2050, 66% of the world's population lives in urban areas [1] and many of these urban areas of these urban areas are modern cities that face ever more challenges and opportunities because of this. Thus, all of us, as citizens and human beings, are increasingly observing an evolution of information technologies exponentially since the beginning of the 21st century using more and more information technologies to communicate and interact with the world around us, we are thus "techno-logical beings."

A smart city uses information and communication technologies to improve the quality and performance of urban services, reduce costs and consume resources, but the most important thing is to engage more efficiently and actively with their citizens [2].

Through the combination of new technologies with the need to make cities more efficient and dynamic, citizen participation becomes vital to make cities more self-sustaining, that is, greater involvement of citizens in data collection and in providing support for specific tasks and the cooperation of human communities at the intellectual level to respond to challenges or to resolve certain more complex contexts [3].

A smart city implies intensive high technology and an advanced city that connects people, information and resources of the city using the new technologies in order to create a sustainable, competitive and innovative city for a recovery of the quality of life through a simple administration and efficient maintenance [4].

In the abstract thought, we can say that a smart city is a smart community, that is, a community where government, business, and residents understand the potential of information technology and make conscious decisions to make technology change their lives in a meaningful and positive way [5].

© Springer Nature Switzerland AG 2019
P. Novais et al. (Eds.): ISAmI 2018, AISC 806, pp. 1–9, 2019.
https://doi.org/10.1007/978-3-030-01746-0_1

It's perfectly clear that the initiatives of an intelligent city are based on the improved functioning of urban performance through the use of data, information and information technologies that provide efficient services to its citizens [6].

Being a Smart City is to be a city open to citizens, businesses, science, innovation and the future. The Guimarães Smart City strategy has framed these different perspectives. The initiatives in Guimarães testify to this. The adhesion, reception, participation of both companies and academia, as well as the citizens and their associations are a manifestation of this.

This paper follows an organic orientation, allied to a thread that demonstrates city profile aligning it with innovation policies. The city presentation section, reveals that Guimarães has a strong momentum for innovation digital and smart city strategy and Industry 4.0.

The innovative and engaged city chapter shows that Guimarães is developing an integrated strategy as a digital and smart sustainable city. The basis for this strategy relies on city engagement and citizen participation concepts.

2   Guimaraes a Smarter City

With a strong focus on the requalification of its territory and heritage, Guimarães is recognized as the birthplace of nationality and plays a central role in the history of the construction of Portugal, is internationally recognized as an example in the rehabilitation and preservation of its Historic Centre, having been distinguished in 2001 as Cultural Heritage of Humanity by UNESCO.

In 2012, Guimarães had the opportunity to show the world its history and culture, after being selected as European Capital of Culture. In the same year Guimarães obtained the Perfect City Award in connectivity and innovation promoted by Siemens and Vision Magazine. In 2013 received another distinction, as European City of Sport, the first in Portugal, a year where, in addition to promoting sports practice, it encouraged its citizens to become healthier, increasing their quality of life. In 2017 Guimarães was award wit ACEPI Navegantes XXI - Best Digital City Award, the Most Sustainable City in Portugal award.

The creation of a Mission Structure for Guimarães Sustainable linked City Hall, Universities, Institutions and people, resulting in an action plan for sustainable development, under the initiative of the Covenant of Mayors, has indelibly marked our future. Today, this is the hallmark of local political and governmental action - that institutions replicate and citizens interiorize on a daily basis – its purpose being to construct a greener and more sustainable territory. In 2017 Guimarães was considered, the most sustainable city in Portugal combining economic development with greening policies. In fact the City's Footprint calculation showed that Guimarães is becoming richer, greener and more sustainable.

With projects like "Guimarães Marca" that was launched to promote local industry internationally, by associating it to cultural heritage and environmental sustainability, Guimarães is becoming one of the most enterprising, innovative and industrial cities in

Portugal, showing a growth rate of around 10% and placing it in has fifth largest national exporter.

With this three vectors in mind, Guimarães has an objective of creating an incubator for the industry, creating a business development center for new entrepreneurs and industries wishing to develop innovative ideas, to work at the intersection between innovation and entrepreneurship by supporting industrial growth based on the added value of innovative ideas, to support technological startups and the industry, to work in the transformation of ideas into products and in the development of products and the scale-up phases. On the other hand, Guimarães industry incubator, wants to contribute to innovation in industry, creating laboratories open to universities and companies, with the objective of developing new ecosystems, helping to create new forms and methodologies for design and development, certification and manufacturing of materials and processes industries.

Energy efficiency measures also takes part of Guimarães vision. With the process of developing a Smart City strategy, to be submitted under H2020, Guimarães desires to leverage energy efficiency projects, like DREAM application that aims reduction of the use of fossil fuels, use of smart grids, efficient public lighting, among others, helping our city to develop and build on its strategic priorities. The Smart City strategy goal is to demonstrate district-wide solutions integrating smart homes and buildings, smart grids, energy storage, electric vehicles and intelligent charging infrastructures as well as the latest generation of ICT platforms based on open specifications. This is accompanied by energy efficiency measures and the use of large amounts of renewables at district level aiming to facilitate a successful transformation towards intelligent, user-oriented and demand-driven infrastructures and city services.

This strategy was designed with the objective of achieving sustainable development in the social, environmental and economic spheres. Thus, it attempts to identify, develop and enable the replication of balanced and integrated solutions. In order to manage the lines of action of the strategy, it was divided into three pillars depending on each other, namely: Structure, movement and environment. Guimarães digital strategy also targets problems that involve citizens, giving the opportunity for their voices to be heard by decision makers.

3 Innovative and Engaged City

3.1 Innovative City

Guimarães has now a strong momentum for innovation and digital & smart strategy. This strategy aims to provide a sophisticated and technological innovation initiatives as a way to establish the common points so that, through them, it can emerge a better governance and dialogue with citizens in which urban reinvention, mobilization, equity, initiative, diversity, integration, attractiveness, investment and growth are the success factors. This allows the creation of policies and regulations, attracting investment, promoting the interest of potential stakeholders for the developing and continuously

revitalizing economic growth. Thus, the objective of achieving inclusion through participation in all spheres of initiative, brings benefits to the city and to improving its
cohesion, involving citizens and making cities more resilient.

In the other hand, the Eco-Innovation and circular economy program demonstrates
the city's ability to reinvent itself even in times of crisis. This was achieved in the last
five years, with the implementation of several innovative projects targeted at the
management of the territory, cooperation and sharing of knowledge along with the
definition of priorities based on sustainability and innovation (Fig. 1).

Fig. 1 Innovation ecosystem

Guimarães was able to prepare infrastructures for the future and to align itself with
the new European policies for the protection of the environment: DREAM H2020
candidacy, Gymnastics Academy, Guimarães for Circular Economy and R&D projects
funded by PORTUGAL 2020 aiming to deal with waste on a large scale trough an
integrated consortium of which the Guimarães Municipality is a part of.

With focus in all entrepreneurs who have an idea or business plan in a rural-based
economic activity and who want to implement it in the county, Guimarães has created
a Rural Base Incubator, which is a support service to early stage startups or companies
especially dedicated to rural-based initiatives such as agriculture, agro industry, forestry
or other services and technologies.

The City Hall promotes an Ecosystem of Innovation and Entrepreneurship, through
the incubation of creative companies and industries in the so-called Creative Labs, or
through the direct management of AvePark-Science and Technology Park, that includes:
Discoveries Centre for Regeneration and Precision Medicine and Spin-off Park of Minho
University among others.

Guimarães is a case study on this by the first joint strategy for preventive conservation
of built cultural heritage within the South-West Europe Heritage care promoted by
Institute for Science and Innovation for Bio-Sustainability.

It promotes knowledge transfer by being a strategic partner in Minho University
center for technology transfer, as well as for the advanced qualification of employment
in the region, setting up several cooperative projects.

In 2014, Guimarães created the Association for Sustainable Development - Landscape Laboratory, in partnership with two Universities. This R&D center, unique in the country – promotes knowledge and innovation, research and scientific dissemination in: Ecology, Hydraulics, Geography and Urban Environment, thus contributing to an integrated action for environmental policies and Sustainable Development.

Today, it is also a center of excellence in environmental education, also being the coordinator of PEGADAS (Education Project for Environmental Sustainability), headquarters the Mission Structure and promotes "Guimarães mais Verde" commitment.

Guimarães innovation ecosystem relies on knowledge as the driver for development. Therefore it is the municipality goal to improve the quality of life and urban environment, attractiveness (talent, resources, ideas, innovation, jobs, and entrepreneurship), openness to innovation, fostering creativeness, quality of public services, security and safety of public realm, intra-community dialog as keystones of success for the city. This can only be possible with an integrated approach supported by a solid strategy namely at the digital level.

Under transformation – at micro scale, every time a city function is transformed or upgraded – value is generated and innovation opportunities are fostered. Generating value from transformation makes cities to be an active (and not passive as in past) generator of economic added-value. Relying on innovation to achieve those urban functions transformations, cities are also providing industry, services and business with new opportunities of offering communities solutions that add even more value. This double added value generates economic growth acceleration and simultaneously consolidates the sustainability of increasing quality of life standards – for all citizens and needs.

For that reason, it has become clearer the importance of making research results more accessible to all, contributing to better and more efficient data, and to innovation in the public sector. With the innovation ecosystem along with a strong digital strategy including ICT tools, networks and media Guimarães will be able to be even more open, global, collaborative, creative and closer to society, which means more reproducibility, more transparency and more impact. This can be achieved with a process based on cooperative work and new ways of diffusing knowledge by using digital technologies and new collaborative tools.

3.2 Engaged City

Guimarães engaged city initiatives intends to plan, organize and implement municipal policies in urban and public space, social and community intervention, education, environment, culture and sport, providing services to citizens. In order to bring citizens closer, Guimarães created a set of digital initiatives [7], pointed towards transparency, simplicity and speed of access to the services provided by the municipality.

Making use of pioneering services, this ecosystem values the proximity relationship with citizens, reinforcing citizen interactivity, granting them tools to have access to the municipal services without having to go to the City Hall building.

Guimarães engaged city objective is to develop innovative technological solutions and digital/social innovations, which tackle challenges faced by the daily difficulties of our citizens.

This framework is based on four important dimensions: social, digital, mobility and environment. This vectors should take benefits from technology, such as efficient and effective public services, innovation, sustainable growth, energy efficiency, environmental strategy, circular economy, optimization and efficient management process, better mobility solutions and better living environments for everyone, addressing and prioritizing it to real community problems and citizen's empowerment.

Ensuring that a digital maturated government is the future, and the use of electronic communications devices, computers and the Internet to provide public services to citizens, this framework intents to use digital interactions between a citizen and government to create a pioneer model that can be broken down into the following initiatives:

Guimarães à Boleia is a free car sharing platform that aims to contribute to a better environment, shaping the future of transportation, enabling users to gain short-term access to transportation on an as-needed basis, allowing citizens significant savings in mobility costs and gain a number of benefits, reducing costs related to transportation, decreasing the amount of time wasted trying to find park and allowing fewer cars to circulate on roads and consequently fewer CO_2 emissions into the atmosphere.

Guimarães Wi-Fi, is a network of internet propagation, offering free access, with user registration, emerging as an innovation for the future, towards citizens and urban innovation. This project goals are to implement a set of state-of-the-art technological solutions related to the Wi-Fi network technologies in an urban context, making them accessible to citizens and tourists. In phase one, 23 hotspots were already implemented, covering the main areas of public interest, in phase two, Guimarães Wi-Fi will implement 20 more hotspots, covering more areas and extending the scope to the 9 villages (Fig. 2).

◻ GUIMARAES WIFI 1.0   ◼ GUIMARAES WIFI 2.0   ● GUIMARAES WIFI 2.0 - Access Points

Fig. 2 Guimarães Wi-Fi coverage area

MyCity platform is a web and mobile service that revolutionizes the notification and occurrence management process, allowing citizens to report the most varied situations related to the city public spaces. This model of civic participation enables citizen involvement in municipality management, allowing it to consult the evolution, follow up and be notified when occurrence status changed.

Biodiversity GO! it's a web and mobile platform, that aims to involve the entire community, both in conservation and in promoting the natural heritage of Guimarães, thus contributing to environment. It aims to engage the entire community, with a citizen-science concept it intends to catalogue and preserve the existing biodiversity helping the construction of the Guimarães Biodiversity Database.

Guimarães Mobitur is a mobile application that accompanies visitors in a visit to the city of Guimarães. With a radar navigation based on the current location, it provides a selection of points of interest and events in an interactive and different way. The new already in progress version will perceive the movement of users in physical space, integrating time-based georeferenced data monitoring framework providing a powerful tool in the decision-making processes of analysis, planning and tourism management [8].

E-paper is a dematerialization of urban operations, procedures and processes platform. E-paper gives citizens the possible to present in digital format, requirements/communications of allotment operations, urbanization processes, construction or demolition authorizations and other relevant licenses/authorizations.

Guimarães 3000, it's the first European project in the permanent elderly monitoring that aims to cover all the isolation situations identified in the "Guimarães 65+" program that aims to reduce sense of isolation, and increase the safety and active solidarity of elderly living alone. Provided with mobile communication and equipped with an SOS system, this device accompanies elderly promoting quality of life and well-being [9] (Fig. 3).

Fig. 3 Guimarães 3000 system

Allparking system is a cooperation project with Guimarães Municipality and Quadrilátero Urbano, and consists on the installation of different parking sensors, which enables the collection of a large amount of information that can be used to improve mobility and quality of service, while at the same time raising the quality of life of citizens.

Participatory budget is a platform that contributes to an informed, active and responsible intervention of citizens in local governance processes, ensuring the participation of citizens and civil society organizations in the decision on the allocation of resources to municipal public policies. In the educational dimension, in participatory budget for schools, students can decide in a democratic process what they want to improve in their school.

4 Conclusions

Guimarães digital transformation supporting strategy, reflects engagement and participation. In overall, integration of citizens, groups or organizations in planning processes and policy decision making. Aspects related to public involvement has an extremely important role in Guimarães strategy. Guimarães happy city approach prioritizes citizens engagement in stages like planning and decision-making, organizing them in mobility-related or public interests areas, including energy, mobility, cultural, educational and environmental or urbanization or other areas in various sub-groups with economic, social and public interests and considering them as stakeholders and giving them the power to participate and affect a project and its implementation.

Specifically in what concerns Guimarães innovative and engaged implementation, some projects faced some difficulties related to the creation of user habits in the use/handling of digital technologies, the civic participation and citizen involvement, others faces problems related to the insurance of robustness and quality of the service or the disclosure of personal information and the possible threat to personal safety.

In order to build and support ongoing and future engagement in the strategic planning and implementation process, our future strategy points will focus on usage of information and communication technologies and other means to improve quality of life, efficiency of urban operation and services, and economic competitiveness. This strategy aims at building a vibrant community with enhanced quality of life, fostering economic growth and face global challenges.

References

1. United Nations: Department of Economic and Social Affairs. World urbanization prospects, the 2014 revision: highlights
2. Giffinger, R., et al.: City-ranking of European medium-sized cities
3. Department for Business Innovation & Skills: SMART CITIES: Background paper. BIS Res. UK, p. 47, October 2013
4. Bakıcı, T., Almirall, E., Wareham, J.: A smart city initiative: the case of Barcelona. J. Knowl. Econ. **4**(2), 135–148 (2013)
5. Hawkins, G.: The smart revolution. AEC Mag. p. 15, September 2014
6. Washburn, D., Sindhu, U.: Helping CIOs understand 'smart city' initiatives. Growth **17**(2), 1–17 (2009)
7. Novais, P., Carneiro, D.: The role of non-intrusive approaches in the development of people-aware systems. Prog. Artif. Intell. **5**(3), 215–220 (2016)

8. Anacleto, R., Figueiredo, L., Almeida, A., Novais, P.: Mobile application to provide personalized sightseeing tours. J. Netw. Comput. Appl. **41**, 56–64 (2014)
9. Costa, A., Novais, P., Simoes, R.: A caregiver support platform within the scope of an ambient assisted living ecosystem. Sensors **14**(3), 5654–5676 (2014)

A Context-Awareness Approach to Tourism and Heritage Routes Generation

Carlos Ramos[1](✉), Goreti Marreiros[1], Constantino Martins[1],
Luiz Faria[1], Luís Conceição[1](✉), Joss Santos[1], Luís Ferreira[1],
Rodrigo Mesquita[1], and Lucas Schwantes Lima[2]

[1] GECAD - Research Group on Engineering and Intelligent Computing
for Innovation and Development, Polytechnic of Porto, Porto, Portugal
{csr,mgt,acm,lef,lmdsc}@isep.ipp.pt
[2] IFSC - Federal Institute of Santa Catarina, Florianópolis, Santa Catarina, Brazil

Abstract. The aims of the TheRoute (Tourism and Heritage Routes including Ambient Intelligence with Visitants' Profile Adaptation and Context Awareness) project is to conduct studies, research and experimentation around the challenge of automatic generation of routes for visitors. The suggested routes fit the profile of visitors and groups of visitors, including aspects like emotion, mood and personality, and be aware of the context (e.g. weather, security). TheRoute is developed according the Ambient Intelligence perspective. At this point of the project execution we have already developed the full system architecture as a System of Systems approach according to an Ambient Intelligence perspective, to allow the best possible performance in the system utilization for the final user. Intelligent route generation uses user preferences for the categories of points of interest, as well as their personality traits.

Keywords: Ambient Intelligence · Tourism and Heritage
Recommender system · User profile and personality · Point of interest

1 Introduction

Several surveys have been published referring research on intelligent tourism recommender systems [1], with some emphasis on mobile devices [2, 3], and the use of context awareness [4].

According [5] travel recommendation systems aim to match the characteristics of tourism and leisure resources and attractions with the user needs. Systems like TOURSPLAN use optimization techniques aimed to define and adapt to the visitant's profile a visit plan combining the most adequate tourism products and point of interest [6–8]. Several other systems have been developed for planning a route, like CT-Planner4 [9], EnoSigTour [10], City Trip Planner [11], CRUZAR [12], Smart City [13], Otium [14], e-Tourism [15]. Social aspects have been included in some recommendation systems like VISIT [16], and iTravel [17].

Several Artificial Intelligence (AI) Techniques have been used in tourism recommendation systems, namely Multi-Agent Systems, Optimization Techniques, Machine Learning, Uncertainty, and Knowledge Representation. TheRoute system is being

© Springer Nature Switzerland AG 2019
P. Novais et al. (Eds.): ISAmI 2018, AISC 806, pp. 10–23, 2019.
https://doi.org/10.1007/978-3-030-01746-0_2

developed having as inspiration the use of Ambient Intelligence (AmI) concept. AmI deals with a new world where computing devices are spread everywhere, allowing the human being to interact in physical world environments in an intelligent and unobtrusive way. These environments should be aware of the needs of people, customizing requirements and forecasting behaviours [18, 19]. Tourism is an excellent scenario for AmI.

The new trends in technology, like the mobile devices (tablets, smartphones) boom on the last years together with positioning systems (GPS, and indoor), sensors, and ubiquitous computing allowed to make real a new vision of AI based on AmI concepts, especially in domains like Tourism.

The remain sections of this paper are organized as follows: Sect. 2 presents the Ambient Intelligence Architecture in a conceptual form, Sect. 3 describes the system's architecture. The Points of Interest and Visitors' Profile modelling are presented in Sect. 4, and in Sect. 5 the algorithms for routes generation are described. In Sect. 6 we present some screenshots and a brief description of the mobile application workflow. Finally, in Sect. 7 we present some conclusions and guidelines that we pretend to follow in the near future.

2   Ambient Intelligence Architecture

Figure 1 illustrates an example of AmI Architecture [19]. The architecture from Fig. 1 involves 4 modules. In this architecture the overall system can be seen as a "system of systems", since the orchestration of several systems is necessary for the overall performance.

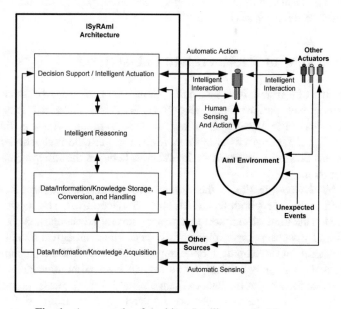

Fig. 1. An example of Ambient Intelligence Architecture

The first module of the architecture has the ability to acquire and join data, information and knowledge. This module has not the responsibility to decide if some part of the data/information/knowledge is correct or complete or affected by uncertainty. Notice that in real life human beings are also affected by these data/information/knowledge problems, and this does not avoid human beings to reason and act in the real world.

Data/Information/Knowledge treatment and fusion will have the responsibility of the second module of the architecture. Using an example of a tour/route, one participant has the responsibility to integrate different visions about some important aspect for the tour/route (opinion from an expert, from the web, etc.).

The Intelligent Reasoning module has the responsibility to perform high-level reasoning processes like Planning, Image and Language Semantics Understanding, Machine Learning, Multi-Agent based Modelling and Simulation, Optimization, Knowledge-based inference, etc. The components of this module are useful for the top module of Decision Support/Intelligent Actuation. This module will use the data/information/knowledge of the two modules previously described, being able to generate more knowledge for the module immediately below.

The top module, Decision Support/Intelligent Actuation, has the responsibility for the direct or indirect actuation on the AmI environment. We claim that some tasks can be assigned to the user, while others can be assigned to the AmI system (by means of agents). In the case of user action, decision support is important. The AmI system suggests actuations and should be able to explain the proposed actuation to the user. It should be able to accept the user changes in the solution, having a kind of "what-if" performance in order to respond to the user showing the implications of the user proposed changes. Another desired characteristic is the ability to learn from the user observation. The Decision Support/Intelligent Actuation module will interact with the other actuators of the AmI environment.

3   General View of TheRoute and Sub-systems

The architecture of a system is its backbone. Thus, if it is not properly structured it will lead to sub-optimal performances, sloppy scaling, and undesired results. Single responsibility and high layering is a key in these systems, which is the goal for theRoute platform. The System of systems (SoS) for theRoute is illustrated in Fig. 2. The goal of this SoS is to represent the interactions between the sub systems as well as enumerating them.

As can be observed in Fig. 2, the system revolves around the Data Access Layer (DAL) system, which acts both as a DAL and a Service Layer. The service layer is a RESTful API. This kind of architectures present several advantages, such as the possibility to build a webservice resource easily accessible through different platforms, particularly useful for developing mobile apps. A RESTful API architecture defines a set of rules and constraints used to define service layer communications. It aims to provide a service focused on the data exchange using HTTP requests. It creates a layer of abstraction that separates server implementations from different client

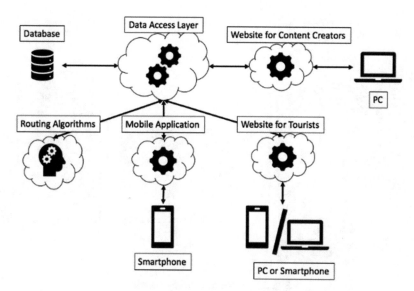

Fig. 2. The SoS for TheRoute exposing all the subsystems

implementations. When we compare with the AmI architecture proposed in [19] it is clear that this part corresponds to the Data/Information/Acquisition module.

The Service Layer role is to offer various services for each Model in Database. By forcing all other applications/websites to go through the DAL we enforce data consistency and concurrence. The DAL is deprived of any Business Logic, that responsibility is entirely given to each system.

From a more abstract perspective, TheRoute system can be seen as an assortment of tools to suggest POI (Points of Interest) and routes recommendations through the interaction of three different actors: TheRoute web platform, content creation web platform, and TheRoute mobile application. Figure 3 shows an overview of the Logical View of TheRoute. According to this logical view, TheRoute web platform works with a centralized server, meaning that tourist tours progress may be registered and consequently monitored. The proposed approach leads to the development of an API based on RESTful principles capable of communicating with the main database. Moreover, TheRoute mobile application and algorithms supports the main philosophy of the project to provide a platform capable of assisting the tourist during the tour. When we compare with the AmI architecture proposed in [19] it is clear that in this part the Data/Information/Acquisition storage, conversion, and handling module is involved.

Considering the centralized architecture of the project, the mobile application requires communication with the server to acquire, display and record data. Additionally, the mobile application is able to work offline.

The Routing algorithms provide the necessary methods to give TheRoute intelligent behaviour. Those algorithms take the input of the user for various parameters (time availability, preferred way of transportation, physical limitations, etc.) and exhibit a Route that will meet all those requirements. When we compare with the AmI architecture proposed in [19] it is clear that in this part the Intelligent reasoning module is involved.

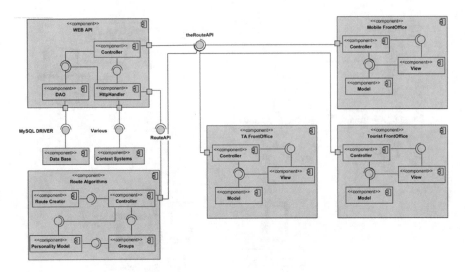

Fig. 3. Global system logical view

The Website for Content creators and management is aimed towards any touristic agent, adding and updating content for the platform. The website has secure authentication, enabling the system administrator to trace any modification to the state of the content. In this case content is anything related to the Points of Interest such as categories, fixed Itineraries, inner details of POI.

As the name suggests, the Website for tourist is the web application on which the tourist will register. Upon registering the tourist needs to fill out a form in which he will answer questions that will enable us to model is personality and cultural tastes in order for them to be used in the Routing Algorithms.

The last system is the Mobile Application. Using it the user will be able to track his itineraries and routes. Other than that, the application, when a route is stared will act as a map and a time tracker: it will update the route during the visit depending on time expended in each POI or in case of abnormal events occur during the trip. If there are either delay or furtherance the route will update to accommodate the time shift. When we compare with the AmI architecture proposed in [19] it is clear that in this part the Decision Support/ Intelligent Actuation module is involved.

4 Modelling Points of Interest and Visitors' Profile

To make a personality-based route, there has to be a connection between the tourist profile and the tourist categories. This is due to the fact that each Point of Interest does not have not a psychological profile, instead they fall within certain tourist categories. Therefore, it was imperative to determine connections between them and a human psychological model. The chosen model was The Big 5 Personality Inventory, the most widely accepted one. It suggests the division of the human psychology into five traits: Openness, Consciousness, Extroversion, Agreeableness and Neuroticism. Openness indicates the person's preference for a variety of experiences and emotions, Consciousness represents his/her tendency for organized environments and overall self-discipline, high Extroversion values signify a high tendency towards open communication and outgoing behavior, Agreeableness reports the person's cooperativeness and Neuroticism how easily he/she develops negative emotions. According to [23] it is plausible to associate the facet of Extraversion with the propension towards nature and/or adventure tourism. Openness has been correlated in the past by [21] with the Artistic facet of the Hollande vocational model, which according to [24] has a predisposition towards galleries, museums and all things related to Art. The discoveries of [25] point towards a strong preference of beach and seaside tourism by the Neuroticism facet, as well as a mild relation towards Consciousness and Openness. The same source reiterates the abundance of results that pointed to an unquestionable preference of individuals with strong agreeableness scores with pre-organized routes. The following image (Fig. 4) depicts the correlations used in the implementation, with yellow as mild connection and red as a strong one.

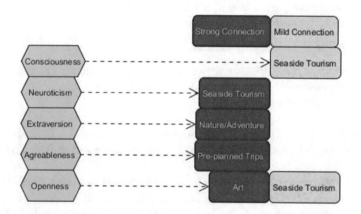

Fig. 4. Association between touristic categories and Big 5 facets

To get the personality of each user, she/he must answer ten questions [26], so the answers will build a psychological profile following the Big 5 Personality Inventory. The questionnaire must be answered at the time of registration, so we can ensure all the users have their psychological profile filled. In addition, we need the user to rate images that correspond to a certain tourist category, so we can build the User Preference Profile. In the end, we ask the user what their limitations are, so that the user can visit all the points of interest recommended.

The Points of Interest, as the core of the user's touristic experience, would require a definition that accurately pinpoints its intrinsic information (such as name, description, touristic categories) and the information that would limit its availability. A Point of Interest is considered to be available after verifying its daily schedule and recommendability towards user defined handicaps. The application would also require keeping tabs on the ever evolving average rating of each Point of Interest and the number of registered visits.

Definition 1: Be it *poi* a Point of Interest such as $(poi = (lat, long, open, cls,$ $name, des, Cat, visual, audio, rat, vst))$, where:

- *lat* is the latitude of the Point of Interest;
- *long* is the Longitude of the Point of Interest;
- *open* that day's opening hour;
- *cls* that day's closing hour;
- *name* the name of the Point of Interest;
- *des* the description of the Point of Interest;
- *Cat* the list of the Point of Interest's categories;
- *visual* a Boolean that indicates if the POI is appropriate to user with visual difficulties;
- *audio* a Boolean that indicates if the Point of Interest is appropriate to user with earing difficulties;
- *rat* average rating spanning $[1.0, 5.0]$;

  vst number of the Point of Interest's visits.

  Each touristic category is associated with its name and the total number of visits that its Points of Interest have accumulated.

Definition 2: Be it *c* a category such as $(c = (cname, cvs))$ where:

- *cname* the name if the category such as *cname* ∈ {Gastronomic Landmark, Historic Landmark, Nature e Exploration, Sports, Seaside Tourism};
- *cvs* total number of category visits.

Definition 2.1: Be it *Cat* a list of categories such as $\{Cat\} = [1, \ldots n]$, where n is the total number of categories; in order words, *Cat* should have at least one category associated.

After her/his registration, the user will be characterized by her/his biographical data (name, gender, nationality, age), handicaps (visual or hearing difficulties) that might compromise the enjoyment of a Point of Interest's experience, a list of touristic interests and a list of Big 5 Personality values.

Definition 3: Be it *user* a user such as $(u = (uname, ugen, unat, uage, uvs, uaud,$ $Int, Pers))$ where:

- *uname* the user's name;
- *ugen* the user's gender;
- *unat* the user's nationality;
- *uage* the user's age;
- *uvs* a Boolean that indicates that the user has visual difficulties;
- *uaud* a Boolean that indicates that user has earing difficulties;
- *Int* a list of categories of interest to the user;
- *Pers* a list of user's Big 5 personality values.

The aforementioned personality values are defined as an association between the Big 5 Personality facets and a numeric value.

Definition 4: Be it f a *Big 5 facet* such as $(f = (fname, fval))$, where:

- *fname* the name of the facet among *fname* $\in \{Extroversion, Neuroticism, Openness,$ *Agreeableness, Consciousness*$\}$;
- *fval* facet value between $[0, 1.0]$;

The connections between the touristic categories and Big 5 Personality Facets that were explained above are an association between the name of the touristic category and respective facet name, including the distinction between strong and mild relationships.

Definition 5: Be it *con* an association such $(con = (y, z))$, being y the name of the category and z is the name of personality facet.

Definition 5.1: Be it *Wcon* a list of associations between mild *The Big 5 personality facets* and touristic categories, as illustrated in Table 1, y_{Wcon} e z_{Wcon} can be defined respectively.

Table 1. Description of mild connections

Touristic category (y_{Wcon})	Big 5 facet (z_{Wcon})
Seaside tourism	Openness to experience
Seaside tourism	Consciousness

Definition 5.2: Be it *Scon* a list of associations between strong *The Big 5 personality facets* and touristic categories, as illustrated in Table 2, y_{Scon} e z_{Scon} can be defined respectively.

Table 2. Description of Strong Connections

Touristic category (y_{Scon})	Big 5 Facet (z_{Scon})
Seaside tourism	Neuroticism
Nature and adventure	Extraversion
Art	Openness to experience

5 Algorithms for Route Generation Context

This section will include the presentation of the algorithms utilized in the personalized route generation. This includes the algorithms for singular person routes.

In the Introduction section, several systems were exemplified, all using different algorithms as a solution to the path finding conundrum. CTPlanner4 [9] for instance, uses genetic algorithms with mutation, cross-over and competition of possibilities. City Trip Planner [11] uses Greedy Randomised Adaptive Search Procedure (or GRASP) and iTravel [17] suggests Points of Interest on the fly, utilizing a collaborative user rating prediction system. TheRoute however, has implemented an adaptation of the

Branch-and-Bound algorithm to formulate the suggested routes. The first step consists on filtering the less than average weighted Points of Interest. Starting with the user defined geographical starting point, the Decision Support Module will look for the five closest available Points of Interest to create five new intermediate routes. On each iteration, the up to twenty-five best possibilities (according to the sum of their user weight) will be selected to proceed unto the next iterations. This process will repeat itself while there is still time to expand them further. When it is not possible to continue, the candidate routes will be validated with resort to the Google Directions API. The best route to be validated will be the one suggested to the user.

The weight is calculated using Formula 1:

$$Weight = \sum_{i=0}^{n} \frac{\frac{POI\ visits}{cat\ visits(i)} * rating}{1.0 - iValue(i)} * Personality\ Modifier \tag{1}$$

The following is the function relative to the *iValue* component of the Point of Interest calculation formula (where *Int* pertains to the list of the user's touristic interests):

$$F_{categorization} = \begin{cases} iValue \rightarrow 0.9, if\ i \in Int \\ \\ iValue \rightarrow 0,\ if\ i \notin Int \end{cases}$$

The following is the function relative to the *Personality Modifier* component of the Point of Interest calculation formula:

$$F_{Personality} = \begin{cases} 1, if\ f \subset Pers \wedge f < 0.5 \cup \{f,i\} \not\subset Wcon \cup \{f,i\} \not\subset Scon \\ 2, if\ f \subset Pers \wedge f > 0.5 \wedge \{f,i\} \subset Wcon \wedge \{f,i\} \not\subset Scon \\ 3, if\ f \subset Pers \wedge f > 0.5 \wedge \{f,i\} \not\subset Wcon \wedge \{f,i\} \subset Scon \end{cases}$$

Example 1: Considering a Point of Interest *poi₁ and user* defined thusly:
$poi_1 = \{0,0,0,0, Dragao, Stadium, Desp, false, false, 4.9, 2000000\}$;
$Desp = \{Sports\}$;

$$Sports = \{Sports, 10000000\};$$
$$user = \{Luís, M, Portuguese, 21, false, false, Int, Pers\};$$
$$Int = \{Sports\};$$
$$Pers = \{0.1, 0.8, 0.4, 0.2, 0.3\};$$

The calculation of the weight of *poi₁* to the user shall be proceed as indicated:

$$Weight = \left(\frac{\frac{2000000}{10000000} * 4.9}{0.1} \right) * 1 = \frac{\left(\frac{1}{5} * 4.9 \right)}{0.1} = \frac{0.98}{0.1} = 9.8$$

Example 2: Considering a Point of Interest poi_2 and user defined thusly:
$$poi_2 = \{0,0,0,0, \text{Beach, Dirty, } Seaside, false, false, 4.0, 50000\}$$
$$Seaside = \{SeasideTourism\};$$

$$Seaside\ Tourism = \{Seaside\ Tourism, 2000000\};$$
$$user = \{Luís, M, Portuguese, 21, false, false, Int, Pers\};$$
$$Int = \{Sports\};$$
$$Pers = \{0.1, 0.8, 0.4, 0.2, 0.3\};$$

The calculation of the weight of poi_2 to the *user* shall be proceed as indicated:

$$Weight = \left(\frac{\frac{50000}{2000000} * 4.0}{1}\right) * 3 = \frac{(0.025 * 4.0)}{1} * 3 = 0.1 * 3 = 0.3$$

Example 3: Considering a Point of Interest poi_3 *and user* defined thus:
$$poi_3 = \{0,0,0,0, \text{Galery, xpto, } Art, true, false, 3.8, 250000\};$$
$$Art = \{Artistic, History\};$$

$$Artistic = \{Art, 1000000\};$$
$$History = \{Historical\ Landmark, 4000000\};$$
$$user = \{Luís, M, Portuguese, 21, false, false, Int, Pers\};$$
$$Int = \{Art\};$$
$$Pers = \{0.1, 0.8, 0.4, 0.2, 0.3\};$$

The calculation of the weight of poi_3 to the *user* shall be proceed as indicated:

$$Weight = \frac{\frac{250000}{1000000}*3.8}{0.1} * 1 + \frac{\frac{250000}{4000000}*3.8}{1} * 1$$
$$= \frac{\frac{1}{4}*3.8}{0.1} + \frac{\frac{1}{16}*3.8}{1} = \frac{0.95}{0.1} + \frac{3.8}{16} = 9.5 + 0.2375 = 9.7275$$

6   TheRoute Mobile Application Interface

TheRoute's interface was developed using the Ionic framework, with the objective of creating a hybrid application, which is, developed using web technologies but built as a native application. Using the native component WebView, (present in every mobile OS), the Apache Cordova platform provides a set of JavaScript APIs that allows the use of the device's resources, (geolocation, camera, contacts list, etc.). The final result is an installable application (an APK, for example) developed as a web app, but with a native look-and-feel and the access to its resources [20].

The application use flow is based in 4 steps: (1) Login/sign up, (2) creation of a route, (3) modifications in the route, and (4) presenting the final route. If it is the user first time in the application, the user fills the fields of first and last name, and her/his birth date. Then, she/he must do the personality test, answering 10 questions with a number in a scale, from 1 (strongly disagree) to 5 (agree strongly) to define her/his big

5 traits (see Fig. 5 – left figure). Big Five Personality Traits model is explained in [21, 22]. The next step is the preferences tests, where he evaluates 6 pictures according to his tastes (family, friends, radical, nature, sightseeing, art/culture) (see Fig. 5 – middle figure). At the end, a chart with the results will be showed, with a brief description of each one of the big 5 traits (see Fig. 5 – right figure). If necessary, these tests can be done again later on.

Fig. 5. User registry and psycologic profile screens

In the page "route requests", the user can choose between requesting a custom route, or pick a thematic route ("ready-to-go route"). If choosing to the "ready-to-go! Route", a list with all the existent thematic routes is shown, ordered by the top rated and most recommended routes first.

If choosing to request a custom route, the user inputs information about her/his trip (date and time, location, mode of transportation, if in group or alone) is considered. The application sends this information to the algorithm of route generation, which returns a unique route especially made to the user (see Fig. 6).

When the user chooses one of his routes, a list with the selected POIs is shown. The user can delete, modify or fix the POIs. The approximated duration and distance are shown in the top. Lastly, a map is shown (Fig. 6), with the route drawn over it with a marker in each POI location, and a small info box showing the next POI. The user can also review the list with all the POIs of the route, where he can see further information about each one.

Fig. 6. Custom route screen

7 Conclusions

TheRoute project intends to conduct studies, research and experimentation around the challenge of automatic generation of routes for visitors. The main goal is the generation of personalized routes regarding tourists' personal preferences and personality traits, in order to present to each tourist, the Points of Interest that he or she will want to visit. At this point of the project execution we have already developed the full system architecture as a System of Systems approach according an Ambient Intelligence perspective, to allow the best possible performance in the system utilization for the final user. Intelligent route generation uses user preferences for the categories of points of interest, as well as their personality traits. To make this possible, the relationships between personality types and tourism categories or styles were studied and with that we developed an algorithm that evaluates each one of the Points of Interest located in a certain city and then generates a route that fits the tourist preferences and needs.

As future work, we will perform a case study with a considerable number of real users (tourists), so that we can have a feedback of the usability and the accuracy of the generated routes for the tourists. On the other hand, we will have results to allow other kind of algorithms, like Machine Learning.

Acknowledgments. We thank the Portuguese Foundation for Science and Technology (FCT), the Operational Programme of North Region of Portugal (PO-Norte), and the FEDER Programme, for the selection and support for the project TheRoute - Tourism and Heritage Routes including Ambient Intelligence with Visitants' Profile Adaptation and Context Awareness (project reference SAICT/023447). We also thank Santander-Totta Bank (BST), from Portugal, and Federal Institute of Santa Catarina (IFSC), from Brazil, for the fellowships to several students involved in TheRoute. Finally we thank all the participants of the different partners of the project, from all the 8 schools of the Polytechnic of Porto (ISEP, ISCAP, ESE, ESMAE, ESTG, ESS,

ESHT, and ESMAD), showing a really multidisciplinary project, and for DouroAzul, an enterprise that is the leader in the sector of river cruises in Portugal, and Polytechnic of Viana do Castelo (IPVC), from Portugal.

References

1. Borras, J., Moreno, A., Valls, A.: Intelligent tourism recommender systems: a survey. Expert Syst. Appl. **41**(16), 7370–7389 (2014)
2. Dickinson, J.E., Ghali, K., Cherrett, T., Speed, C., Davies, N., Norgate, S.: Tourism and the smartphone app: capabilities, emerging practice and scope in the travel domain. Curr. Issues Tour. **7**(1), 84–101 (2014)
3. Gavalas, D., Konstantopoulos, C., Mastakas, K., Pantziou, G.: Mobile recommender systems in tourism. J. Netw. Comput. Appl. **39**, 319–333 (2014)
4. Lamsfus, C., Wang, D., Alzua-Sorzabal, A., Xiang, Z.: Going mobile defining context for on-the-go travelers. J. Travel. Res. **54**(6), 691–701 (2015)
5. Ricci, F.: Travel recommender systems. IEEE Intell. Syst. **17**(6), 55–57 (2002)
6. Coelho, B., Martins, C., Almeida, A.: Adaptive tourism modeling and socialization system. In: Proceedings of the International Conference on Computational Science and Engineering, vol. 4, pp. 645–652 (2009)
7. Luz, N., Anacleto, R., Martins, C., Almeida, A., Lucas, J.P.: Lightweight tourism recommendation. Int. J. Web Eng. Technol. **9**(2), 106–123 (2013)
8. Lucas, J.P., Luz, N., Moreno, M.N., Anacleto, R., Almeoida, A., Martins, C.: A hybrid recommendation approach for a tourism system. Expert Syst. Appl. **40**(9), 3532–3550 (2013)
9. Kurata, Y., Hara, T.: CT-planner4: Toward a more user-friendly interactive day-tour planner. Information and communication technologies in tourism, pp. 73–86. Springer International Publishing (2014)
10. Borràs, J., Moreno, A., Valls, A., Ferré, M., Ciurana, E., Salvat, J., Anton-Clavé, S.: Uso de técnicas de inteligencia artificial para hacer recomendaciones enoturísticas personalizadas en la Provincia de Tarragona. In: Proceedings of the Congreso Nacional de Turismo y Tecnologías de la Información y las Comunicaciones, pp. 217–230 (2012)
11. Vansteenwegen, P., Souffriau, W., Berghe, G.V., Van Oudheusden, D.: The city trip planner: an expert system for tourists. Expert Syst. Appl. **38**(6), 6540–6546 (2011)
12. Mínguez, I., Berrueta, D., Polo, L.: CRUZAR: An Application of Semantic. Cases on Semantic Interoperability for Information Systems Integration. In: Practices and Applications, no. 255 (2009)
13. Luberg, A., Tammet, T., Järv, P.: Smart city: A rule-based tourist recommendation system. In: Law, R., Fuchs, M., Ricci, F. (eds.) Information and communication technologies in tourism. Springer (2011)
14. Montejo-Ráez, A., Perea-Ortega, J.M., García-Cumbreras, M.A., Martínez-Santiago, F.: Otium: a web based planner for tourism and leisure. Expert Syst. Appl. **38**(8), 10085–10093 (2011)
15. Sebastia, L., Garcia, I., Onaindia, E., Guzman, C.: e-Tourism: a tourist recommendation and planning application. Int. J. Artif. Intell. Tools **18**(5), 717–738 (2009)
16. Meehan, K., Lunney, T., Curran, K., McCaughey, A.: Context-aware intelligent recommendation system for tourism. In: Proceedings of the IEEE International Conference on Pervasive Computing and Communications, pp. 328–331 (2013)

17. Yang, W.S., Hwang, S.Y.: iTravel: a recommender system in mobile peer-to-peer environment. J. Syst. Softw. **86**(1), 12–20 (2013)
18. Ramos, C., Augusto, J.C., Shapiro, D.: Ambient intelligence—the next step for artificial intelligence. IEEE Intell. Syst. **23**(2), 15–18 (2008)
19. Ramos, C.: An architecture for ambient intelligent environments. In: Proceedings of the 3rd Symposium of Ubiquitous Computing and Ambient Intelligence, pp. 30–38. Springer, Berlin, Heidelberg (2009)
20. Brostowe, J.: What is a Hybrid Mobile App?", Telerik Developer Network (2015). https://developer.telerik.com/featured/what-is-a-hybrid-mobile-app/. Accessed 13 Dec 2017
21. Larson, L.M., Rottinghaus, P.J., Borgen, F.H.: Meta-analyses of big six interests and big five personality factors. J. Vocat. Behav. **61**(2), 217–239 (2002)
22. Santos, R., Marreiros, G., Ramos, C., Neves, J., Bulas-Cruz, J.: Personality, emotion, and mood in agent-based model for group decision making. IEEE Intell. Syst. **26**(6), 58–66 (2011)
23. Kahle, L.R., Matsuura, Y., Stinson, J.: Personality and personal values in travel destination. In: Ha, Y.-U., Yi, Y. (eds.) AP – Asia Pacific Advances in Consumer Research, vol. 6, p. 311 (2005)
24. Frew, E., Shaw, R.N.: The relationship between personality, gender, and tourism behavior. Tour. Manag. **20**, 193–202 (1999)
25. Neidhardt, J., et al.: Eliciting the users' unknown preferences. In: Proceedings of the 8th ACM Conference on Recommender Systems, pp. 309–312 (2014). http://dl.acm.org/citation.cfm?doid=2645710.2645767
26. Rammstedt, B., John, O.P.: Measuring personality in one minute or less: a 10-item short version of the big five inventory in English and German. J. Res. Pers. **41**, 203–212 (2007)

User-Guided System to Generate Spanish Popular Music

María Navarro-Cáceres[1]([⊠]), Matilde Olarte-Martínez[1], F. Amílcar Cardoso[2], and Pedro Martins[2]

[1] University of Salamanca, Patio de Escuelas, 37001 Salamanca, Spain
maria90@usal.es
[2] CISUC, Department of Informatics Engineering, University of Coimbra,
Rua Sílvio Lima, Pólo II da Universidade de Coimbra, 3030 Coimbra, Portugal

Abstract. The automatic generation of music is an emerging field of research that has attracted wide attention in Computer Science. Additionally, the interaction between users and machines is nowadays very present in our daily lives, and influences fields such as Economy, Sports or Arts. Following this approach, this work develops an intelligent system that generates melodies based on Spanish popular music and some indications of the users through an interface. The system creates a melody by learning from the corpus selected through a Markov model, which is also influenced by the users' preferences. Several experiments were carried out to evaluate the musical quality and the usefulness of the system to interact with the user and generate music. The results of the evaluation shows that the proposal is able to generate music adapted to the style standards of Spanish popular music and to the users' indications.

Keywords: Computational Creativity · Melody generation
Popular songs

1 Introduction

The different computational advances in the field of Artificial Intelligence that have occurred in recent years have attracted the attention of researchers of multiple origins and motivations, creating innovative fields that unite apparently disparate concepts such as Artificial Intelligence and Art. From an interdisciplinary research field that sits at the intersection of the areas of AI, Psychology, Cognitive science, Linguistics, Anthropology and other human-centered sciences, the area of Computational Creativity (CC) was born. CC can be defined as a "philosophy, science and engineering of computational systems which, by taking on particular responsibilities, exhibit behaviors that unbiased observers would deem to be creative" [4].

The area of Computational Creativity has recently been significantly developed with the entry of companies as important as Google, with projects such as DeepDream, which uses a convolutional neural network that transforms images

© Springer Nature Switzerland AG 2019
P. Novais et al. (Eds.): ISAmI 2018, AISC 806, pp. 24–32, 2019.
https://doi.org/10.1007/978-3-030-01746-0_3

or, more recently, Magenta [3], the equivalent for the generation of music. Although many different techniques have been applied to generate music automatically, statistical methods were a preferable tool due to its ability to learn from the context or the corpus. The deep analysis performed in [11] about different statistical models, is particularly interesting. We should also remark that Markov Models are used very commonly to generate different kinds of music [8–10].

However, to the best of our knowledge, the initiatives have usually focused on the generation of music of a rather tonal character and, to date, there has been no study addressing the generation of Spanish popular music. This genre of music differs from the classical music in many aspects, including the sonority, the sounds disposition or the rhythmic formulas used. The automatic generation of this kind of music also depends on multiple factors that are intrinsically connected, such as the representation of the tonality, the melodies and the rhythm.

Additionally, more and more humans work in collaboration with other humans and/or computational entities, both directly and indirectly, to meet a series of objectives that, individually, would be impossible. Currently, the support of new technologies are key to generating a collaborative organization in the educational or business framework. This new paradigm of virtual collaboration through electronic devices has great potential in areas as diverse as Home automation, Energy saving, Medicine or even Visual Arts or Music.

There have been several approaches in which the interaction of the users is essential for the achievement of music [2,7,8,10]. However, many of them are not focused on users who are not familiar with musical education, making hard for them to generate melody in a semi-automatic way. Drawing on this phenomenon, a user-guided melody generation is proposed, with the goal of bringing the Spanish popular music to those ones without a specific musical education.

For this purpose, new melodies are generated from original Spanish popular songs, extracted from multiple sources of popular music. After the analysis, encoding and storage of relevant features, original melodies are used as a training corpus. Among the different learning models that can be applied to generate music from a previous corpus [1], Markov Models (MMs) have been selected due to their successful application in other related works [9–12]. MMs are trained in a corpus and then used to generate a new melody that fits the style of popular songs. However this melody generation is always guided by the users, who can influence the final result through a hardware device. For the purposes of this work, a mouse is selected as the tool to guide the melody generation. The user interacts in a screen moving the mouse in two axes: horizontal axis to control the rhythm of the melody, and the vertical axis to control the pitches.

Once the melodies are generated, a listening test was developed to evaluate the musical quality according to the Spanish popular music standards and to collect the users' opinion about the usefulness of the system to interact with them and generate music.

The remainder of this paper is structured as follows. Section 2 details the generation process of music and the interaction with the users. Section 3 gives an analysis about the evaluation of the system developed. Finally, Sect. 4 describes the final conclusions and future work.

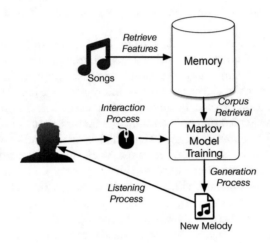

Fig. 1. Overview of music generation flow.

2   System Description

Figure 1 gives an overview of the overall system to create popular Spanish music.

The system retrieves the musical data from popular music songs. Consequently, it is provided with a memory to store different melodies that belong to the popular music style. This corpus has been manually analyzed and retrieved from song books and include a wide variety of popular music.

Initially, the system selects the collected melodies and learns a model for generating a new composition. For the purposes of this paper, Markov Models (MM) were used as the learning algorithm. The Markov Models are influenced by the users' indications through the position of the mouse in the interface of the system.

2.1   Music Generation

The platform intends to generate music based on the features of the popular music, such as the rhythm or the sonority.

In order to generate music according to the popular music (or popular songs) standards, a set of musical sources has been retrieved. The sources should reflect the most common sonorities of the popular music. In order to make the process easier, the songs are extracted from songbooks, recordings or digital scores.

Folk songs in Spain are as varied as its regions. However, generally it is very difficult to understand the origin of the sonority in popular music, since this type of music has been in constant evolution due to its characteristic grammar and its transmission from generation to generation orally. However, analyzing many of the melodies collected over the years by different ethnomusicologists and in different songs, we can draw some conclusions in this regard that allow us to discern the popular music, selecting those ones that better represent the popular musical features according to [6].

The sources should be digitalized to represent the information about rhythm and sonority. To identify the popular music we should not only analyze the particular duration or pitch, but also the duration of the musical phrases, the degrees in which the melody reposes (notes with a long duration), and the particular cadences. Unlike the classical music, in popular music the harmonic tension and the use of the chords degree are not particularly relevant, as it does not follow harmonic rules; they are only used according to the melodic course. Drawing on these properties and also inspired by the concept of viewpoints exposed in [12], the following features of the popular songs were selected:

- Pitch: Musical note
- Duration: Rhythmic formula of one note
- Degree: Position of the note within the musical scale
- First in bar: Boolean value which indicates if the note is the first in a bar or not.
- Time Signature: Number that represents the time signature of the melody
- Musical Phrase: It represents the position of a note in a musical phrase.

Currently, there is no standard format that only addresses these features. However, there are a wide number of songs already encoded as MIDI (Musical Instrument Digital Interface) files [5], due to its availability throughout the network, the low difficulty in creating such files based on digital scores, and its structure, which allows easy access to notes and durations. The files do not contain the sounds. Instead, they include instructions that allow the reconstruction of the song by using a sequencer and a synthesizer that work with MIDI specifications. Therefore, the files are quite light since they allow to encode a complete song in a few hundred lines. The mathematical data inside these files along with a manual analysis have been used to encode the features above.

Once the files are available, the next step is to extract the necessary information for the project: notes, durations, bars, time signature, etc. These data are considered the training set for the learning model. In the first experiments, we combine different musical features and check which ones works better with the Markov Models. The Markov model determines the possible transitions for each state and the initial probabilities of each one.

Consequently, the position of the mouse must be translated into a note and a duration. To do so, the mouse works as an indicator with two dimensions, and the user can move it through the screen provided. In this sense, the Y axis was set to indicate the reference note (higher or lower pitches), since we intuitively associate "climbing" with higher notes and "lowering" with lower notes. Likewise, the X

axis indicates a reference duration. On the Y axis, 0 represents silence (lowest position of the mouse; while 2 reference octaves are represented, from the lowest to the highest. On the X axis, 1 represents the shortest note (sixteenth note) and 16 represents the longest note (round). These delimitations of the space of reference facilitate the training process of the model.

In this case, the probability $P(t)$ of each note t_i to be selected as the next note in the melody is a linear combination between the probability $P_M(t_i)$ given by the Markov Model and the position of the mouse $P_D(t_i)$ given by the user:

$$P(t) = k \cdot P_D(t_i) + (1 - k) * P_M(t_i), \tag{1}$$

where k is the weight of each probability and is empirically set to 0.55.

Once we selected the notes of the melody, the music generated by the system will be encoded in MIDI format to use a standard synthesizer that could be incorporated in the own computer. We used a standard piano as the synthesis sound due to its versatility.

3   Results and Discussion

A system which generates popular songs is built. For the generation of the music, 180 popular songs were selected, 85 dance melodies and 95 work songs. All of them make use of similar rhythm patterns, with time signature of 2/4, 3/4 and 6/8. Each song consisted of 3 or 4 musical phrases with similar length, and uses songs with the two different sonorities, meaning we divided the corpus in Frigian mode with possible modifications of this mode in its evolution to E minor and Eolian mode with possible modifications to evolve to A minor.

The melodies have been encoded according to the Sect. 2 and saved in an Excel file with all the properties. These features were the corpus for the Markov Model training. The memory of the MM, meaning the number of estates that it can remember for the future generation of the melody was empirically set to 4.

During the generation process, each iteration of the system consists of adding of a new note in the melody assisted by the mouse position and the MM, and it is iterated until the user decides to stop. Figure 2 shows the interactive screen of the user while he is generating the melody.

The upper buttons indicate to start or stop the generation process. The interaction interface is divided into sections in order to frame the pitches and durations of the musical compositions. The cell where the mouse is over is painted in green to highlight the musical reference in which the system is currently working to generate music and rhythm.

The evaluation of this system is twofold. On the one hand, we aim to validate the musical results, meaning the melodies generated can follow the style of the Spanish popular music. On the other hand, we aim to validate the usefulness and the interactivity of the overall system to generate such kind of music.

In order to assess the quality of the music, a listening test was designed in which 10 melodies generated by the system were chosen. There were 5 melodies of Eolian mode and 5 melodies of Frigian mode. Additionally, 4 melodies have a

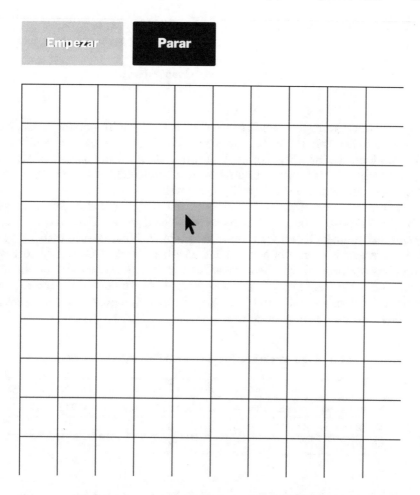

Fig. 2. Interface of music generation system.

time signature of 6/8, 3 of 3/4 and 3 of 2/4. 10 expert users in ethnomusicology were selected and asked if they think the melodies generated follows the standards of popular music according to the sonority and the rhythm. They should evaluate the quality in a scale from 1 to 5, where 1 means Very poorly and 5 means Perfectly.

For such categorical data, we use the chi-square test to prove that the results reflect the quality of the melodies according to the subjective ratings given by the listeners. It is important to note that we can get very subjective ratings since all listeners can have different interpretations of the musical pieces and different musical tastes. Thus, we considered the analysis of the median M_e and the mode M_o a useful step.

Table 1 collects statistical values calculated from the data, namely the p-value for the chi-value χ^2 the median M_e and the mode M_d. The statistical analysis

Table 1. Statistics resulting from the listening test results.

χ^2	M_e	M_o
$6.8381e - 04$	Good	Good

suggests that the system captures the style of the Spanish popular melodies. The majority of listeners think the values are "Good" (according to the M_d) and the median indicates at least half of the ratings obtained are scored as "Good" or even better. The statistical results lead us to conclude that the fitness function captures perceptual musical quality quite well.

The second part of the test consists of validating the usefulness of the system. For this purpose, a total of 8 users were selected to test the system. Each user generated 10 melodies and then were asked for their experience with the system. They answered a questionnaire about whether the system is easy to use, the interface conforms to the real movements of the device, as well as the overall score for the system and possible suggestions. All the questions could be rated from 1 ("Completely disagree") to 5 ("Completely agree"). Table 2 shows the mean scores for all these questions.

Table 2. Final statistics after the users finished testing the system.

	Easy to use	Interface	Control quality	Overall ratings
Mean ratings	4.22 ± 0.86	3.23 ± 1.53	4.09 ± 0.94	4.01 ± 0.79
Mode	4	3	3	3
Median	4	3	3	3

The Table 2 shows that the general satisfaction degree is quite high, with a mean of 4.01 and a mode and median values of 3. The users consider the system to be very easy to use even for people without any musical training (4.22, and a median a mode of 4), although the interface could be improved (3.23, with mode and median of 3). Some users have suggested the addition of a complete score with notes and rhythms instead of only listening to the final sounds and see the score at the end of the generation process.

4   Conclusions

This paper presents an intelligent system to compose melodies using a common device such as a mouse to control the duration and the pitch of the generated notes. The melodies adapt to user preferences through the indications given by the position of the mouse in an interface. As a first step, the proposed approach retrieves a set of MIDI files from which some musical features are extracted. A Markov model is then trained with the data of the collection of music, and the transition probabilities of this model are modified according to the control device to generate a melody that respects these "controlled" probabilities.

To deploy the system, an application has been developed that could be described as a controllable intelligent sequencer, since on the one hand it learns to generate sequences from sets of examples and on the other it admits the direct intervention of the user to guide the process of generation of the melody.

The results of the different experiments carried out emphasize the importance of user preferences in the melody generation. However, despite the users' indications, we do not avoid to follow the standards of the popular music in the generation process. In fact, the user has an essential role to guide the generation process and adapt the melody to his preferences through a mouse.

To improve the interaction of the users, we would like to analyze the incorporation of more standard devices, such as a keyboard or a joystick, to indicate a more accurate feature related to pitch and rhythm duration. In order to improve the automatic generation of music, it is needed to analyze the melodies with new properties, including a more general view of the composition. In popular music, the vocal songs are really important for ethnomusicologist studies. Consequently, a deeper analysis of popular songs and a study to incorporate popular lyrics to the music generated will be addressed in a future work.

Acknowledgments. This work was supported by the Spanish Ministry of Economy and FEDER funds. Project SURF: Intelligent System for integrated and sustainable management of urban fleets TIN2015-65515-C4-3-R.

References

1. Ajoodha, R., Klein, R., Jakovljevic, M.: Using statistical models and evolutionary algorithms in algorithmic music composition. In: Encyclopedia of Information Science and Technology, 3rd edn., pp. 6050–6062. IGI Global (2015)
2. Bergsland, A., Wechsler, R.: Composing interactive dance pieces for the motion-composer, a device for persons with disabilities. In: Proceedings of the International Conference on New Interfaces for Musical Expression, pp. 20–23 (2015)
3. Casella, P., Paiva, A.: MAgentA: an architecture for real time automatic composition of background music. In: Intelligent Virtual Agents, pp. 224–232 (2001)
4. Colton, S., Wiggins, G.A.: Computational creativity: the final frontier? Front. Artif. Intell. Appl. **242**, 21–26 (2012)
5. Jungleib, S.: General Midi. AR Editions, Inc. (1996)
6. Manzano, M.: El folklore musical en españa, hoy. Boletín Informativo de la Fundación Juan March, vol. 204, pp. 3–18 (1990)
7. Moreira, J., Roy, P., Pachet, F.: Virtualband: interacting with stylistically consistent agents. In: ISMIR, pp. 341–346 (2013)
8. Pachet, F.: The continuator: musical interaction with style. J. New Music. Res. **32**(3), 333–341 (2003)
9. Pachet, F., Roy, P.: Markov constraints: steerable generation of Markov sequences. Constraints **16**(2), 148–172 (2011)
10. Papadopoulos, A., Roy, P., Pachet, F.: Assisted lead sheet composition using flow-composer. In: International Conference on Principles and Practice of Constraint Programming, pp. 769–785. Springer (2016)

11. Whorley, R.P., Conklin, D.: Music generation from statistical models of harmony. J. New Music. Res. **45**(2), 160–183 (2016)
12. Whorley, R.P., Wiggins, G.A., Rhodes, C., Pearce, M.T.: Multiple viewpoint systems: time complexity and the construction of domains for complex musical viewpoints in the harmonization problem. J. New Music. Res. **42**(3), 237–266 (2013)

Getting Residents Closer to Public Institutions Through Gamification

Manuel Rodrigues[1]([✉]), Vasco Monteiro[2], Paulo Novais[2], and César Analide[2]

[1] CIICESI, School of Technology and Management – Polytechnic of Porto Felgueiras,
Felgueiras, Portugal
[2] Informatics Department/Computer Science and Technology Center, University of Minho,
Braga, Portugal
vmnokk@gmail.com, {pjon,analide}@di.uminho.pt

Abstract. With the huge growth of mobile device users, in recent years, the need
and the opportunity to create new digital services and platforms has arisen. These
platforms and services not only make people's daily life easier, but also facilitate
and improve communication between institutions and people. Also, new ways of
achieving the intended goals are being developed and tested. Gamification is an
example, where institutions and people's communication is encouraged through
the offering of incentives/rewards that potentiate involvement with that particular
institution. In practice, institutions offer rewards to participants who perform
predetermined tasks, for recommendation, dissemination, evaluation or greater
involvement of people with that particular institution. The concept of smart cities
is also getting huge attention nowadays. Making a city "smart" is emerging as a
strategy to mitigate the problems generated by the urban population growth and
rapid urbanization. In this work, we propose a digital solution, in the form of a
mobile application, which has as its main goal to improve city hall public services
and people's communication, bringing them closer. This is achieved using gamification techniques that aim to engage residents with city hall services. It is
provided a report system that enables residents to communicate to the city hall
services, some issues regarding their town that they wish to be solved, such as
broken structures, clogged sewers, among others. The proposed system also has
a lore that leads to extra motivation to complete missions, be part of player gath-
erings and events, and meet new people and to better explore the cities' wonders.
An admin platform for the maintenance and administration of the system is also
proposed, to better help keeping the application's content fresh and updated,
allowing for a better user experience for the population. The proposed system is
being prepared for testing in real environments, the simulation results, as showed
in this work, indicate very promising results towards the achievement of the
proposed goals.

Keywords: Smart Cities · Mobile computing · Computational creativity

1 Introduction

We live in an era in which a large percentage of the population, regardless of age, has
contact with the most diverse technologies, and in particular, it is rare to find those who,

P. Novais et al. (Eds.): ISAmI 2018, AISC 806, pp. 33–39, 2019.
https://doi.org/10.1007/978-3-030-01746-0_4

on current days, do not have access to a computer and/or to mobile devices such as tablets or smartphones. Coupled with this growing trend, and in particular the need for people to be in constant contact with the rest of the world, there has been a huge growth of applications that are now within reach of these devices. Taking advantage of this growth, there is an opportunity to conceive new ways of bringing citizens closer to public institutions, throughout technology. Alongside all this, and with the fast growth in the field of intelligent systems and artificial intelligence, the increasingly fashionable concept of Smart Cities appeared naturally, often framed with the concepts of e-Participation [1] and e-Government [2], which characterize the transition from existing, and sometimes somewhat outdated, processes towards digital ones. Specifically, the concept of e-Government is the opportunity, driven by necessity, to bring government products and services to citizens, taking advantage of the evolution of technology and of their own technologies to shorten the distance between them and the government and try to return some transparency to some of these services provided by the state. It can be compared to electronic commerce in the way that it is intended to automate state service processes, making them available in any device, anywhere, twenty-four hours a day, and every day. The main purpose of this concept is to make the work of public institutions significantly more effective and efficient, increasing, as mentioned above, transparency, but also the capacity to respond and manage resources of each of these institutions. However, the growth of this trend is largely conditioned by the decision of those responsible for public institutions. On the other hand, we need to take into account the acceptance of "the customers", the citizens. It is not new that the human being is unenthusiastic to change [3]. The challenge is then to propose and develop a platform that will allow public institutions of cities to create events or challenges for their citizens to participate and have fun while taking care of their own city. It is intended that a City Hall, for instance, will be able to get closer to all its citizens, including those who live on the outskirts of the city and that, in the case of users, they feel important, heard and involved in the matters that concern their city functioning. With this, these institutions gain the possibility of studying the support that their citizens may give to certain initiatives, such as the construction of something new or an alteration of something existing, to perceive which paths can be traversed in the direction of the future as a species of study of daily and constant market, allied to a tool of promotion not only of the pre-populated city and its values as well as its commercial, cultural, artistic and leisure spaces.

2   Background Issues

To accomplish what we propose to do in this work, several technologies and concepts need to be addressed in order to establish a solid background that will allow us to propose and validate this new approach. All these steps will be briefly mentioned next.

2.1   Gamification

Gamification is a relatively informal term to describe the use of game elements, such as rewards, classifications and missions, in systems or environments without any relation to games, with the aim of improving the user experience and also their degree of involvement and motivation [4]. Despite the implementation of new games and with new ideas appearing every day, the concept itself is more than a hundred years old. Among the many companies that use this technique, there were a number of companies that created playability platforms such as Bunchball [5], which in 2007 was the first to offer game engines or components and Badgeville a company which started in 2010. Large names in the market, Microsoft, IBM and Deloitte, among others, began to apply gamification to various applications and processes. Massachusetts Institute of Technology has a program called Education Arcade whose aim is to explore the natural interest of the people in the games and transform the process of education in a way that motivates students [6]. In [7] some behaviors that are associated with gaming, are found to be also necessary at school for instance. Many advantages are recognized to gaming, as in [8–10]. All these being said, gamification arises as a promising field to be used in several contexts, particularly in this work associated with the concept of Smart Cities.

2.2 Smart Cities

More than half of the World's population now lives in urban areas. This shift to a primarily urban population is expected to continue for the next couple of decades. Many new issues regarding how we live in those cities and how they are organized are then in order. In the last two decades, the concept of "smart city" has become more and more popular in scientific literature and international policies. Although there is an increase in frequency of use of the phrase "smart city", there is still not a clear and unfailing understanding of the concept among specialists and academia. Only a limited number of studies investigated and began to systematically consider questions related to this new urban phenomenon of smart cities. The concept of a smart city is still emerging, defining and conceptualizing it is still in progress [11, 12]. In [13] several smart cities definitions are enumerated. As a good starting point we can use the following statement: "The buzz concept of being clever, smart, skillful, creative, networked, connected, and competitive becomes a key ingredient of knowledge-based urban development and hence of a smart city [14].

2.3 Used Technologies

Nowadays, a huge variety of technologies are available, that enable us to develop our intend work. The choice or preference of one to another, must be driven by many factors, such as the specific goals of the work, development timings and availability for instance. For the web server, Flask (a python microframework) was chosen along with a PostgreSQL database, after a carefully study of the ones available, because it complies with all the conditions necessary for the development of the project. In particular, it was chosen the PostgreSQL version 9.5 that allows to have JSON type fields to store dynamic structures. For the mobile application, after evaluating several alternatives Cordova [15] was chosen as the development framework. With Cordova we wore able to develop the entire application using web technologies (HTML, CSS and JavaScript) and use the same code for multiple platforms such as Android, iOS and Windows Mobile. Cordova offers a minimum support for the device's sensors, especially the GPS which is a huge part of the application.

3   The Proposed Platform

As stated before, the main goal of this work is to get residents engaged in tasks related to their city, throughout the use of gamification techniques, allowing participants to explore and knowing better their city, as well as having goals to achieve and contending to win prizes.

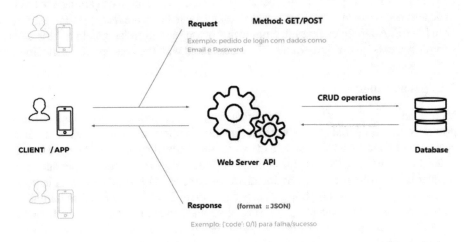

Fig. 1. Platform architecture

The proposed platform will revolve around two concepts: competition and reward. A virtual world was conceived where society is governed by a council of tribes who, together, democratically dictate the course of the city. This council will meet once in a while and the level of influence each tribe has on the decisions taken will vary with the support it receives from the population each month. Each of these tribes represents a different philosophy of the city. Although the tribes are not necessarily rivals, they will compete for attention, respect and dedication of the citizens. This project requires the establishment of communications between a mobile application and a central web server that is the provider of all the necessary information for the correct operation of the application. This client-server communication is carried out through HTTP requests, the application trigger actions that, in turn, get a response by the server. The platform architecture is detailed in Fig. 1. Users are encouraged to report anomalous situations they encounter in their city using the mobile app (Fig. 2).

Also, they are challenged to attend to events and fulfill challenges, placed in the platform by the systems administrator (Fig. 3), that intent to motivate users to participate in city events.

Fig. 2. Reporting anomalous situations

Fig. 3. New event

Throughout this process, users can get rewards that they can trade in the application's store for the available products, and can with their vote, rate events and participate in city hall decisions.

4   Results

This platform was tested in a controlled environment, using volunteers that installed the mobile application in their smartphones. For a relative short period of time, they wore encouraged to participate in the study, attending to "virtual events", reporting "virtual anomalous situations" and participating in "virtual challenges". The collected data of these interactions showed very promising results and are still being processed.

5   Conclusions and Future Work

With this work, we intended to provide a platform to allow citizens to interact with public services. To do so, and to motivate users, gamification was used. The application components, the administration area that controls it and the web server that supports and feeds both, are implemented in a sufficiently modular way to allow the system to be applied to any city in the world. It is only necessary to fill in initial information that one wants to use to fully set up information for the desired city. The platform makes it possible for the entity responsible for the system to retrieve information on the trends, complaints and wishes of the population. Speaking of interest and utility, the idea presented with this work is meant to encourage citizens to leave their homes and to get to know their city better, building a bridge between the fastidious process of interacting with public institutions and the playful aspect of participating in events and having fun, with one goal in mind: the rewards. It is intended to apply these concepts and this platform in a real situation, in order to validate all the assumptions that wore made with the preliminary tests, which showed encouraging results.

Acknowledgements. This work has been supported by COMPETE: POCI-01-0145-FEDER-007043 and FCT – Fundação para a Ciência e Tecnologia within the Project Scope: UID/CEC/00319/2013.

References

1. Sharma, G., Kharel, P.: E-participation concept and Web 2.0 in E-government. Gen. Sci. Res. **3**(1), 1–4 (2015)
2. Gil-García, J.R., Pardo, T.A.: E-government success factors: mapping practical tools to theoretical foundations. Gov. Inf. Q. **22**(2), 187–216 (2005)
3. Kanter, R.M.: Ten reasons people resist change. Harv. Bus. Rev. (2012). https://hbr.org/2012/09/ten-reasons-people-resist-chang
4. Deterding, S., Sicart, M., Nacke, L., O'Hara, K., Dixon, D.: Gamification using game-design elements in non-gaming contexts. In: CHIEA 2011/CHI 2011 Extended Abstracts on Human Factors in Computing Systems, pp. 2425–2428 (2011)

5. Bunchball, The engagement and performance solution, powered by gamification. http://www.bunchball.com/
6. https://education.mit.edu/research/
7. de Freitas, S., Maharg, P.: Digital Games and Learning, pp. 152–171 (2011)
8. Granic, I., Lobel, A., Engels, R.C.: The benefits of playing video games. Am. Psychol. 69(1), 66 (2014)
9. Virvou, M., Katsionis, G., Manos, K.: Combining software games with education: evaluation of its educational effectiveness. J. Educ. Technol. Soc. 8(2), 54–65 (2005)
10. Prensky, M.: Don't Bother Me, Mom, I'm Learning!: How Computer and Video Games are Preparing Your Kids for 21st Century Success and How You Can Help!. Paragon House, St. Paul (2006)
11. Boulton, A., Brunn, S.D., Devriendt, L.: 18 cyberinfrastructures and 'smart' world cities: physical, human and soft infrastructures. Int. Handb. Glob. World Cities, 198 (2011)
12. Hollands, R.G.: Will the real smart city please stand up? Intelligent, progressive or entrepreneurial? City 12(3), 303–320 (2008)
13. Albino, V., Berardi, U., Dangelico, R.: Smart cities: definitions, dimensions, performance, and initiatives. J. Urban Technol. 22(1), 3–21 (2015)
14. Dirks, S., Gurdgiev, C., Keeling, M.: Smarter Cities for Smarter Growth: How Cities Can Optimize Their Systems for the Talent-Based Economy. IBM Global Business Services, Somers (2010)
15. https://cordova.apache.org/

Monitoring Mental Stress Through Mouse Behaviour and Decision-Making Patterns

Filipe Gonçalves[1(✉)], Davide Carneiro[1,2], José Pêgo[3,4], and Paulo Novais[1]

[1] Algoritmi Research Centre/Department of Informatics,
University of Minho, Braga, Portugal
fgoncalves@algoritmi.uminho.pt, {dcarneiro,pjon}@di.uminho.pt
[2] CIICESI, ESTG, Polytechnic Institute of Porto, Felgueiras, Portugal
[3] School of Health Sciences, Life and Health Sciences Research Institute (ICVS),
University of Minho, Braga, Portugal
jmpego@ecsaude.uminho.pt
[4] ICVS/3B's - PT Government Associate Laboratory, Braga/Guimarães, Portugal

Abstract. More and more technological advances offer new paradigms for training, allowing novel forms of teaching and learning to be devised. A widely accepted prediction is that computing will move to the background, weaving itself into the fabric of our everyday living spaces and projecting the human user into the foreground. This forecast turns out to be an opportunity for human-computer interaction as a way to monitor and assess the user's stress levels during high-risk tasks. The main effects of stress are increased physiological arousal, somatic complaints, mood disturbances (anxiety, fear and anger) and diminished quality of working life (e.g. reduced job satisfaction). To mitigate these problems, it is necessary to detect stressful users and apply coping measures to manage stress. Human-computer interaction could be improved by having machines naturally monitor their users' stress, in a non-invasive and non-intrusive way. This article discusses the development of a random forest classifier with the goal of enabling the assessment of high school students' stress during academic exams, through the analysis of mouse behaviour and decision-making patterns.

Keywords: Stress monitoring · Human-computer interaction
Performance assessment · Machine learning

1 Introduction

Stress is nowadays a well-known and studied phenomenon, especially due to its significant influence in many aspects of our existence, including our health, well-being, physiological state, social relationships, among many others. It could be stated that stress (or the lack of it) influences virtually all our decisions and consequently our existence. One of the most interesting aspects of stress is that its effects are two-fold in that they can be positive or negative, depending on factors such as intensity, context, duration or consequences [6]. For instance,

P. Novais et al. (Eds.): ISAmI 2018, AISC 806, pp. 40–47, 2019.
https://doi.org/10.1007/978-3-030-01746-0_5

a certain amount of stress in the participants of a brainstorming meeting is desirable in that it will promote activity and engagement levels and promote the generation of ideas. Such meetings are generally short thus an increased level of stress will not have negative effects on the individuals' health. However, if this increased level of stress becomes regular in the individuals' work routine, negative health consequences (and others) will start to emerge.

Key challenges in this context are understanding how each individual is affected by stress, how each individual copes or the magnitude and valence of the effects of stressors. In this work we contribute to the challenge of understanding the effects of stress on an individual. In the past, we have studied this phenomenon based on behavioural features extracted from the individual's interaction with the computer, following the idea that stressed individuals interact differently with the computer when under stress and that these differences are consistent [3].

In this paper we improve this prediction model with information regarding two new aspects: (1) several individual decision-making metrics and (2) the individual's stress profile. Decision-making metrics are extracted from the user's interaction with the computer, as happens with the behavioural features. The individual stress profile, on the other hand, is based on the students' answer to the Perceived Stress Scale (PSS) questionnaire [1]. The PSS is a psychological instrument for measuring the perception of stress. It is a measure of the degree to which situations in one's life are appraised as stressful. Thus, this new model incorporates the behaviour of the user, the decision-making patterns and the stress profile. To validate this work, the proposed system was used in the School of Medicine of the University of Minho, Portugal. Specifically, we collected data from 101 students participating in computer-based exams, which are particularly stressful moments in a student's life, especially in the field of Medicine. The developed model can be used to predict the user's experienced stress level during demanding tasks, through the analysis of human-computer interaction, decision-making patterns and past PSS scores.

This paper is organised as follows: Sect. 2 defines the individual's calibration processes required for the classifier to better predict the user's stress; Behavioural biometrics features and decision-making patterns analysed and applied are described in Sect. 3; Sect. 4 describes the set of steps/precautions taken for obtaining the input from the selected population, taking into account their environment and routines; Sect. 5 presents an analysis of the trained random forest classifier to predict the PSS score value based on the selected features recorded during an exam; finally Sect. 6 presents conclusions about the work developed so far and future work considerations.

2   User Profile

In the present, there are three popular instruments for measuring perceived stress: the Stress Appraisal Measure (SAM), the Impact of Event Scale (IES) and the Perceived Stress Scale (PSS) [1,4,9]. Among these, PSS is the most widely used, namely in studies assessing the stressfulness of events, physical and psychiatric diseases and stress management programs. It is a measure of the degree to which situations in one's life are appraised as stressful. Items were designed to tap into how unpredictable, uncontrollable, and overloaded respondents find their lives. The scale also includes a number of direct queries about current levels of experienced stress.

The PSS was designed for use in community samples with at least a junior high school education. The items are easy to understand, and the response alternatives are simple to grasp. Higher PSS scores are associated with greater vulnerability to stressful life-events. For this research, the PSS model used was the PSS modified 13-item, described in [8], applied to a group of high school medical students of the University of Minho. Medical education is perceived as being stressful, where studies show that relatively high mean levels of distress are experienced by both male and female medical students [2,7]. The most common sources of stress are related to academic and psychosocial concerns, such as "high parental expectations", "frequency of examinations", "vastness of academic curriculum", "sleeping difficulties", "worrying about the future", "loneliness", "becoming a doctor", "performance in periodic examinations" are the most frequently and severely occurring sources of stress [10]. Thus there is the need for improving these future professional's performance while encouraging psychological well-being and stress management programs for these individuals, which spend a notable part of their life in stressful environments. Stress management skills and supportive environments alongside mental health promotion strategies are necessary and useful to develop community health through school health programs.

In order for the system to correctly predict the stress levels of an individual, a calibrator capable of supporting the classifier was necessary. Indeed, each student has a different stress management/coping profile, which leads to drastic inter-student variations in the PSS score. As such, an individual analysis is carried out for each individual, in order to determine their profile, by reviewing their past PSS scores (designated as PSS profile). In the 13-Item PSS, individual scores can range from 0 to 52 with higher scores indicating higher perceived stress:

- Scores ranging from 0–17 are considered low stress;
- Scores ranging from 18–34 are considered moderate stress;
- Scores ranging from 35–52 are considered high perceived stress.

Stress student profile calibration is done by processing three different features extracted from the individual's past PSS responses: PSS Maximum Score, Minimum Score and Average Score. These features are calculated for each student, by analysing their past PSS results. In the current case study, data from a total of 1143 PSS questionnaires was used, in which each of the students responded to 1 to 5 questionnaires over different moments in time. Students with less then 3 questionnaires were removed, since these cases would present a low amount of information to build an individual PSS profile. This resulted in data from 101 students. All questionnaires were applied before an exam, on the same day, as a way to assess the stress perception of each student in periods in which stress levels are higher.

3   Features Selection

The architecture of the system developed by our team is detailed in [3], and is composed by two biometric behaviour analysis modules: one which focuses in the study of mouse behaviour and another in the decision-making patterns.

One of the main aims of this research group was to build a non-invasive and non-intrusive system capable of monitoring and assessing the stress levels of an individual through the analysis of these biometric features. Unlike many types of physical biometrics, behavioural biometrics can often be gathered with existing hardware, needing only software for analysis. That capacity makes behavioural biometrics simpler and less costly to implement, presenting the opportunity to develop new and innovative technologies which can be used to improve the well-being of the users. In this study, we will only review and detail the attributes selected and used in the created classifier.

The listed features (plus the features defined in Sect. 2) were chosen to be later applied in the classifier model (given the complexity of the Distance of the Mouse to the Straight Line and Excess of Distance Between Clicks mouse features, Fig. 1 demonstrates a graphical example with a small explanation):

- *Exam Duration (ED):* This feature presents the duration between the moment the individual logged in to start the exam until its completion, represented by the logout event (unit in milliseconds);
- *Exam Total Questions (ETQ):* The feature quantifies the total number of questions of the exam (unit in integer);
- *Number of Decisions Made (NDM):* This feature quantifies the number of answers selected by each individual during an exam (unit in integer);
- *Time Double Click (TDC):* The time span between two consecutive mouse clicks events when its latency is smaller than 200 ms, i.e. the duration of a double click. Similarly to other features, a shorter double click time represents increased interaction performance (unit in milliseconds);
- *Distance of the Mouse to the Straight Line (DMSL):* This feature quantifies the sum of the successive distances of the mouse to the straight line defined by two consecutive clicks (unit in pixels). More specifically, it measures the distance between all the points of the path travelled by the mouse, and the closest point in a straight line (that represents the shortest path) between the coordinates of the two clicks;
- *Excess of Distance Between Clicks (EDBQ):* This feature measures the excess of distance that the mouse travelled between each two consecutive mouse clicks (unit in pixels);

The selection of the features was carried out based on the Recursive Feature Elimination method, where a random forest algorithm is used on each iteration to evaluate the model. The algorithm is configured to explore all possible subsets

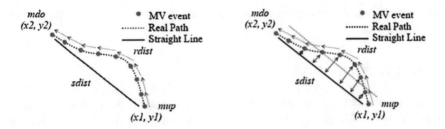

Fig. 1. Graphical visualisation of some of the behavioural features used.

of the attributes and identify those attributes that are and aren't required to build an accurate model to predict the PSS score of a student in an exam. Figure 2 shows the Root Mean Square Error (RMSE) variation when exploring the different possible subset combinations, in a total of 33 features. This value is the lowest when the number of features is 9 (with a RMSE value of 4.5). For this selection, 10-fold cross-validation was used to validate the results.

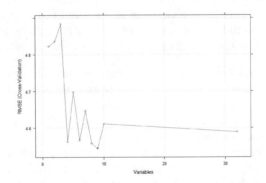

Fig. 2. Plot of RMSE variation results from the Recursive Feature Elimination algorithm, where the RMSE value is lowest for 9 features. For this feature selection analysis, mouse behaviour features, decision-making patterns and PSS profile metrics were taken into account.

4   Study Design

To determine the levels of perceived stress of a group of individuals, based on their mouse behaviour performance and decision making behaviour performance, data was collected from the participation of a group of medical students in high-stakes computer-based exams. In these exams, students are indicated to their seats, and at the designated time they log in the exam platform using their personal credentials and the exam begins. The participation in the data-collection process does not imply any change in the student's routine, and all monitored metrics are calculated through background processes, making the collection data process completely transparent from the student's point of view, just like a normal routine exam. These exams consists mostly of single-best-answer multiple choice questions, where the students only use the mouse as an interaction means. When the exams end, students are allowed to leave the room.

A PSS questionnaire is applied every month before the exam in which the students indicate their perception of stress, allowing the analysis of the evolution of each student's stress levels during their academic course.

All processes of analysis and acquisition of data were properly explained and authorised by the medical students voluntarily, where the individual's identification remains confidential (an unique id key is assigned to each student). Through the use of the id key, it is possible for the research group to aggregate different features with relevant data (e.g. biometric behavioural features, salivary cortisol, PSS scores, etc.). The case study considers a group of 101 medical students of the 1st, 2nd and 3rd years, which are monitored to study the effect of stress/anxiety in the performance of high demand tasks, such as the execution of exams.

5   Results

After the calibration, dataset preparation and feature selection process, the next step was to train a classifier to study the possibility of predicting the PSS score of students during the exam, based on the student's interaction patterns.

Random forest predictors were used to this end, with 500 trees and a leaf size of 7. Random forest predictors use an ensemble of decision trees to predict the intended value. Each decision tree is trained on a random subset of the training set and only uses a random subset of the features. The holdout method [5] was employed to build the training and testing samples. This method suggests that the available data should be randomly split into two disjoint subsets for a single train-test experiment. One group is used for the task of training while the other is held for the task of testing. In this specific work, the training set holds 70% (100 cases) of the data while the test set holds 30% (43 cases) of the remaining cases.

Initially, all possible 9 predictors which could be found at each split with the best performance in the random forest model were used. Based on these models, the number of Error of all Trees fitted and the number of Mean Squared Test Error were calculated and compared for each random number of possible predictors. Figure 3(a) shows the results from this analysis, where it was concluded that the optimal number of features is 7. Moreover, the error was lowest when the number of decision trees was around 500, as shown in Fig. 3(b).

Given these results, we proceed to the optimised classifier model analysis. Figure 4(a) depicts a plot of the actual values versus the predicted ones, for the test set. The correlation value between the actual and the predicted values presented a result of 0.851. Additionally, Fig. 4(b) shows the importance of each feature in the developed classifier, emphasising the relevance of the features related to the user's past stress profile.

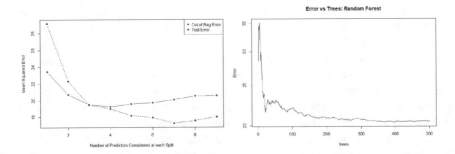

Fig. 3. (a) Out of Bag Sample Errors vs Error on Test set when trying all possible 9 predictors which can be found at each split with 500 trees; (b) Plot of Error vs Number of Trees in Random Forest method.

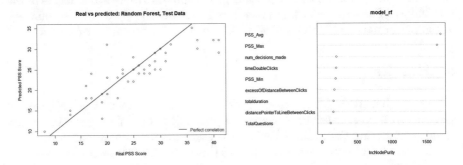

Fig. 4. (a) Real PSS Score vs Predicted PSS Score values. Predicted values obtained through the use of Random Decision Forest Model, presenting a coefficient correlation between predicted and real values of 0.851; (b) Relevance of selected features sorted by descending order.

6   Conclusions and Future Work

This work allows drawing some interesting conclusions about students and their behaviour during exams. First of all, it shows that stress actually influences students' interaction patterns with the computer during high-stake exams. This could open the door to the development of new approaches for assessing and managing stress, especially in the context of superior education. It also shows that it is possible to train classifiers that can carry out this task. Moreover, it is the research group's conviction that this kind of approach can be extended to other domains, namely the workplace. Nonetheless, this calls for the carrying out of new studies since the characteristics of these milieus are imminently different.

There are, however, some issues still to address. The first is to understand why some students improve their general performance with stress while others do not. Knowing which factors influence this might allow higher education institutions to implement individualised coping strategies with the aim to mitigate negative stress effects. It is also our aim to improve the training of the classifiers through the use of clustering techniques, where we will try to identify groups of students that behave alike. The main advantages that we expect from this are: we will be able to train group classifiers that have performances similar to the ones trained for individual students; students who are participating for the first time, for whom there is still not a model trained, can be assigned to a known group with a similar behavioural pattern, thus using the model of that group for classifying his own behaviour; in addition, we intend to arrange an alternative method of automatic calibration without the need to analyse the student's past PSS profile values (e.g. through the differentiation between an user's behaviour profile of the mouse and decision-making patterns with the set of metrics recorded during a present exam).

Acknowledgements. This work is part-funded by ERDF–European Regional Development Fund and by National Funds through the FCT–Portuguese Foundation for Science and Technology within project NORTE-01-0247-FEDER-017832. The work of Filipe Gonçalves is supported by a FCT grant with the reference ICVS-BI-2016-005.

References

1. Cohen, S., Kamarck, T., Mermelstein, R.: A global measure of perceived stress. J. Health Soc. Behav. **24**(4), 385–396 (1983)
2. Dahlin, M., Joneborg, N., Runeson, B.: Stress and depression among medical students: a cross-sectional study. Med. Educ. **39**(6), 594–604 (2005)
3. Gonçalves, F., Carneiro, D., Novais, P., Pêgo, J.: EUStress: a human behaviour analysis system for monitoring and assessing stress during exams, pp. 137–147. Springer, Cham (2018). https://doi.org/10.1007/978-3-319-66379-1_13
4. Horowitz, M., Wilner, N., Alvarez, W.: Impact of event scale: a measure of subjective stress. Psychosom. Med. **41**(3), 209–218 (1979)
5. Kohavi, R., et al.: A study of cross-validation and bootstrap for accuracy estimation and model selection. In: IJCAI, vol. 14, Montreal, Canada, pp. 1137–1145 (1995)
6. Kranner, I., Minibayeva, F.V., Beckett, R.P., Seal, C.E.: What is stress? Concepts, definitions and applications in seed science. New Phytol. **188**(3), 655–673 (2010)
7. Lloyd, C., Gartrell, N.K.: Psychiatric symptoms in medical students. Compr. Psychiatry **25**(6), 552–565 (1984)
8. Pais Ribeiro, J., Marques, T.: A avaliação do stresse: a propósito de um estudo de adaptação da escala de percepção de stresse. Psicol. Saúde Doenças **10**(2), 237–248 (2009)
9. Peacock, E.J., Wong, P.T.P.: The stress appraisal measure (SAM): a multidimensional approach to cognitive appraisal. Stress and Health **6**(3), 227–236 (1990)
10. Shah, M., Hasan, S., Malik, S., Sreeramareddy, C.T.: Perceived stress, sources and severity of stress among medical undergraduates in a Pakistani medical school. BMC Med. Educ. **10**(1), 2 (2010)

Help Cateter: An Application for Mobile Device About Central Venous Catheters Handling

Agnes Peruzzo Innocente[(✉)], Eduardo Pooch, and Sílvio César Cazella

Federal University of Health Sciences of Porto Alegre, Rio Grande do Sul,
Rua Sarmento Leite 245, Porto Alegre, RS 90050-170, Brazil
{agnesp,silvioc}@ufcspa.edu.br, eduardopooch@gmail.com

Abstract. At the same time, it is necessary to constantly improve competences related to nursing care, there is a shortage of tools used for permanent education of the professionals involved in direct assistance of patients with central venous catheters, a prolific field for the development and using of based solutions in mobility. This study presents an application for mobile devices in which the contents can be accessed regarding the correct handling of central venous catheters. The prototype has been evaluated by healthcare workers, presenting promising results regarding the use of mobile applications for permanent education of nurses.

Keywords: Catheters · Applications · e-Learning · e-Health

1 Introduction

Nowadays, the use of mobile devices in work processes is very present, a practice that can be seen even within the health services, whereas this sector, lacks new technological alternatives in the search for improving their practices [5]. The use of mobile devices linked to health care allows its professionals to be more efficient, bringing benefits to the staff, but, mainly, qualifying the care offered to the patients.

As a result of the growing use of mobile devices linked to health care contrasts the tools used for bedside consultations, medical-patient monitoring, health education, and others. Thus, mobile applications play important roles in direct patient care and in clinical decision making, as well as in teaching new professionals, since they can be used for permanent education of nurses.

The using of smartphones and tablets runs very fast, reaching about 57.7 million units in 2014. This fact is due to the constant search for a technological upgrade, access to mobile internet, and by the increasing interest in using services offered through applications [6].

According to [7] telephone companies have sent 344.3 million smartphones for sale worldwide in the first term of 2017, a 3.4% increase over compared to 2016, which shows the continuous increase acquisition of these devices. The most used operating system is still Android, reaching 85% of the smartphones sold.

Mobile phones used by nurses promote the resolution of problems in daily practice, provides a reflexive practice, and teaching situations. These facts are consequences of

P. Novais et al. (Eds.): ISAmI 2018, AISC 806, pp. 48–55, 2019.
https://doi.org/10.1007/978-3-030-01746-0_6

the mobility offered by using these devices and the possibility of access to information at any time and place [11].

Mobile devices play a very important role in permanent education of health professionals, supporting the acquisition of knowledge free of barriers [9]. The possibility of information access, even in distant places where people would not have access to knowledge acquisition easily, is one of the main points that have helped to strengthen the use of mobile applications for the education of health professionals [4].

In health, the number of intravenous solutions administered in hospitalized patients has increased. At the same time, the use of Central Venous Catheters (CVC) has been becoming more common as a fundamental part of a therapeutic plan of these patients. Linked to this, it is necessary to reduce the complications related to the use of these devices, especially the Bloodstream Infections related to the use of CVC (BSI/CVC), responsible for 60% of nosocomial infections, which increase mortality, length of hospital stay, and consequently the costs for health care [1].

Considering the topics described so far, the objective of this article is to report the development and evaluation of a prototype of mobile application on the handling of CVC for permanent education of nurses. This paper is structured as follows: the section two presents the method of the study. The prototype of the developed application is presented in the third section. The evaluation of the prototype is presented in section four. Finally, in section five, the conclusions of the research is described.

2 Applications for m-Learning

The educational process has undergone important changes over the years in search of the incorporation of methods that facilitate the teaching and the assimilation of the contents. In this context, the use of digital technologies [8], specifically mobile applications, was incorporated as a means of providing the user with the control of their own learning, in addition to providing them with the flexibility to search/acquire knowledge [3].

The learning process through the use of mobile devices and the wireless network is known as m-learning or mobile-learning. From the use of this modality of teaching, mobility, convenience, and accessibility are provided to all who seek to acquire knowledge [13].

Specifically, in nursing, we highlight the benefits already mentioned above, the improvement of practices [2], the approximation of theory with reality and also the reflexive process that the use of these tools instigates [10], being this one of the purposes of continuing education in health.

3 Method

This is an applied research, that refers to a research driven by the need for knowledge for immediate application of results. Its approach has been qualitative and quantitative. The research methods which guided this study has been bibliographic research and prototyping. This work has involved the participation of a professional in the area of

Information Technology, who helped with the development of the application prototype software. The construction of the prototype has been based on the care referred to by the participants of this study and has been followed by guidelines issued by the Center for Disease Control and Prevention (CDC) [12] and the Agência Nacional de Vigilância Sanitária [1].

Twenty nurses have performed the evaluation of the application prototype, among them nursing assistants, professors, and residents. This evaluation included the application of two questionnaires, the first one to characterize the participants, and the second, to evaluate the prototype. Qualitative and quantitative analysis techniques were used to analyze the data.

4   Help Cateter Prototype

When a patient carries a CVC there is specific care to keep the functionality of it and complications resulting from its use are avoided. Many healthcare institutions have documents to guide this care, known as Standard Operating Procedures (SOPs) to describe the care to be performed. Most of the time, these documents are available for consulting via local network or they are printed in the sections. This consultation, as reported by the participants, is wasted time, or demands the availability of computers for accessing, which affects the agility and productivity of the nurses. Thus, it has been proposed to develop a prototype of an Android OS app, which was called Help Cateter. This tool was developed based on the CDC Infection Prevention Guidelines [12] and the Anvisa publication [1] about the same topic.

4.1   Help Cateter Features

Help Cateter has been developed based on the care for CVC management issued by CDC and Anvisa, which were selected and organized based on the clinical experience of one of the authors of this study.

The content definition frames CVC management in any setting, either in home care or intensive care units. It was tried to maintain, as faithful as possible, the information collected from the consulted literature.

The application has a main menu with six items, which are: learning about the topic/theme; BSI/CVC; catheter evaluation; other information; educating your team; and Quiz.

In the item "learning about the topic/theme", it has tried to present the main precautions to be taken in CVC handling. It has been subdivided into topics: general recommendations; catheters; stabilization; covering; scrub the hub; flushing catheter; syringes for handling; infusion systems; and heparinization, or salinization.

The second topic approaches the pathophysiology of BSI/CVC, and the problems related to them.

In the section "catheter evaluation", there are the possible characteristics of the CVC regarding the bandage entirety, bandage expiration date, and infusion systems, and the users can identify the best description of the catheter under evaluation.

The fourth section of Help Cateter provides additional information of CVC management, such as: length of stay; infusions and via used to administer solutions; preparation of medicines; phlebitis; blood collections; nursing care; trichotomy; and tube with luer lock with antimicrobial substances.

The fifth section approaches issues related to the education of the health professionals to handle the CVC and the need for guidance and education for the patients and their families to maintain the catheter.

The sixth topic has two tests, from which it is possible to perform a review of all content covered in the application. It proposes objective questions, in which the user checks the answer that he/she considers correct, and at the end of the test, the score is shown on the screen.

The Fig. 1 presents some interfaces of the prototype, they help nurses on looking for CVC handling content. The Fig. 1 shows in (a) the main interface of Help Cateter menu; (b) interface of the menu according to the topic: "learning about the theme - catheters".

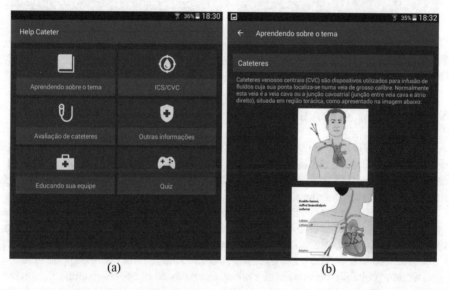

(a)   (b)

Fig. 1. (a) Interface of Help Cateter main menu; (b) Interface of the menu according to the topic: "learning about the theme - catheters".

The Fig. 2 presents in (a) interface of the menu according to the topic "learning about the theme - coverings", (b) questionnaire interface to test the knowledge you have gotten after using the prototype.

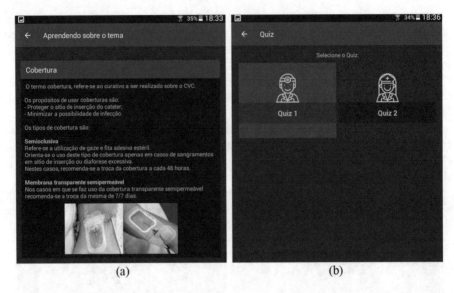

(a)   (b)

Fig. 2. (a) Interface of the menu according to the topic "learning about the theme - coverings"; (b) Interface of the Quiz.

5 Prototype Evaluation

A questionnaire was used evaluate the prototype. The questionnaire comprises: (a) 28 questions, each one could be scored from 5-1, where "5" means "excellent" and "1" means "bad"; and (b) four open questions. The maximum score in the evaluation of each item is 140 and the minimum, 28.

The 28 questions that evaluate the prototype can be divided into two groups. The questions from one to twelve have the objective of ergonomic evaluation of the prototype. The items for evaluation that relate to this criteria are: organization, interfaces, content and technical. The other questions evaluate the usability of the application.

The non-probabilistic sample of evaluators consisted of twenty evaluators, and this sample was composed by nurses, among them: nursing assistants with two or more years of experience in CVC management, professors of the area of health and nursing residents.

At the time of the invitation to participate in this research, the participants answered a characterization questionnaire. The objective was to investigate the participants' familiarity with the use of applications on mobile devices, such as smartphones and tablets. With the answers, it has been concluded that all of them have already used some kind of application on their everyday activities, including to obtain and to disseminate knowledge.

To present the quantitative results, the values equivalent to the averages reached have been used, and the variation from 1 to 5 could be used.

According to the questionnaire of evaluation, the first group of questions, which refers to the ergonomic evaluation of the developed tool, it has scored among "very good" and "excellent". In evaluating the organization of Help Cateter application proto-type, the overall average score was 4,73; the interface question reached an average of 4,45 points; the evaluation of the contents that were in the prototype and its interrelation with the origin area had an average of 4,75; and for the technical part, which refers to the data structure, security, privacy and functioning of the application, the average was 4,6 points.

Any question had scored 1 or 2, which correspond to the concepts of "bad" and "regular" in the evaluation of ergonomics criteria. Likewise, there were no items that have received maximum marks, although most participants rated the application proto-type as "excellent". The feature of "screen appearance" was the only one that presented the highest percentage of evaluations of 4 points (very good), having a total of 45% of the evaluations, the others were divided into 40% with "excellent" and 15% with "good".

On the ergonomic evaluation, there was a significant association between domain intensity of mobile phones and/or tablets with the questions "screen appearance" and "logical structure of the data", and the more technological domain they have, their scores were higher, too (rs = 0,472, p = 0,035 and rs = 0,539, p = 0,014).

The second group of questions, which evaluated the usability of Help Cateter appli-cation, had an average score of 4,57. The item "the user accesses the application easily" was the only one that has received score 2 (regular) by one of the participants. Another evaluated item that was scored with 3 (good) by 4 of the participants was "the application screens are clear, easy to read and interpret", followed by "the menu is visible and easy to use" and "information exchange between user and application".

Questions about the application computational requirements, such as "application with appropriate growth limit", "memory requirements do not prevent it from running" and "required operating system is available", have not been evaluated by three of the evalua-tors (15%), who felt themselves limited in their computational knowledge to evaluate.

The open questions tried to verify the opinion of the participants regarding the use of the prototype as a tool for permanent education, and they unanimously gave positive responses to these questions. Some comments were made according to the use of the prototype in the permanent education of nurses (Table 1).

Table 1. Contributions to the evaluators in relation to Help Cateter for permanent education.

Participant	Contribution
E4	*Excellent initiative to bring knowledge since we have been inserted in a technological environment. Ease and practicality to use in any place. The application came to aggregate, update and improve the professionals' knowledge*
E9	*A very good option for teaching development mainly for many professionals who do not have time to take courses after time and workplace*
E14	*It brings technical relevance, safe and reliable information source. It looks excellent for permanent education or tool for use in the classroom*
E20	*It will have fundamental importance for nurses and other professionals, because it is quick and easy to access and get information, it can be used both as an educational tool for the professional in training and for the ones who are at the bedside and need a quick consultation on the subject. I hope that this application can be applied to various health professionals as an educational tool*

6 Final Remarks

The increasing use of mobile devices makes them possible allies to work in several areas. In the health area, the use of mobile technologies is highlighted by facilitating access to information at any time and place. The Help Cateter application confirms the availability of using mobile applications for permanent education of nurses and brings all the content they need for CVC handling.

All evaluators have agreed the prototype facilitates and expedites the knowledge acquisition. The application or use of Help Cateter can, therefore, be used for permanent education of nurses offering information based on scientific evidence, and according to the results found in this study, the use of Help Cateter may provide the teaching of nursing undergraduates, as well as being used as a support tool in the clinical decision at the bedside, because it provides information about CVC management and brings the steps to be followed in the care of these devices.

After all the research, in particular, the massive acceptance obtained by the prototype, the objective of this work has been reached. In the same way, it has been also perceived that the solution presented here - the application -, is part of a real need for the health professionals.

To conclude the considerations about this research, it is possible to identify gaps for future studies: (a) to expand the target audience, covering all the professionals who handle the CVC, such as the inclusion of licensed practical nurse and surgeons; (b) authenticate the prototype; and (c) create versions for Windows Phone and iOS platforms.

References

1. Anvisa: Agência Nacional de Vigilância Sanitária. Medidas de Prevenção de Infecção Relacionada à Assistência à Saúde. Brasília (2017)
2. Barra, D.C.C., Sasso, G.T.M.D., Almeida, S.R.W.: Usability of computerized nursing process from the ICNP in intensive care units. Rev. Esc. Enferm. USP **49**(2), 0326–0334 (2015)
3. Barros, W.C.T.S.: Aplicativo móvel para aprendizagem da avaliação do nível de consciência em adultos (OMAC). Universidade Federal de Santa Catarina (2015)
4. Campos, K.F.C., Brant, P.B.O., Randow, R., Guerra, V.A.: Educação permanente: avanços, desafios para a gestão em saúde no Brasil. Investigação Qualitativa em Saúde **1**(5), 1276–1285 (2016)
5. Cazella, S.C., Feyh, R., Ben, A.J.: A decision support system for medical mobile devices based on clinical guidelines for tuberculosis. In: Ambient Intelligence – Software and Applications of the 5th International Symposium (2014)
6. Galinari, R., Cervieri Junior, O., Teixeira Junior, J.R., Rawet, E.L.: Comércio eletrônico, tecnologias móveis e mídias sociais no Brasil. BNDES Setorial **41**, 135–180 (2015)
7. IDC: Smartphone OS Market Share, 2017 Q1, According to IDC. Framingham (2017). https://www.idc.com/promo/smartphone-market-share/os
8. Lahti, M., Hätönen, H., Välimäki, M.: Impact of E-learning on nurses' and student nurses knowledge, skills, and satisfaction: a systematic review and meta-analysis. Int. J. Nurs. Stud. **51**, 136–149 (2014)

9. Mercês, J.M.R., Redeiro, M.M.P.: A importância dos dispositivos móveis como estratégia para a formação e desenvolvimento de profissionais de saúde. In: Congresso ABED, Rio de Janeiro (2016)

10. Pereira, F. G. F., Silva, D. V., Sousa, L. M. O., Frota, N. M.: Construção de um aplicativo digital para o ensino de sinais vitais. Rev Gaúcha Enferm. **37** (2016)

11. Pimmer, C., Brysiewicz, P., Linxen, S., Walters, F., Chipps, J., UrsGröhbiel, U.: Informal mobile learning in nurse education and practice in remote areas - a case study from rural South Africa. Nurse Educ. Today **34**(11), 1398–1404 (2014)

12. O'Grady, N.P., Alexander, M., Dellinger, E.P., Gerberding, J.L., Heard, S., Maki, D.G., Masur, H., McCormick, R.D., Mermel, L.A., Pearson, M.L., Raad, I.I., Randolph, A. Weinstein, R.A.: Guidelines for the prevention of intravascular catheter-related infections. Clin. Infect. Dis. (2011). https://www.cdc.gov/hai/pdfs/bsi-guidelines-2011.pdf

13. Saccol, A., Schlemmer, E., Barbosa, J.: M-learning e u-learning: novas perspectivas das aprendizagens móvel e ubíqua, 1st edn. Pearson Prentice Hall, São Paulo (2011)

Distributed Management of Traffic Intersections

Cesar L. Gonzalez[1,4(✉)], Jorge L. Zapotecatl[2,3], J. M. Alberola[1], V. Julian[1], and C. Gershenson[2,3]

[1] D. Sistemas Informáticos y Computación, Universitat Politècnica de València, Valencia, Spain
cegonpin@upvnet.upv.es, {jalberola,vinglada}@dsic.upv.es
[2] D. de Ciencias de la Computación, I. de Investigaciones en Matemáticas Aplicadas y en Sistemas, Universidad Nacional Autónoma de Mexico, Mexico City, Mexico
jzapotecatl@gmail.com, cgg@unam.mx
[3] Centro de Ciencias de la Complejidad, Universidad Nacional Autónoma de Mexico, Mexico City, Mexico
[4] Universidad ECCI, Bogotá, Colombia

Abstract. The development of autonomous vehicles has the potential to considerably improve traffic management on urban zones. The coordination of autonomous vehicles at intersections is a trending problem. In this area of research, several approaches have been proposed using centralized solutions. However, centralized systems for traffic coordination have a limited fault-tolerance and can only be optimal if these systems never fail. This paper proposes a distributed coordination management system for intersections of autonomous vehicles through the employment of some established rules to be followed by vehicles. To validate our proposal, we show experiments to compare our approach with centralized approaches. Simulations have been made using a cellular automaton traffic model with different traffic densities.

Keywords: Autonomous intersection management
Vehicle coordination · Self-organization · Multiagent systems

1 Introduction

Due to the increase of the number of vehicles, traffic management in cities has become increasingly complex. This has resulted in more interactions and communications requirements, such as: vehicle to vehicle (V2V), vehicles to infrastructure (V2I), and vehicles to signals and traffic management devices (V2X) [6]. If we consider autonomous vehicles, the number of traffic signals or vehicles that share the road with others is enormous. With all of them, different types of vehicles interactions are needed.

One of the critical points for traffic management is intersection control. Intersections between streets has had different ways of management and control to

© Springer Nature Switzerland AG 2019
P. Novais et al. (Eds.): ISAmI 2018, AISC 806, pp. 56–64, 2019.
https://doi.org/10.1007/978-3-030-01746-0_7

avoid collisions and maximize flow. Thus, each vehicle that finds a conflict in a lane on an intersection has a time and a space of priority to cross, while other vehicles wait for their priority. Current technologies have an important role to play on intersection management and also over general traffic management: e.g., intelligent traffic lights, sensors, wireless communications (V2V, V2I), GPS, automation, etc. [6].

Although technologies can help intersection management, there are still challenges to be addressed to improve vehicle crossing at intersections. These challenges are focused on decreasing the delay time or increasing the flow of vehicles, that will result in lower energy consumption and increased sustainability.

Several approaches have proposed alternatives to solve this problem. In the literature, we can find centralized solutions to the problem of traffic intersection management [1,3,7,10]. In these approaches, there is a unique control system that is the responsible of communicating with all the actors at an intersection (vehicles, signals, traffic lights, etc.). This centralized control system takes the decision about who must cross first and which vehicles must keep waiting. Though existing approaches have reported the benefits of using a centralized traffic management system, problems can arise when conditions in the environment are very changeable and devices fail.

Our approach proposes a locally distributed solution for efficient traffic management of autonomous vehicles at intersections. In this sense, we establish behavior rules to be followed by vehicles. These rules establish negotiations on priority when crossing at intersections. The proposed approach has been tested with different levels of traffic and it has been compared with other centralized solutions. Simulations have been made using a simulator software developed at UNAM [11,12].

The rest of the paper is structured as follows. In Sect. 2, we present the definition of the model for distributed traffic management. In Sect. 3, we compare the performance of our model with other approaches. Finally, in Sect. 4, we summarize the main conclusions of this work and present some future research lines.

2 Distributed Intersection Management (DIM)

In this section, we present the Distributed Intersection Management (DIM) system to provide autonomous vehicles with the capacity to negotiate and manage crossings at intersections. This system is aimed at being scalable and flexible as well as achieving similar levels of efficiency than a centralized system. The DIM model is composed by three parts: the traffic flow model, the autonomous vehicle model, and behavioral roles.

2.1 Traffic Flow

The traffic flow model of DIM is based on the LAI [8] model for large traffic networks simulation. LAI is a model for traffic flow that captures the drivers's reactions in a real environment. We use this model to understand the behavior of the traffic while preserving safety on the road.

The LAI model allows us to represent the interactions of the vehicles on a shared lane and direction. This model defines the following three main rules in order to represent the behavior of a vehicle:

- A vehicle a_i can accelerate as long as exists a distance D_{acc} between this vehicle and the vehicle that comes before a_{i+1}.
- A vehicle a_i keeps its velocity as long as exists a distance $D_{keep} < D_{acc}$ between this vehicle and the vehicle that comes before a_{i+1}.
- A vehicle a_i has to decrease its velocity if exists a distance $D_{brake} < D_{keep}$ between this vehicle and the vehicle that comes before a_{i+1}.

The above three rules provide the mechanism to maintain safe distances among the vehicles, guaranteeing safe driving. As long as safe distances exists between a vehicle and its predecessor, collisions will be avoided between these vehicles.

The LAI model defines three equations to calculate safe distances according to the above rules [8]. These equations are incorporated into the DIM model to describe the dynamics of the vehicles on the same trajectory and lane. In addition, we based our distributed model on the Gershenson centralized negotiation model [2,4,5] for the design of our distributed rules for autonomous vehicles.

2.2 Autonomous Vehicles

We assume a group of agents $A = a_0, ..., a_n$ that represent autonomous vehicles moving through the different streets of a city. Each vehicle a_i includes sensors to detect other vehicles that are inside an area. Each vehicle is also provided with a wireless communication system to send messages and request information to other vehicles.

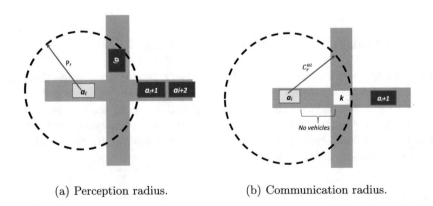

(a) Perception radius.   (b) Communication radius.

Fig. 1. Example of the perception radius and the communication radius.

To represent this, an autonomous vehicle a_i defines two radii: the perception radius and the communication radius. The perception radius P_r defines a detection area inside which, other autonomous vehicles are detected by the sensors of a_i. This radius simulates LIDAR[1] sensors (see Fig. 1(a)).

The communication radius C_r defines a communication area inside which, other autonomous vehicles receive messages sent by a_i. Messages can be delivered to specific receivers or can be broadcasted to any receiver inside this area (see Fig. 1(b)).

2.3 Behavioral Roles

An autonomous vehicle can play two different roles: Follower and Negotiator. The role played by an autonomous vehicle depends of information that receives and the actions the vehicle can take. This is similar to the approach already proposed in the context of automated highway systems in the 1990s [9].

The follower role (represented as F_v) is played by autonomous vehicles that are moving just behind another vehicle. At the beginning of the execution, every autonomous vehicle has associated this role. An autonomous vehicle a_i plays F_v if it detects another vehicle a_{i+1} driving before it, inside the detection area defined by P_r. In this situation, a_i has the goal of keeping its safe distance with a_{i+1} (see Fig. 2(a)).

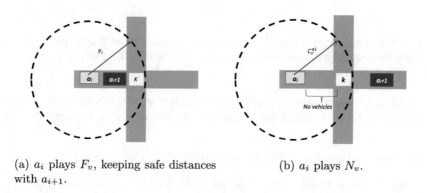

(a) a_i plays F_v, keeping safe distances with a_{i+1}.

(b) a_i plays N_v.

Fig. 2. Examples of the roles played by a vehicle

A vehicle a_i playing F_v is able to detect the distance with respect to the vehicle that comes before a_{i+1}. Taking into account its safe distance, it could decide to increase, to keep or to decrease its velocity according to the above commented LAI model rules.

---

[1] https://news.voyage.auto/an-introduction-to-lidar-the-key-self-driving-car-sensor-a7e405590cff.

The negotiator role (represented as N_v) is played by autonomous vehicles that do not detect other vehicles inside their communication areas and before the next intersection k (see Fig. 2(b)).

When a vehicle a_i starts playing role N_v, this vehicle broadcasts a message with information of its position and velocity with respect to intersection k. If a vehicle a_i playing role N_v intersects its $C_r^{a_i}$ (communication radius of agent a_i) with the $C_r^{a_j}$ (communication radius of agent a_j) of another agent a_j playing role N_v in a conflict way, they must share the information of velocity and positions in order to negotiate who should be the first to cross at the intersection (see Fig. 3).

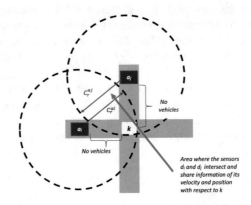

Fig. 3. Vehicle a_i playing role N_v sharing information with vehicle a_j playing role N_v in a conflict way intersecting their communication radius ($C_r^{a_i}$ and $C_r^{a_j}$).

Finally, an agent playing role N_v leaves this role, when it enters in the intersection and shares its messages to the new vehicle with role N_v behind it.

2.4   Negotiation Between Autonomous Vehicles

Each autonomous vehicle must follow a set of rules in order to do the negotiations among the different vehicles trying to cross the intersection. As a consequence, they obtain a priority to cross that they should fulfill contributing cooperatively to achieve the expected behavior of the system.

Due to the limits of this paper only the two most representative rules are described below.

Avoid Intersection Blocked. If an agent a_i playing role N_v goes to an intersection k, and detects first in its $C_r^{a_i}$ or in its P_r another autonomous vehicle a_{i+1} that has crossed the intersection k but still is in a distance e with respect the intersection, the vehicle a_i playing N_v must begin to decrease its speed before

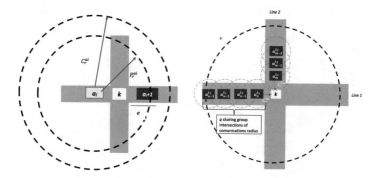

(a) a_i will avoid to cross on intersection k if $a_i + 1$ is in e.

(b) The queue of vehicles in lane 1 has priority to cross over vehicles in lane 2 according to the rule *Avoid intersection blocked*.

Fig. 4. Examples of rules

the intersection. The vehicle a_i will avoid to cross the intersection k until distance e will be free. While distance e isn't free the vehicle a_i playing N_v will come to stop before the intersection (see Fig. 4(a)).

If there exists two conflict lanes L_1 and L_2 with an agent in each line $a_n^{L_1}$ and $a_m^{L_2}$ inside the distance e after the intersection k, then the rule about *avoid intersection blocked* will be executed iteratively until any lane inside distance e will be free. If both lanes will be free at the same time, then the vehicle playing role N_v that has been waiting more time is who reaches the priority to cross.

Reach Priority to Cross. In this approach we give a higher priority in order to cross to convoys or groups of autonomous vehicles that are in the same lane. According to this, an agent $a_n^{L_1}$ playing role N_v in lane 1 reaches priority to cross over the rest of lanes in conflict, if the quantity of autonomous vehicles behind to it (q) (e.g. $a_{n-1}^{L_1}$, $a_{n-2}^{L_1}$, $a_{n-3}^{L_1}$, ...) is the higher respect the rest of the lanes in conflict. To calculate this q we introduce a threshold ϵ which indicates the distance limit of a queue of vehicles in the same lane before an intersection (see Fig. 4(b)).

3 Experiments

In this section, we show the experiments using our DIM model in the simulator developed by Zapotécatl [11], which is based on cellular automata. This tool simulates the dynamic of vehicular traffic in cities composed by streets and intersections. This simulator is developed following the rules of LAIE's[2] model [5, 11, 13]

---

[2] The LAIE's model is an extension of the LAI model, which introduces conflict ways but maintaining the same dynamic model.

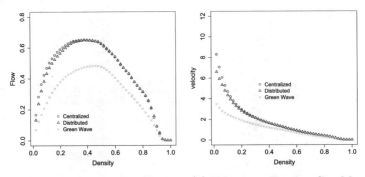

(a) Flow vs Density. City Manhattan style of 100 intersections.   (b) Velocity vs Density. City Manhattan style of 100 intersections.

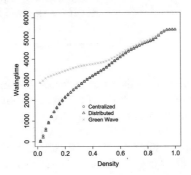

(c) WaitingTime vs Density. City Manhattan style of 100 intersections.

Fig. 5. Results of experimentation

We compare the performance of our DIM model with two other traffic intersection management systems. The first system (Green Wave) is a traditional approach in which traffic lights are the responsible of setting the priority in each intersection. In this approach, the traffic light switches between green and red light every period of time, giving priority to the vehicles located in the line with green light. The second system (Centralized) is the self-organizing proposal developed by [2, 4, 5].

The experiments evaluate the performance of the three systems in a Manhattan-style grid with 100 intersections. We start from a traffic density of 0.02 and we increase this density until reaching 1 (that means a collapse where no vehicle is moving because all spaces are occupied). Each density was repeated 20 times with different initial random positions of vehicles.

Figure 5(a) shows the performance of the three systems in terms of traffic flow, which can be calculated by multiplying velocity by density. As it can be observed, the behavior is similar for low traffic densities. However, as the density increases, the behavior of the Green Wave system is significantly lower than the other two approaches. In this sense, the maximum flow achieved by Green Wave is 0.48 for density 0.5, while the performance of the other two systems achieves a traffic flow of 0.65, maintaining similar values for density ranging from 0.2 to 0.6. As it can be observed, the performance of the DIM system is quite similar to the Centralized approach.

Figure 5(b) shows the performance of the systems in terms of the average velocity reached on intersections per vehicle during the simulation. It can be observed that the performance of the Green Wave is again lower than the other systems for density values from 0.02 to 0.5 (since each vehicle stops while the traffic light is red). For densities greater than 0.5, the performance of the three systems is quite similar. This is caused due to as the density increase, the average velocity tends to decrease until the city is collapsed by vehicles and the velocity reaches 0.

Figure 5(c) shows the performance of the three systems in terms of the average waiting time on intersections during the simulation. Similar to the previous figures, the performance of the Green Wave system is the worst. In this case, the maximum difference appears with the lowest density values. This is because vehicles stop several times during the execution because some traffic lights are red. In contrast, DIM and the Centralized system show very short waiting times with low densities, This is due to the rules for dynamically changing traffic lights in the case Centralized approach and to the reactive negotiation in intersections in the case of the DIM system. For densities greater than 0.5, the performance of the three systems becomes similar since the traffic tends to collapse the city.

4   Conclusions

In this paper, we proposed the DIM model for distributed management of traffic intersections. The performance of DIM is quite similar to other centralized adaptive approaches such as the one proposed Gershenson. At the same time, our proposal outperforms a conventional traffic system control like Green Wave in terms of velocity, waiting time and traffic flow. In our proposal, each autonomous vehicle that reaches an interaction coordinates with the rest of vehicles for crossing safely and efficiently.

The coordination of autonomous vehicles in DIM does not need a central control for management [3]. Therefore, this distributed system is more scalable since there is not any centralized manager that could become a bottleneck. What is more, DIM is much tolerant to changes in the conditions in the environment and device failures.

With regard to adaptive centralized systems like the one proposed by Gershenson, the DIM model requires less hardware and road infrastructure to manage traffic. Due to the roles defined for the vehicles, the negotiation rules are considered to cross intersections in a safe way without obstructing the intersections.

For future work, it is planned to introduce communication failures into the system to measure its robustness and operability performance compared to other systems. Additionally, we plan to include in our model the possibility of considering multiple lanes and directions.

References

1. Bazzan, A.L., Klügl, F.: A review on agent-based technology for traffic and transportation. Knowl. Eng. Rev. **29**(3), 375–403 (2014)
2. Cools, S.-B., Gershenson, C., D'Hooghe, B.: Self-organizing traffic lights: a realistic simulation. In: Advances in Applied Self-organizing Systems, pp. 45–55. Springer (2013)
3. Dresner, K., Stone, P.: A multiagent approach to autonomous intersection management. J. Artif. Intell. Res. **31**, 591–656 (2008)
4. Gershenson, C.: Self-organizing traffic lights. arXiv preprint arXiv:nlin/0411066 (2004)
5. Gershenson, C., Rosenblueth, D.A.: Self-organizing traffic lights at multiple-street intersections. Complexity **17**(4), 23–39 (2012)
6. Gregor, D., Toral, S., Ariza, T., Barrero, F., Gregor, R., Rodas, J., Arzamendia, M.: A methodology for structured ontology construction applied to intelligent transportation systems. Comput. Stand. Interfaces **47**, 108–119 (2016)
7. Guo, D., Li, Z., Song, J., Zhang, Y.: A study on the framework of urban traffic control system. In: Proceedings of the IEEE Conference on Intelligent Transportation Systems, vol. 1, pp. 842–846. IEEE (2003)
8. Lárraga, M., Alvarez-Icaza, L.: Cellular automaton model for traffic flow based on safe driving policies and human reactions. Physica A Stat. Mech. Appl. **389**(23), 5425–5438 (2010)
9. Li, P., Alvarez, L., Horowitz, R.: Ahs safe control laws for platoon leaders. IEEE Trans. Control Syst. Technol. **5**(6), 614–628 (1997)
10. Wu, J., Abbas-Turki, A., El Moudni, A.: Cooperative driving: an ant colony system for autonomous intersection management. Appl. Intell. **37**(2), 207–222 (2012)
11. Zapotecatl, J.L.: Qttrafficlights (2014). https://github.com/Zapotecatl/Traffic-Light
12. Zapotecatl, J.L., Rosenblueth, D.A., Gershenson, C.: Deliberative self-organizing traffic lights with elementary cellular automata. In: Complexity 2017 (2017)
13. Zubillaga, D., Cruz, G., Aguilar, L.D., Zapotécatl, J., Fernández, N., Aguilar, J., Rosenblueth, D.A., Gershenson, C.: Measuring the complexity of self-organizing traffic lights. Entropy **16**(5), 2384–2407 (2014)

Public Building Energy Efficiency - An IoT Approach

Catarina Santos, Joao C. Ferreira$^{(\boxtimes)}$, Vasco Rato, and Ricardo Resende

Instituto Universitário de Lisboa (ISCTE-IUL), ISTAR-IUL, Lisbon, Portugal
{cmnss,jcafa,vnpmr,jrpre}@iscte-iul.pt

Abstract. Buildings play an important role in energy consumption, mainly in the operation phase. Current development on IoT allows implementing sustainable actions in building towards savings, identify consumption patterns and relate consumption with space usage. Comfort parameters can be defined, and a set of services can be implemented toward the goals of saving energy and water. This approach can be replicated in most buildings and considerable savings can be achieved thus contributing to a more sustainable world without negative impact on building users' comfort.

Keywords: IoT · Sensors · Sustainability · Building efficiency · Energy

1 Introduction

Electric power grids in Europe - and worldwide - are gaining intelligence and becoming "smart grids". The increase electricity consumption in developed countries, caused by a larger number of more powerful and diversified power-connected devices, creates consumption peaks, which lead to the need of integrating new ways to produce, distribute and consume energy with more efficiency. Also considering a constant increase in fuel prices, threats of global warming, implications of carbon and other emissions from traditional fuels, there is a growing interest in improving energy efficiency. One of the most important elements in ensuring energy efficiency is energy management and monitoring. Energy monitoring is an energy efficiency technique based on the standard management axiom stating that "you cannot improve what you cannot measure". It implies the necessity of measurements and data organisation [1]. But measuring is just the first part of the journey. There is also a need to transform collected data into correlated and usable information using a sustainable, well designed, and upgradable energy efficiency monitoring system [2].

Effective energy management requires chronological knowledge of both the relevant energy uses and the main influencing factors such as operational requirements (e.g. production data) and environmental data (e.g. external temperature, humidity, etc.). This activity concerns all types of energy (electricity, gas, steam, chilled water, compressed air, etc.) [2]. Some important questions are to determine which parameters should be monitored, define the optimal number and position of meters, choose the suitable frequency of collecting data (annually, monthly, daily, hourly or less). It is essential to

© Springer Nature Switzerland AG 2019
P. Novais et al. (Eds.): ISAmI 2018, AISC 806, pp. 65–72, 2019.
https://doi.org/10.1007/978-3-030-01746-0_8

identify main, independent, factors to reduce the number of monitored parameters. Creating a suitable database is essential to analyse energy use of buildings properly [1].

Sustainability initiatives at university level falls into three categories:

(1) research-based sustainability - there is a proliferation of masters and doctorate's courses adopting the environmental angle on traditional disciplines, from environmental economics to climate modelling;
(2) operational-based sustainability on the university itself. The focus is the reduction of deleterious environmental effects, cutting carbon and energy bills. Less common, but still important, is the role universities have in contributing to their local environment - socially, culturally, economically and ecologically;
(3) "Universities of Sustainability", where the focus is on the education of environmentally and socially responsible citizens, on improving course curricula to ensure that courses include useful contents develop skills for a world altered by climate change and post-peak oil [3].

These levels are not independent. For example, sustainability research (level 1) will be converted into content for sustainability courses (level 3). Campus energy or water saving efforts (level 2) must involve the population thus educating them (level 3).

Several universities around the world are working on making their campuses sustainable, and one of the aspects is smart energy management - which is the main focus of this work, therefore aligned with the second level. Energy waste in various space types, such as teaching auditoriums, working areas (offices, laboratories, computer rooms, etc.) or residential buildings (dormitories) can be found [4]. The energy and environmental impact of universities could be considerably reduced by applying organisational, technological and energy optimisation measures [5, 6]. Actions can be taken and aimed at improving the production, distribution and consumption of energy within the campus, to increase buildings energy performance, improve energy management and educate people about efficient and sustainable energy use [7].

2   Energy Efficiency Project

ISCTE-IUL has a global community of ca. 10000 people, of which 9234 students from undergraduate, master, PhD and postgraduate programs. In 2017 it had a budget of 38.5 M€ of which 54% were self-generated. Four main buildings compose the campus: (1) Building Sedas Nunes (also known as Building I); (2) Building II; (3) Ala Autónoma; and (4) Building INDEG. These buildings are 20 to 40 years old and have a total gross built area of 48,500 m². ISCTE-IUL also has a multi-sports field, a parking lot and an off-campus student residence. In 2017, ISCTE-IUL started a Strategic Program on Sustainability. A formal sustainability specific organizational structure is managed by the Director of Sustainability to implement several projects, namely: (1) Campus Operations, such as water, energy and waste management; (2) Core Activities, like research and education; and (3) Outreach to the community, meaning in this context, activities to connect to society and share knowledge and expertise.

The objective of the university is to become the most sustainable university in Portugal. Under the Sustainability Program, the work described in this paper is focused on the Energy Efficiency Project which includes four topics: (1) Replacement or improvements of the HVAC systems; (2) Upgrade of electric lighting; (3) Installation of photovoltaic panels; and (4) Refurbishment of the Sedas Nunes building's roof to improve thermal insulation. The university believes there will be an average saving of one third on the energy consumption and CO_2 emissions. To accomplish these, there is a need to study energy efficiency and monitor energy consumption – the main focus of this work. What makes IoT interesting is the ability to save ISCTE-IUL a considerable amount of financial resources by optimizing processes. This is possible through the installation of sensors and the respective data analysis which, in turn, allows to take decisions on building operations and influence user behaviour.

3   Related Work

There is a considerable number of theses and research on IoT related to energy efficiency. In Portugal, there are already several smart grids and consumption control pilot-projects, such as, (1) in 2010 Galp company started the development of an energy management system pilot – SmartGalp – which monitors energy consumption through a platform that interacts with domestic users, from electricity and natural gas, to fuel. The installation of devices in houses or cars of the end users allows to follow the effective consumption and establish reduction goals to save on the energy bills. Through the monitoring of results, the company verified that this tool allows for effective savings, being able to reduce up to 8% in the energy consumption [8]. EDP electricity company launched in 2011 the InovCity Project, a program where it is possible to be energetically efficient [9]. Within the scope of this project, Évora became the first Iberian metropolis to test a new way of thinking about electricity production and distribution. The first stage focused on the automation of electric grid management to reduce operating costs. With the smart grid, any citizen can know in real time its energy consumption [10]. The project had a very positive impact regarding energy efficiency since 60% savings in electricity costs were attained with the implementation of LED and AI technology. Parque Escolar, a public company in charge of modernising Portuguese public school buildings, implemented a system that allows it to track and control energy consumption on all of installed equipment. This technology is already implemented in several schools' buildings in Lisbon controlling air conditioning, lighting and even IT devices. The pilot-project reduced the use of energy consumed in IT in 25% - including computers, IP phones, wireless access points or video cameras. This system is complemented with an easy to read information presentation that has become a teaching tool in schools encouraging "green" individual behaviour [11]. This is an example of what ISCTE-IUL university may achieve through integrating institutional strategic goals with researchers and students' cooperative work. By 2020, the prediction is that IoT will be a trillion-dollar industry in selling solutions [12]. Specifically dedicated to smart campus, there are already several companies creating custom-made solutions for the university campus market, such as Huawei [13] and Cisco [14].

Many universities throughout the world have set in motion projects aimed at achieving a smart campus. Most also created labs to work specifically in smart environments. An example is the European Commission-funded project aiming at the development of services and applications supported by a data gathering platform that integrates real-time information systems and intelligent energy management systems that drive a bi-directional learning process. The user learns how to interact with the building, and the building learns how to interact with the user in a more energy efficient way [15]. This was applied to chosen universities in Lisbon, Helsinki, Luleå and Milan. This project reached 30% in energy savings through use of ICT, Living Lab methodologies and gamification to promote user behaviour transformation on public building users [16].

4   Proposal

We have developed a Sensor Network to create smart environments: temperature sensors control the classroom temperature, and BLE beacons track user movement and emit sustainability-related information.

Figure 1 shows our vision for the problem. The sensors installed on the campus provide data to a central cloud server, where information is manipulated towards the desirable goal. A service-based approach is used to provide flexibility and allow the reuse of algorithms towards knowledge extraction. The proposed architecture is composed of four layers: (1) **Data layer**, which comprises data collection from installed sensors; (2) **Information layer,** where data is manipulated towards achieving desirable information, based on data mining algorithms (out of the scope of this paper); (3) **Knowledge layer**, where this information can be used for campus management, and specific functional roles act on infrastructure and systems to optimise operating conditions; and (4) **Services layer,** which feeds main applications in a service-based approach,

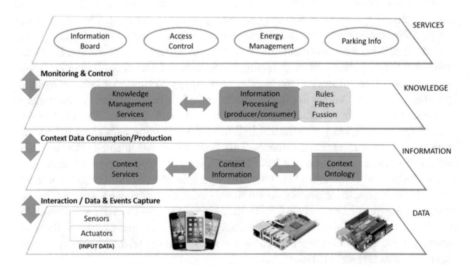

Fig. 1. A general overview of the proposed architecture of our IoT smart campus system.

where information can be incorporated in the related service. For example, the info about the number of empty spaces at the parking facilities can be used to increase the number of persons using them.

From the gathered (big) data, patterns can be extracted and analysed. It is thus possible to make predictions about the physical or social phenomena being observed. The task of identifying patterns from big data is related to the application domain and oriented to a specific usage. In a university, the information to be collected and analysed has the main objective of allowing an improved operations management that leads to savings and therefore to more a sustainable performance. Nonetheless, in this type of institution, the ability to have data and detect patterns should also be related to research and teaching goals. In fact, these big data sets are also a major opportunity in the search for models relating several levels of information: external environmental conditions (temperature, relative humidity, solar radiation, wind speed, noise pollution, and air quality); internal environmental conditions (air temperature, radiant temperature, relative humidity, air displacement, noise level and air quality); time, date and location-related occupancy patterns and rate; resource consumption and waste and emissions generation.

4.1 Data Layer

This layer is mainly composed of a sensor array on a Lora communication network linked to a cloud IBM server, Bluemix. The following sensors are installed: (1) Electricity measurements - Current sensor: YHDC SCT013-000, current transformer, 100A: 50mA and a Receiver - Raspberry Pi 3 Model B + LoRa Module; (2) Temperature and relative humidity measurements based on Texas instruments CC2650STK; and (3) Beacons – Bluetooth Beacon from Estimote which emit data through Bluetooth that is received by a smartphone app linked to a cloud-based backend that calculates users' position as they move through the campus. Sensors are installed in classrooms; data is being sent through Lora network using the MQTT protocol to publish messages to the IBM server.

Sensors are calibrated and specifically used to collect data such as electricity consumption and temperature. Lora network was installed in the university, allowing the sensors (if connected to a hardware device with Lora technology) to transmit the data captured by the sensors. A gateway which receives this data will send it to other similar gateways if needed until the data arrives at a central server which manages the whole network and communicates with the internet [17].

4.2 Information and Knowledge Layers

Data collected from the sensors will be stored in a database within Bluemix platform; then, through IoT services, we can correlate information and create knowledge from the raw data. Stored data can also be interpreted to identify patterns and create reports. We use the Bluemix platform, from IBM, which provides templates to overview collected data. Rules and alerts based on sensor data in the platform better monitor all the variables. An example for electricity consumption is the identification of residual consumption in empty classrooms leading to corrective actions towards its elimination, or consumption can be correlated with room occupancy and external temperature.

4.3   Service Layer

Based on extracted data and knowledge, diversity of automatic actions can be implemented based on a service approach. Heating and cooling are activated based on sensor temperature data correlated with the presence of users in each space. Light intensity can be controlled based on luminosity information and presence. Water flow in bathrooms can be correlated with human presence. These services perform actions based on sensor input using node red platform easily. Manual inputs available from mobile devices can complete these actions [18].

We develop services oriented to room comfort temperature control because there is a connection between environmental temperature and cognitive performance [19]. Higher room temperatures can increase heartbeat to above 100 beats per minute. On a higher cardiac frequency, students end up consuming more calories diminishing their cognitive performance. Our service used input pre-defined temperature and, based on external weather conditions (exterior temperature), adapts the interior conditions to these pre-defined values. In winter, the temperature comfort should be around 18, while during the summer it should be around 26. Several factors need to be considered to actually achieve a comfortable environment, such as the number of students in the classroom - if there is a high number of students within the room, temperature will be higher due to more internal gains, and therefore we need to adapt the heating and cooling system; the insulation of the building will also affect the temperature, and since in ISCTE-IUL we have buildings with different construction materials, there is a possibility of studying systems operations more suitable to each type of external envelope; also due to the university having different buildings, which have rooms with a great diversity in spatial orientations, we can study the impact of room' solar orientation, and adjust the systems operations accordingly. With this, we see it is fundamental to regulate temperature and thermal comfort, so classrooms provide the conditions for students to learn in a comfortable environment.

Available information regarding external climate conditions can be used to predict near-future thermal comfort constraints. Further correlating this set of data with sensors-collected data provides very useful management information to predict future needs regarding heating and cooling. It is, therefore, possible to better manage the relationship between energy supply and demand taking better advantage of renewable energy produced on-site.

5   Results

Experience at ISCTE-IUL shows a significant saving potential. For the last five years, the learning management system included online information about room occupancy based on class schedules. By manually analysing this information on a weekly basis, it was possible to prepare custom-made routines to supply the systems management contractor so that HVAC systems were activated based on actual occupancy predictions. This process led to energy savings of 12%, based on actual consumption determined through energy invoice analysis.

With the implementation of this new sensor-based automatically processed information collection, we foresee a considerable improvement in how the energy-consuming systems are managed. It will be possible to improve occupancy rate-based systems activation at two scales further: the space scale, fine-tuning where heating or cooling should be supplied; and timescale, reducing weekly-based definitions to daily-based information. This is possible putting together the real-time low-frequency data collection with an integrated digital management system.

Also, detailed information on the facilities, such as the building geometry, wall, floor and roof composition, room door and window type and size and room size, type, identification, functions, etc. are stored in Building Information Model (BIM) models. BIM models are 3D descriptions of buildings which associate information with the geometry of the building and its contents. ISCTE-IUL's facility management office has been developing and maintaining a BIM model which is being used to feed maps, room listings and locations. This model has been linked to the academic management system to gather and display information such as room capacity, office occupancy and other parameters. Its visualisation capabilities are used to represent sensor location and results, provide info on which to base thermo-hygrometric simulations and to display gathered data in a geo-referenced, visually rich environment. This visualization platform is important when insights on the occupants and buildings systems response are sought for. Spaces occupancy, building materials, solar orientation and other factors can easily be understood, supporting data interpretation.

6   Conclusions

This work describes ISCTE-IUL approach towards building energy efficiency services where context information from locally installed sensors can be manipulated to identify consumption patterns and later implement actions in a service basis to save energy or water in a building. Usage of external information, like local maps, building materials, external weather conditions and room occupancy can be used to improve further saving actions. In spite of this work being a local dedicated approach to our campus, the service basis approach allows an easy deployment to other cases. In the near future, we will also add to this IoT platform a gamification approach to encourage users in saving energy and water.

References

1. Sretenovic, A.: Analysis of energy use at the university campus. Ph.D. thesis, Norwegian University of Science and Technology, Trondheim (2013)
2. Mongiovi, L.G., Cristaldi, L., Tironi, E., Bua, F., Liziero, M., Frattini, G., Martirano, L.: Architectural criteria for a distributed energy monitoring system. In: 2017 IEEE International Conference on Environment and Electrical Eng and 2017 IEEE Industrial and Commercial Power Systems Europe (EEEIC/I&CPS Europe), Milan, Italy, pp. 1–6. IEEE (2017)

3. Leal Filho, W.: Sustainability at Universities: Opportunities, Challenges, and Trends, 1st edn. P. Lang, UK (2009)

4. Wong, N.H., Kardinal Jusuf, S., Aung La Win, A., Kyaw Thu, H., Syatia Negara, T., Xuchao, W.: Environmental study of the impact of greenery in an institutional campus in the tropics. Build. Environ. **42**(8), 2949–2970 (2007)

5. Kolokotsa, D., Gobakis, K., Papantoniou, S., Georgatou, C., Kampelis, N., Kalaitzakis, K., Santamouris, M.: Development of a web based energy management system for university campuses: the CAMP-IT platform. Energy Build. **123**, 119–135 (2016)

6. Venetoulis, J.: Assessing the ecological impact of a university. IJSHE Int. J. Sustain. High. Educ. **2**(2), 180–196 (2001)

7. Mattoni, B., Pagliaro, F., Corona, G., Ponzo, V., Bisegna, F., Gugliermetti, F., Quintero-Nunez, M.: A matrix approach to identify and choose efficient strategies to develop the smart campus. In: 2016 IEEE 16th International Conference on Environment and Electrical Engineering (EEEIC), Florence, Italy, pp. 1–6. IEEE (2016)

8. SmartGalp – Galp. http://www.galpenergia.com/EN/sustainability/Innovation-research-technology/Paginas/Smart-Galp.aspx. Accessed 4 Jan 2018

9. Évora Inovcity – EDP. https://www.edp.com/en/stories/evora-inovcity

10. Gouveia, C., Rua, D., Soares, F.J., Moreira, C., Matos, P.G., Lopes, J.A.P.: Development and implementation of Portuguese smart distribution system. Electr. Power Syst. Res. **120**, 150–162 (2015)

11. Mendes, R.C.: Smart City Oeiras: O rumo a um concelho verdadeiramente inteligente. Ph.D. thesis, Universidade Atlântica, Lisbon (2016)

12. Columbus, L.: Roundup of IoT Forecasts and Market Estimates. Forbes (2016)

13. Smart Campus Solution for Education—Huawei solutions. http://e.huawei.com/en/solutions/industries/education/higher-education/smart-campus. Accessed 5 Jan 2018

14. Smart Campus Technology - Connected Campus. https://www.cisco.com/c/en/us/solutions/industries/education/connected-campus.html

15. European Commission: CORDIS: Projects and Results: SMART CAMPUS - Building-User Learning Interaction for Energy Efficiency. http://cordis.europa.eu/project/rcn/191915_en.html. Accessed 5 Jan 2018

16. EC - funded Smart Campus project achieves 30% in energy savings. https://www.metering.com/ec-funded-smart-campus-project-achieves-30-in-energy-savings/. Accessed 5 Jan 2018

17. Raza, U., Kulkarni, P., Sooriyabandara, M.: Low power wide area networks: an overview (2016)

18. Node Red. https://nodered.org. Accessed 6 Jan 2018

19. Taylor, L., Watkins, S.L., Marshall, H., Dascombe, B.J., Foster, J.: The impact of different environmental conditions on cognitive function: a focused review. Front. Physiol. **6**, 372 (2015)

Visual Recognition of Gestures in a Meeting to Detect When Documents Being Talked About Are Missing

Hugo Tovar Lopez and John Dowell[✉]

University College London, London, WC1, UK
j.dowell@cs.ucl.ac.uk

Abstract. Meetings frequently involve discussion of documents and can be significantly affected if a document is absent. An agent system capable of spontaneously retrieving a document at the point it is needed would have to judge whether a meeting is talking about a particular document and whether that document is already present. We report the exploratory application of agent techniques for making these two judgements. To obtain examples from which an agent system can learn, we first conducted a study of participants making these judgements with video recordings of meetings. We then show that interactions between hands and paper documents in meetings can be used to recognise when a document being talked about is not to hand. The work demonstrates the potential for multimodal agent systems using these techniques to learn to perform specific, discourse-level tasks during meetings.

Keywords: Smart meeting room · Multimodal agent · Meetings video
Object recognition · Hand tracking · Gesture recognition

1   Introduction

Having the relevant documents to hand is key to whether meetings are properly informed and even whether the right decisions are made. Yet it is a common experience of attending meetings that a particular document being discussed or cited is not always present, perhaps because it could not have been anticipated that the discussion would take a certain turn or that a document would become so important to an agenda item. The same is true regardless of whether the meeting is co-located or distributed and whether documents are digital or paper. This absence of a needed document is one of the factors that can conspire against the success of meetings [12]; consider, for example, when a person at a meeting makes a critical or controversial claim citing their recall of a particular document.

Meetings and the activities they support have a domain of known documents such as minutes of previous meetings and documents such as reports, contracts, proposals etc. We contrast these kinds of document with unknown but relevant documents that might be gathered opportunistically, for example during an ideas generating activity

© Springer Nature Switzerland AG 2019
P. Novais et al. (Eds.): ISAmI 2018, AISC 806, pp. 73–85, 2019.
https://doi.org/10.1007/978-3-030-01746-0_9

[19]. We are focusing on the known and particular documents that a meeting might discuss.

Having the right documents to hand in a meeting was traditionally a task for administrator assistants, both by anticipating what documents would be needed and by following the discussion in a meeting and proactively retrieving a relevant document. Nowadays we seek to facilitate meetings and mitigate counter-acting factors with tools, but with the demand that their benefits must outweigh any downsides such as distraction effects [12]. Moreover the concept of the smart meeting room has been advanced as a space possessing ambient intelligence capable of functions like automatic summarization and detection of decision and action items [4]. We can envisage the future smart meeting room having an agent system that spontaneously retrieves a document being discussed which is absent.

An agent system with this capability would need to be able to judge whether a meeting is Talking_About_a_Document (the 'TAD' judgement) and whether that Document_is_Not_There (the 'DNT' judgement). It would retrieve the absent document or the most likely documents from a digital repository, using search terms from the dialogue proximal to where discussion of a document was detected. This agent system is most similar to the implicit querying, just-in-time information retrieval system for meetings investigated by Popescu-Belis and colleagues [24]. Their proposed system monitors conversations for pre-defined keywords informed by topic-modelling and uses these terms to search local repositories and the web for the most relevant documents (including unknown documents) and presents a continuously updating list of search results to the meeting participants. By contrast, the system we envisage uses multiple modalities to monitor for discussion of known documents and only offers documents that the meeting doesn't already have.

To assess how amenable are the TAD and DNT judgement tasks for an agent system to perform, we can first ask how well do people do them? Since both involve cognitive and socio-cultural aspects, we can expect people to not always agree. As well as providing a performance baseline, such an enquiry would be able to identify the sources of information that a person uses and which an agent system might also exploit. We first report a study in which participants watched videos of meetings and performed both the TAD and DNT judgement tasks. The videos were selected from the AMI corpus of meetings videos [21]. Participants' judgements were used as the ground truth annotations of the videos to train agent techniques to make the same judgements.

An agent system making DNT judgements should base its judgements primarily on the visible activity and behaviours of participants in the meeting. Visual data can also be used for making TAD judgements. We report our exploratory application of computer vision techniques that learn to make DNT and TAD judgements. Visual object recognition and tracking are used to train neural network based classifiers. With selected videos from the AMI corpus, the classifiers are trained first to recognize and track hands and paper documents, second to recognize types of interaction between hands and paper, and third to make the TAD and DNT judgements. Other applications of computer vision techniques to meetings have invariably concerned analysis of visual focus of attention of participants [3] or social signals [30].

2   People Performing the TAD and DNT Judgement Tasks

To understand how and how well people perform the TAD ('Talking-About-a-Document') and DNT ('Document-Not-There') judgements, we performed a study with ten participants watching videos of meetings. We wanted to establish how consistently would different people make the same judgements, what were the high level cues they were using, and the predicted value for the meeting of being given that document. Further, we wanted a dataset of meetings videos annotated with 'expert' judgements that would provide a ground truth for exploring the application of agent system techniques.

Ten AMI videos were selected, each approximately half a minute long and with varying views from multiple cameras (Fig. 1), together representing a variety of meeting situations such as presentations and discussions. Each participant is recorded by two microphones and also available is a transcript of the meeting that is punctuated but not grammatically improved; it also includes non-word vocal sounds. There were explicit and implicit references to documents, discourse without reference to documents, explicit and implied statements about the presence of a document, and visual evidence for the presence and absence of documents (i.e. printed documents, projection screen, laptops).

Fig. 1.   Example views from multiple cameras for two AMI meetings.

Multi-choice questions about the TAD and DNT judgements were created through preliminary trials with 'observer' participants. Ten of these participants watched each of the videos with accompanying transcript and answered the open ended questions (a) 'Do you think this specific fragment is referring to a document*? (*As a document, consider any physical or electronic media containing text such as paper sheet, book, file, email or website)', and (b) 'If there is reference to a document, is this document currently present?' Participants were recruited through crowdsourcing and carried out the tasks online. Their responses were merged to produce a preliminary list of multi-choice answers. For example, responses to question (a) included: '(i) Yes: It has been explicitly mentioned. (ii) Yes: They are talking about information that seems to be present in a document', etc. Responses to question (b) included: '(i) Yes: The document has been explicitly mentioned as currently present; (ii) Yes: The document is displayed on a personal screen, (iii) No: The speaker refers to a document in the past tense', etc. A new group of ten observer participants watched the videos and answered the multi-choice questions with the option of providing additional answers to the questions. Very few additional answer categories were submitted. A final panel of ten 'test' participants were recruited again by crowdsourcing, all having experience of attending meetings. They were shown the videos with transcripts and asked to answer each of the questions with the aggregated multi-choice answers.

The group of ten test participants were 74.4% consistent in their TAD judgements concerning whether a document was mentioned (calculated by dividing the number of agreements by the number of judgements). In the DNT judgements concerning whether the document was not there in the meeting, their agreement was 74.2%. However, as the percentage of overall agreement (POA) doesn't take chance into account, we are interested in calculating the kappa coefficient [10] as a chance-adjusted measure of agreement, which can take a value in the range of -1 (i.e. perfect disagreement) to 1 (i.e. perfect agreement), where zero is the chance probability. In the situation of having multiple raters not restricted to assign a certain number of cases to each category, free-marginal multi-rater kappa (κ_{free}) [26] is recommended [7]. For TAD judgements, the inter-rater reliability κ_{free} is 0.488, while for DNT judgements κ_{free} is 0.484. A positive κ_{free} coefficient means there is a level of agreement above chance and in this case it equates to a moderate agreement level [18]. Even though the overall agreement on each judgement was substantial, the judgements are independent, in other words, a positive judgement in the TAD task does not make a positive judgement in the DNT task more likely.

The high-level cues used by participants to make their judgements were elicited from their answers to the multichoice questions. Participants were found to rely on the dialogue in 78% of TAD judgements about whether the meeting was talking about a document whereas DNT judgements about whether the document was absent relied evenly on both auditory and visual cues.

The study confirms that observers of a meeting can judge whether a document is being discussed and whether it is present. It confirms that they are able to do this with an acceptable level of agreement which therefore warrants using their video annotations as ground truth for training an agent system to perform this task. The study also provides insight into the cues people use to make the TAD and DNT judgements. It confirms that people use both visual and verbal evidence for performing the tasks and combine the judgements they make with each. Our exploration of the use of multimodal agent techniques to performing the tasks use the same cues, although the transcribed speech is only used for the TAD task. We now report that application of text analysis methods.

3   Computer Vision Method for the TAD and DNT Judgement Tasks

Our participants used the views of meetings in the videos, and in particular the views of documents being manipulated by hands, to judge whether the meeting was talking about a document (the TAD task) and whether that document was not there in the meeting (the DNT task).

We now describe the exploratory development of trained computer vision methods for performing these same judgement tasks. 'Automatic' Human-Object Interaction analysis has not been pursued previously in relation to meetings analysis, other than recognising if participants are looking at a whiteboard or a shared screen as part of dealing with the 'visual focus of attention' problem. Generally, Human-Object Interaction recognition relies on the static pose of both human and object to infer if there is an interaction between them. Gupta et al extended this approach to analyses involving motion [16]. For example, they trained a model by supervised learning to recognize from videos whether people were drinking or making a phone call in interaction with different objects including a cup and a phone. The almost perfect results they achieved were unsurprising given the diversity of objects and interactions they studied.

The meetings agent detects participants' hands and paper documents, tracking their movements, then recognizing those movements as particular kinds of interaction between hands and paper. The association of those interactions with the two judgement tasks, TAD and DNT, can then be learnt from the ground truth annotations obtained from our observer participants.

3.1 Object Detection and Tracking

Detection of hands and paper documents in the videos has been implemented with C++ and OpenCV, an open source computer vision library that includes standard image processing techniques and analysis methods. Once hands and paper have been detected for each frame, they are tracked across frames. The tracking copes both with new hands and papers appearing and with existing ones disappearing due to occlusion.

To assess the effectiveness of the tracker, an epoch (16 s long) from the IB4010 meeting video was used. Annotation of the ground truth location for each object was carried out by a volunteer using the Viper video annotation tool [11]. Figure 2 shows a typical frame containing several labelled hands and paper documents, including one hand that is currently occluded. The trajectory of each hand as determined by the tracker is shown. The performance of the tracker in locating hand #1 is shown in Fig. 3 in terms of the relative displacement of the hand (measured in pixels) in each successive frame. The ground truth annotated data are also shown for comparison. Although the automatic tracked data has already been smoothed to remove false peaks, it still remains highly noisy. This error should be significantly reduced by implementing more advanced tracking techniques [32].

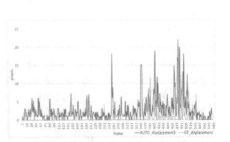

Fig. 2. Hands and paper tracking. Red lines show the trajectory of each hand. The red line on the lower left denotes a hand that is currently occluded but that has been tracked in previous frames.

Fig. 3. Tracking of hand #1 over selected meeting epoch. Displacement of hand in each successive frame (pixels) from automatic tracking (AUTO) and manual annotation (Ground Truth GT).

Assessment of the tracker's performance in terms of the CLEAR MOT (Multiple Object Tracking) Metrics [6] is given in Table 1. The MOTP value for the hands and paper tracker is very high. The miss rate refers to objects not detected and is acceptable for this tracker, particularly so for paper. The false positive rate (FPR) expresses how many of the detected objects were incorrect and is high for the hand+paper tracker, a consequence of visual elements sharing the same pixel colours as hands and paper.

Table 1. Evaluation of the tracker for hands and paper: MOTP (Multiple Object Tracking Precision); Miss rate (objects not detected). FPR (False Positive Rate); Mismatches (object confusion during tracking); MOTA (Multiple Object Tracking Accuracy).

Object	MOTP (pixels)	Miss rate (%)	FPR (%)	Mismatches	MOTA (%)
Hand	9.5	10.4	31.8	0	57.8
Paper	5.6	12.1	32.1	0	55.75

This rate can therefore be improved if shape detection techniques [23] and template matching are used [17]. Mismatches count the ratio of objects incorrectly matched during tracking. The perfect mismatch values for the tracker confirm the assumption that tracked hands and documents will be in a similar position in consecutive frames. The multiple object tracking accuracy (MOTA) metric is an aggregation over the miss rate, FPR and mismatches and conveys the tracker's overall strengths. The indifferent MOTA scores are due to the high FPRs and the impact of occlusions on the miss rate.

3.2   Hand+Paper Interaction Meta-Features

Being able to track hands and papers in a meeting is a preliminary to automatically recognizing types of interactions between hands and paper. We assume there are few different types of such interactions, that they have a set of distinctive meta-features, and that an agent system can be trained to recognize them. We further assume that those interaction types associate systematically with the two judgements (TAD and DNT) and that an agent system can learn to make those judgements using recognized hand-paper interactions.

Three short videos (each of approximately 20 s) were selected from the IB4010 meeting of the AMI Corpus (hereafter, Episodes A, B and C) and visually inspected for hand+paper interactions. Three interaction types were apparent in these episodes:

- GRABBING_STATIC: a static hand holding paper;
- GRABBING_MOVING: a moving hand holding paper;
- POINTING_READING: a finger points and touches a paper during reading.

Three meta-features were hypothesized as discriminating between the three interaction types

- Intersected area between hand and paper: the percentage of overlap between both objects, 0% being no intersection and 100% complete intersection.
- Inter-distance between hand and paper, the distance between both objects' bounding box centers, given in pixels.
- Self-displacement of hand and paper: the number of pixels the center of an object's bounding box moved from the previous to the current frame.

To investigate the regularity of these meta-features with the interaction types, the three video episodes were first annotated for: (i) the location of one relevant hand and one paper; (ii) the type of interaction between the hand and paper from the three categories; (iii) epochs where participants are talking about a document; (iv) epochs where the discussed document is available. Annotation of (i) and (ii) was carried out by a single volunteer whilst annotations of (iii) and (iv) were carried out by 10 crowdsourced volunteers because of the greater degree of individual judgement involved. An annotated video for Episode A is available at: (url omitted for draft anonymization).

A hand+paper dyad from each video episode is analysed in relation to the interaction types and meta-features. Figure 4 shows (a) the area of overlap, and (b) the inter-distance for one hand+paper dyad in Episode A. Both ground truth annotated inter-distance and the inter-distances generated automatically by the hands+papers tracker are shown. Four instances of interactions of two different types (POINTING_READING occurring three times, GRABBING_MOVING occurring once) are mapped in the figure. Similar charts were generated for the other two meta-features of intersected area and self-displacement.

3.3   Developing a Hand+Paper Interactions Classifier

The question raised by the assaying of interaction meta-features from the detection and tracking of hands and papers, is whether there is a systematic relationship of those spatial data with the different interaction types that is amenable to machine learning. For example, in Fig. 4 is it possible that a consistent pattern in inter-distance variation is associated characteristically with POINTING_READING which a classifier could be trained to identify?

We trained a neural network to classify every instance of interaction from the three videos of meetings episodes over every combination of hand and paper dyads from the set of hands and the set of papers. The neural network was trained with the annotated data and tested separately with the annotated spatial data and the data generated by our hands+papers tracker. We selected one hand+paper dyad from Episode A and an epoch consisting of 587 frames. Each frame was manually annotated for four meta-features (i.e. hand displacement h_d, paper displacement p_d, intersected area between hand and paper $(h, p)_a$, and the inter-distance $(h, p)_d$ between them). Hand+paper interactions were manually annotated; GRABBING_MOVING and POINTING_READING interactions but not GRABBING_STATIC interactions were identified in the epoch.

The data were randomly grouped as 70% for training, 15% for validation and 15% for testing. Table 2 shows the evaluation results from testing with both annotated and automatically tracked data. For the annotated data, the accuracy of classification is 97.7% for both types of interaction occurring in the selected epoch. The sensitivity rates are very high and the False Positive Rates extremely low. Accuracy deteriorates markedly when the automatically generated tracking data is used. This is consistent with the mixed precision (MOTP) and accuracy (MOTA) evaluation results obtained for the tracker, confirming the need to introduce more sophisticated image analysis techniques into the tracker to more accurately detect hands and paper. Nevertheless the potential for the classifier to recognise hand+paper interactions is clear from the performance with the annotated hand and paper location data.

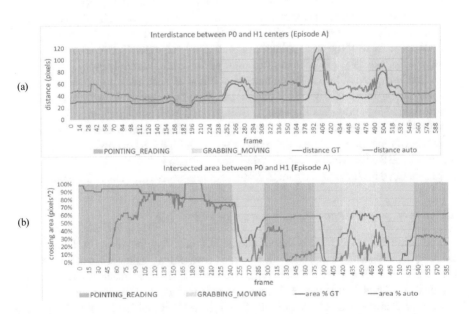

Fig. 4. Plots of (a) intersected area and (b) inter-distance between a Hand and Paper dyad on a succession of frames for both manually annotated (GT) and tracker generated (auto) data. Annotated Hand+Paper interactions (POINTING-READING; GRABBING_MOVING) are overlaid.

Table 2. Evaluation results for the hand+paper (h, p) interactions classifiers. Results for manually annotated data and data produced by the trackers are shown. Inputs are hand and paper (h, p) dyad spatial features.

Classifier	Input (h, p)	Accuracy %	Sensitivity %	FPR %
GRABBING_ MOVING	Annotation	97.7	97.6	0.0
POINTING_READING	Annotation	97.7	100.0	3.3
(GRABBING_ STATIC)	Annotation	n/a	n/a	n/a
GRABBING_ MOVING	Tracker	60.1	95.3	82.6
POINTING_READING	Tracker	55.2	44.2	1.7
(GRABBING_STATIC)	Tracker	n/a	n/a	n/a

3.4   Developing Two Classifiers for the TAD and DNT Judgements

The ability to recognise hand+paper interactions in a meeting implies that these interactions can be used for the two judgement tasks, TAD and DNT that together represent the decision of whether a meeting is talking about a document that is not to hand. Two neural network classifiers were trained, each performing one of these tasks. The classifiers were trained with the same 587 frame epoch as the interactions classifier and trained with the same selection of 70% of the data. Each classifier is capable of single-layered and of hierarchical recognition approaches [2], see Table 3.

Table 3.   Abstraction layers for recognition and classification of hand-paper interactions for the TAD and DNT judgement tasks.

Layer	Analysis	Judgement	Features to extract
High level	TAD & DNT	Document talked about? Document not there?	Output
Mid-level	Hand and paper interactions	What type of interactions are there between hands and papers?	Intersected area, inter-distance, and self-displacement
Low level	Hand and paper detection and tracking	Are hands and papers visible and where are they moving?	Bounding box positions and areas
Input	n/a	n/a	Video frames (pixels)

- Single-layer approach: Uses hand and paper dyad (h, p) spatial features as input: hand displacement hd, paper displacement pd, intersected area between hand and paper (h, p)a, and the inter-distance (h, p)d between them. Outputs a binary judgement (high-level) for TAD and another for DNT.
- Hierarchical approach: hand+paper interactions from the interactions classifier and uses these to output a binary judgement (high level) for the TAD and DNT tasks.

The evaluation results for the two classifiers are shown in Table 4. For the TAD and DNT judgements with the annotated hand and paper spatial data, the single-layer approach obtains a higher accuracy (94.3% for TAD and 93.2% for DNT) than the hierarchical (70.5% and 81.8%), although for DNT this includes a false positive rate of 20%, in contrast with the 0% for the hierarchical. When the automatically tracked hand and paper data rather than the annotated hand and paper spatial data are used with the single layer classifier, the accuracy again deteriorates to just above chance as a consequence of the tracker's poor accuracy.

Table 4. TAD classifier and DNT classifier evaluation results with manually annotated data and automatically tracked data.

Classifier	Input	Accuracy %	Sensitivity %	FPR %
TAD (single-layer)	(h, p) spatial annotation	94.3	93.9	5.5
DNT (single-layer)	(h, p) spatial annotation	93.2	97.1	20.0
TAD (hierarchical)	h+p interactions	70.5	66.7	27.9
DNT (hierarchical)	h+p interactions	81.8	81.8	0.0
TAD (single-layer)	(h, p) spatial tracker	62.2	50.1	2.7
DNT (single-layer)	(h, p) spatial tracker	68.8	76.3	68.4

4   Discussion

People observing meetings can be reasonably consistent in their judgements about whether a document is being discussed and whether that document is present. This was the finding from our study with observer participants who made the TAD (is the meeting Talking-About-a-Document?) and DNT (is the Document-Not-There?) judgements. They used both the spoken discourse and the visible behaviour of people manipulating paper documents in the meetings to make these judgements.

We have also shown that there are distinct types of spatial interaction between hands and paper documents in meeting situations. The spatial meta-features we used to distinguish these interactions have been confirmed as good descriptors for classifying interactions. The interactions we identified are POINTING_READING, GRABBING_MOVING, GRABBING STATIC and also NO-INTERACTION. The classifier we trained with annotated AMI meetings recordings has shown that it is possible to classify hand+paper interactions with an accuracy of 97.7% when using the manually annotated object tracking.

The results demonstrate that visually tracked hands and papers in a meeting can be used for the TAD and DNT judgements. A discrete classifier was trained for each judgement. Each was both single layer and hierarchical, capable of making the judgements directly from training with the hand and paper spatial data, and hierarchically with input of recognised hand+paper interactions. Evaluation found that the single layer accuracy exceeds the hierarchical (94.3% and 93.2% for TAD and DNT respectively, against 70.5% and 81.8% for the hierarchical model). These results were obtained with a subset of the AMI corpus and need to be replicated by re-training with a larger data set.

The classifiers' accuracies fell markedly with the automatically tracked data. The limitations of the tracker were evident in its inconsistent MOTP and MOTA evaluation scores, particularly its false positives rate. More sophisticated tracking algorithms [25] can be directly substituted into the tracker to replace the simple gaussian distribution mechanism. Optimising the tracker was a subsidiary aim in this work, our priority was to show that tracking of hands and paper documents is possible and that it can be exploited by the higher level judgement tasks. This is evidenced in the comparison of TAD and DNT judgements with the manually annotated and automatically tracked hands and paper.

The superiority of the single-layered approach is in part due to the interaction classifier discarding spatial feature information that is not relevant for its own purposes but may be useful for the subsequent TAD and DNT judgements. Since we can obtain high accuracy results for the TAD and DNT judgements using a single-layered approach (i.e. using the tracking information alone), it could be argued that there is no need to recognise hand and paper interactions. However, identifying the interactions enables explanation of the meeting participants' behaviour with documents. Moreover, the hand+paper interactions could be used as input to the text analysis modality as an intermediate analysis product to improve its modality specific judgements and to support fusion of modalities more extensively.

This study reveals how such a modality-specific technique can be exploited by an agent system [22]. The results indicate that the methods are usable and with development would be deployable for spontaneous retrieval of documents. An agent system using multiple modalities for the same judgements would achieve a higher performance by combining the judgements. That fusion process could occur earlier, with intermediate analysis products, or later with decision outcomes. The capability of the techniques revealed in this study would therefore improve not only with the modality specific improvements described, but also through their eventual fusion.

An agent system spontaneously retrieving an absent document for a meeting would need to decide whether and when that document should be presented. For example, alerting a meeting when a speaker has finished speaking, using a visual cue to indicate which document is being offered and perhaps which meeting topic this relates to, might make the distraction or interruption more acceptable [20]. This sort of judgement is often performed exquisitely well by people and it is feasible that an agent system would be capable of making this judgement too, given the capability we have found for judging whether a meeting is discussing an absent document.

References

1. Aberdeen, J., Burger, J., Connolly, D., Roberts, S., Vilain, M.: Mitre-bedford: description of the alembic system as used for MUC-5. In: 5th Conference on Message Understanding Association for Computational Linguistics, pp. 137–146 (1993)
2. Aggarwal, J.K., Ryoo, M.S.: Human activity analysis: a review. ACM Comput. Surv. (CSUR) 43(3), 16 (2011)
3. Ba, S.O., Odobez, J.-M.: Multiperson visual focus of attention from head pose and meeting contextual cues. IEEE Trans. Pattern Anal. Mach. Intell. 33(1), 101–116 (2011)
4. Banerjee, S., Rudnicky, A.I.: Segmenting meetings into agenda items by extracting implicit supervision from human note-taking. In: Intelligent User Interfaces IUI 2007, pp. 151–159 (2007)
5. Benjelloun, O., Garcia-Molina, H., Menestrina, D., Su, Q., Whang, S.E., Widom, J.: Swoosh: a generic approach to entity resolution. VLDB J. Int. J. Very Large Data Bases 18(1), 255–276 (2009)

6. Bernardin, K., Stiefelhagen, R.: Evaluating multiple object tracking performance: the clear mot metrics. EURASIP J. Image Video Process. **1**, 1–10 (2008)

7. Brennan, R.L., Prediger, D.J.: Coefficient kappa: Some uses, misuses, and alternatives. Educ. Psychol. Meas. **41**(3), 687–699 (1981)

8. Chen, Z.Q., Kalashnikov, D.V., Mehrotra, S.: Exploiting context analysis for combining multiple entity resolution systems. In: ACM SIGMOD/PODS 2009 Conference, pp. 207–218 (2009)

9. Chinchor, N., Robinson, P.: MUC-7 named entity task definition. In: Proceedings of the 7th Conference on Message Understanding, vol. 29 (1997)

10. Cohen, J.: A coefficient of agreement for nominal scales. Educ. Psychosoc. Meas. **20**, 37–46 (1960)

11. Doermann, D., Mihalcik, D.: Tools and techniques for video performance evaluation. In: Proceedings of 15th International Conference on Pattern Recognition, pp. 167–170 (2000)

12. Ehlen, P., Purver, M., Niekrasz, J., Lee, K., Peters, S.: Meeting adjourned: off-line learning interfaces for automatic meeting understanding. In: Intelligent User Interfaces IUI 2008, pp. 276–284. ACM (2008)

13. Galliano, S., Gravier, G., Chaubard, L.: The ester 2 evaluation campaign for the rich transcription of french radio broadcasts. In: Interspeech, pp. 2583–2586 (2009)

14. Garofolo, J.S., Laprun, C., Michel, M., Stanford, V.M., Tabassi, E.: The NIST meeting room pilot corpus. In: LREC Citeseer (2004)

15. Grouin, C., Rosset, S., Zweigenbaum, P., Fort, K., Galibert, O., Quintard, L.: Proposal for an extension of traditional named entities: from guidelines to evaluation, an overview. In: Proceedings of the 5th Linguistic Annotation Workshop Association for Computational Linguistics, pp. 92–100 (2011)

16. Gupta, A., Kembhavi, A., Davis, L.S.: Observing human-object interactions: using spatial and functional compatibility for recognition. IEEE Trans. Pattern Anal. Mach. Intell. **31**(10), 1775–1789 (2009). https://doi.org/10.1109/Tpami.2009.83

17. Kulkarni, S., Manoj, H., David, S., Madumbu, V., Kumar, Y.S.: Robust hand gesture recognition system using motion templates. In: 2011 11th International Conference on ITS Telecommunications (ITST), pp. 431–435. IEEE (2011)

18. Landis, J.R., Koch, G.G.: The measurement of observer agreement for categorical data. Biometrics, 159–174 (1977)

19. Li, N., Dillenbourg, P.: Designing conversation-context recommendation display to support opportunistic search in meetings. In: Proceedings of the 11th International Conference on Mobile and Ubiquitous Multimedia, p. 12. ACM (2012)

20. Lopez-Tovar, H., Charalambous, A., Dowell, J.: Managing smartphone interruptions through adaptive modes and modulation of notifications. In: Proceedings of the 20th International Conference on Intelligent User Interfaces, pp. 296–299. ACM (2015)

21. Mccowan, I., Carletta, J., Kraaij, W., Ashby, S., Bourban, S., Flynn, M., Guillemot, M., Hain, T., Kadlec, J., Karaiskos, V.: The AMI meeting corpus. In: Proceedings of the 5th International Conference on Methods and Techniques in Behavioral Research (2005)

22. Mccowan, I., Gatica-Perez, D., Bengio, S., Lathoud, G., Barnard, M., Zhang, D.: Automatic analysis of multimodal group actions in meetings. IEEE Trans. Pattern Anal. Mach. Intell. **27**(3), 305–317 (2005)

23. Ong, E.-J., Bowden, R.: A boosted classifier tree for hand shape detection. In: Proceedings of Sixth IEEE International Conference on Automatic Face and Gesture Recognition, pp. 889–894. IEEE (2004)

24. Popescu-Belis, A., Yazdani, M., Nanchen, A., Garner, P.N.: A just-in-time document retrieval system for dialogues or monologues. In: Proceedings of the SIGDIAL 2011 Conference Association for Computational Linguistics, pp. 350–352 (2011)
25. Pulford, G.W.: Taxonomy of multiple target tracking methods. IEE Proc. Radar Sonar Navig. **152**(5), 291–304 (2005). https://doi.org/10.1049/ip-rsn:20045064
26. Randolph, J.J.: Free-marginal multirater kappa (multirater k [free]): an alternative to fleiss' fixed-marginal multirater kappa (2005). Online submission
27. Talburt, J.R.: Entity Resolution and Information Quality, pp. 1–235 (2011)
28. Tur, G., Stolcke, A., Voss, L., Peters, S., Hakkani-Tur, D., Dowding, J., Favre, B., Fernandez, R., Frampton, M., Frandsen, M., Frederickson, C., Graciarena, M., Kintzing, D., Leveque, K., Mason, S., Niekrasz, J., Purver, M., Riedhammer, K., Shriberg, E., Tien, J., Vergyri, D., Yang, F.: The calo meeting assistant system. IEEE Trans. Audio Speech Lang. Process. **18**(6), 1601–1611 (2010). https://doi.org/10.1109/Tasl.2009.2038810
29. Van Leeuwen, D.A., Huijbregts, M.: The ami speaker diarization system for NIST RT06s meeting data. In: International Workshop on Machine Learning for Multimodal Interaction, pp. 371–384. Springer (2006)
30. Vinciarelli, A., Pantic, M., Bourlard, H.: Social signal processing: survey of an emerging domain. Image Vis. Comput. **27**(12), 1743–1759 (2009)
31. Wilcock, G.: Introduction to linguistic annotation and text analytics. Synth. Lect. Hum. Lang. Technol. **2**(1), 1–159 (2009)
32. Yilmaz, A., Javed, O., Shah, M.: Object tracking: a survey. ACM Comput. Surv. **38**, 4 (2006)

EUD4SH: A EUD Model
for the Smart Home

Danilo Caivano[1], Fabio Cassano[1(✉)], Daniela Fogli[2], and Antonio Piccinno[1]

[1] Department of Computer Science, University of Bari, Bari, Italy
{danilo.caivano,fabio.cassano1,antonio.piccinno}@uniba.it
[2] Department of Information Engineering, University of Brescia, Brescia, Italy
daniela.fogli@unibs.it

Abstract. The increasing availability of low-cost smart devices is bringing them to be used more and more in the smart home. However, the development of a smart home environment requires to take into account several aspects. First of all, designers must consider the end user (namely the person that actually uses the smart home), and not only the technology, at the center of any intervention. Another important aspect is the interaction between the smart home and the appliances that are already deployed in the same smart environment. Moreover, most of smart home solutions are static and do not allow end users to customize them according to their real needs and preferences. Finally, not all end users may possess the necessary knowledge and skills to customize a smart home, but someone else, such as an adult child or a caregiver, may be called on to carry out this task for them. In this paper, we analyze all these aspects that can influence the development and evolution of a smart home. We then propose a model supporting developers and software engineers to deploy and evaluate a smart home solution that adopts end-user development techniques. It is based on the International Classification of Functioning scale, which is used to characterize the person that is mainly going to live in the smart home and define a solution suitable to his/her needs.

1 Introduction

Smart homes are usually conceived for supporting people in their daily activity, such as managing home security, energy consumption, and comfort. Most of the devices available in the mass market and adopted in smart homes exploit the Internet of Things (IoT) technology. IoT allows small micro-controllers equipped with sensors to be connected to the Net. These appliances are usually adopted to monitor the environment (for example: the closed circuit camera, the smoke sensor, the power consumption absorption, etc.), and share the produced information with the house owner. Other devices can autonomously take a decision to facilitate the user in his/her daily activity (for example: the thermostat, the light sensor, etc.). Smart devices can also be used by the smart home system to send alerts or take decisions, as well as to automate some actions.

© Springer Nature Switzerland AG 2019
P. Novais et al. (Eds.): ISAmI 2018, AISC 806, pp. 86–93, 2019.
https://doi.org/10.1007/978-3-030-01746-0_10

However, common commercial solutions for the smart home do not provide users with functionalities to modify or customize the behavior of their smart devices. To achieve a good "orchestration" of smart devices, an external software, like IFTTT, Atooma or Tasker, is usually required. Furthermore, different user studies have shown the difficulties encountered by smart systems endowed with Artificial Intelligence (AI) algorithms when they try to address the needs of multiple users in the same home [1].

Last but not least, big companies commonly advertise that their smart home solutions may be used by young or middle-age people. Indeed, people who are working outside everyday can take advantage from a little automation of their home; however, they are not the only people who may improve their life by using a smart home. Elderly, disabled and non-computer-literal people might gain profits from such an environment, even though it often provides them with functionalities not easily accessible and modifiable. In fact, these functionalities are usually fixed by the smart home manufacturer and do not satisfy the real needs of those specific users.

In this paper we propose an End-User Development (EUD) model, called EUD4SH (EUD for Smart Home), to help developers, software engineers and home inhabitants to define, adapt and evolve a smart home environment. In particular, the EUD4SH model allows taking constantly into account the changing requirements of the inhabitants, by keeping attention to their evolving needs and health status.

The paper is organized as follows: in Sect. 2, related works are discussed; in Sect. 3 the International Classification of Functioning (ICF) schema is explained; in Sect. 4, it is described how the ICF can influence the development of a smart home; in Sect. 5, the EUD4SH model is proposed; finally, in Sect. 6 discussion and future works are presented.

2   Related Works

Defining what is the best smart home technology to be used for helping people live alone is an important issue for developers and software engineers. However, smart home inhabitants are heterogeneous: there may be young, middle age, older, disabled, or people with limited knowledge of hardware/software technology. Some proposed works consider the smart home as a part of the rehabilitation process of a patient [2]. Others propose the smart home technology for supporting lonely people (in particular older people with impairments) to better manage their lives [3]. On one hand, it is positive to let these people be independent in their smart home environment; on the other hand, many aspects about their health must be considered [4]. Some researchers are trying to fill the gap between the technology and the smart home, by considering the users at the center of the design process [5]. In our work, we are making a step further, by considering a continuum process of evolution of the smart home according to the user's needs [6].

IoT technology allows pervasive devices to be connected on the Net. In this way, personal data can be collected and analyzed by remote machines in order to elicit knowledge about the home and its inhabitants [7]. Sometimes, AI techniques are adopted, but they are not sufficient to satisfy users' requirements. For example, the Nest thermostat is one of the most discussed (and controversial) appliance in this field [1].

To let people "have the power" to modify and customize their smart home, a new approach is required. EUD is currently well known in the Human-Computer Interaction field [8]. Recently, EUD has been defined as an approach that "encompasses methods, techniques, methodologies, situations, and socio-technical environments that allow end users to act as professionals in those domains in which they are not professionals" [9]. It has been adopted not only to support creation and tailoring of software artifacts, but also in the Ambient Intelligence field [10]. It is also useful to support smart home inhabitants to cooperate and solve conflict situations. For example, "gamification" has been proposed to allow them to creatively solve real problems by getting rewards for their participation in smart home management [11].

3   The ICF Model

Focusing on different people with different abilities, it is important to define a way to uniquely characterize persons living in a smart home environment. In medicine, the ICF schema [12] is widely used to objectively define the complete physical and psychological status of a given patient. Thanks to its broad spectrum of analyzing topics, this schema can be applied to any person, including those presenting some kind of disability and sane people.

By compiling a list of detailed information about the person's status, a score of its health can be generated [13]. This compilation can be cyclically repeated to monitor the evolution of the patient's health status. The ICF model is represented with the white boxes reported in Fig. 1. Each white box represents a part of the person's life: from the social life to the mobility or any physical impairment. The connection between white boxes denote the sharing of one or more parameters. Each box refers to an objective assessment about:

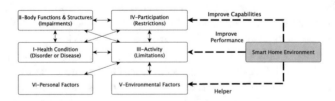

Fig. 1. The white boxes represent the original ICF schema, the gray one is the modification proposed

  I: "Health Condition": it represents the health status of the person;
 II: "Body functions and Structures": it represents the ability of the person to move (with or without assistance) and if he/she is able to use the arms or the legs;
III: "Activity": it involves the ability of the person to perform daily tasks and communicate with other people;
 IV: "Participation": it represents the quality of the family members or the caregivers to the person's life;
  V: "Environmental Factors": it comprehends all the information related to the place where the person lives;
 VI: "Personal Factors": it involves all the personal characteristics (age, level of instruction etc.).

The I: box contains a brief health information about the patient, while the VI: one contains some general information (age, sex, address etc.). The II: box, describes the physiological functions of body systems (including psychological functions) and is divided into two sections. The first one includes the body functions and is expressed using a qualifier representing the "Extent of impairments" (problems in a body function as a significant deviation or loss) with a scale from 0 (No impairment) to 4 (Complete impairment); the second one represents the body structures and expresses the "Nature of the changes". The III: and IV: boxes are described using two qualifiers. "Activity" is the execution of a task or action by an individual. "Participation" is involvement in a life situation. While "Activity (limitations)" are difficulties an individual may have in executing activities, "Participation (restrictions)" are problems an individual may have in being involved in life situations. The former represents the person's capacity and indicates the Extent of Activity limitations. The latter represents the person's performances and indicates the Extent of Participation Restriction. Both are evaluated on a scale from 0 (No difficulty) to 4 (Complete difficulty). The V box makes up the physical, social and attitudinal environment in which people live and conduct their lives. It is represented with one qualifier and a scale from −4 (Complete barriers) to 4 (Complete facilitator). The value 0 represents both a "No Barrier" and a "No facilitator". A barrier is a part of the environment that prevents the person's movement in the environment, while the facilitator is something that helps or supports it.

4   Relationships Between ICF and Smart Home

The usage of the ICF scale to determine the people's ability to live in the smart home is a well-known research topic [14]. By comparing the ICF model and the capabilities of a smart home environment, we realized that some ICF sections (or parameters) may be influenced by the smart home. The smart home environment is represented in gray in Fig. 1, while the dotted lines denote its major influence on the ICF parts. A detailed representation on how the smart home can improve the values of the white ICF boxes is shown in Fig. 2. The "Activity (Limitations)" and "Participation (Restriction)" are considered as a unique entity, according to the WHO rules. The white boxes represent the "Activity limitations and Participation restriction" details. The labels on each line (if present) represent how the smart home can influence people's life, by supporting them in the activities. The smart home environment can act as helper in the "Environmental Factors" by facilitating the end user's life in the house. There are many technologies that, when used in the smart home environment, can improve the quality of the life of the inhabitants [15]. For example, using IoT devices, the activity performance can be improved and, thanks to the technology such as the social network connectivity, participation capabilities can be improved as well [16]. The smart home environment improves the person's "Learning and Applying knowledge" by proposing mini-games (such as brain training games) to complete on the smart-TV, or on devices like the smartphone or tablet. With the same idea, the "General Task and Demands" is supported with task-oriented games (or tasks) like: "Take the flour from the sideboard". The "Communication" value is improved by letting the person talk with bots, if no other person is available, like Google Assistant, Siri, or Amazon Alexa. With the aid in preparing (or buying) food, the "Domestic Life" area is improved. The "Interpersonal Interaction and Relationship" is improved by stimulating the person to get out home and interact with other people like neighbors. The same solution can be adopted to increase "Mobility". To improve the "Social Life", the smart home may support the free time management (e.g.: communicate social events and find a way to move there). Finally, the person's self care is supported by scheduling eating and drinking tasks. As to the environmental factors, "Support and Relationship" are improved by letting health-related people (such as caregivers or physicians) become more closer to the person's assistance needs. The technology (chats, visual call, etc.) improves the "Products and Technology" aspects of the person's life. By spending more time with the person (e.g.: helping to edit the behavior of the smart home), friends, relatives and family members may affect the "Attitudes" value. Finally, the home environment is improved by automating the smart home behavior, having the best environmental condition in the house (e.g.: lower and rise up the blinds in order to have the best light conditions).

5   A Model for Smart Home Development

In this work we aim at investigating how to support people in improving the ICF values through their smart home environment. In particular, we propose the EUD4SH model, in which user's needs and capabilities (assessed by the ICF model) and the smart home (supported by the IoT devices) meet and play together a key role of supporting people with EUD techniques. In other words, through EUD techniques, the end users (i.e., home inhabitants) or their caregivers (physicians, therapists, adult children, etc.) are allowed to set and manage the smart home in order to adapt it (and thus its smart devices) to the changing requirements. The model is presented as a cycle shown in Fig. 3. The left node in the cycle represents the "ICF (assessment)" of the inhabitants of a smart home environment. This step of the cycle is needed to objectively evaluate the

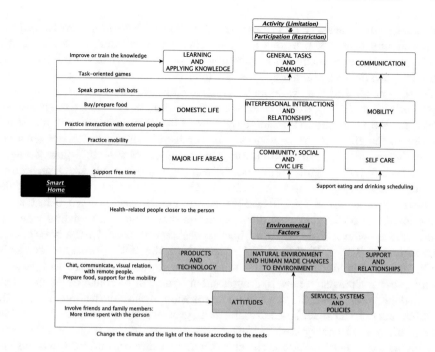

Fig. 2. The detailed influence area of the Smart Home on the ICF parts.

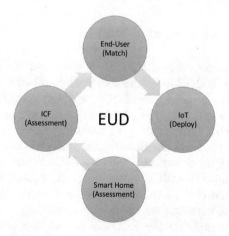

Fig. 3. The EUD4SH IoT model cycle

physical/psychological/health status of the inhabitants in order to better define the smart home to support them. The second node of the cycle is represented by the "End-User (Match)". This step is aimed at defining the best support type for the inhabitants according to individual's ICF values (coming from the previous node). The support is not always a static solution (the "one size fits all" idea), but it needs to be tailored to the end user's needs that evolve in time. Data coming from the previous node are then collected and "deployed" in the "IoT (deploy)" node. This node encompasses all the information related to the used and available technology. This is the step in which the relations between the users' needs and the available IoT technologies are studied and defined. The last node, the "Smart Home (assessment)" represents the smart environment, implemented using the IoT technology defined in the previous step. The orchestration of the different smart devices is also defined in this step. Finally, since the smart home influences the inhabitants, this node affects the "ICF (assessment)" node, thus starting again the cycle. In the proposed model, EUD techniques are needed in each of the four nodes: in the first node to possibly help also non-technical expert assess ICF values, customize the user's choice, select which functionalities of the deployed devices can be used, and finally define where to place the specific IoT devices and how to orchestrate them in the smart home environment. The EUD4SH model reflects how the ICF assessment cyclically influences and is influenced by the smart home definition. In case the person's health status changes, a new ICF assessment can then be reflected in the smart home deployment, resulting thus in an adaptation of the smart home itself to the new changes. EUD techniques are aimed at granting the possibility to consider the user's needs in every step of the described cycle. In this way, the inhabitant is able to customize and adapt the smart home to support and improve the ICF values.

6 Conclusions and Future Works

In this work we have proposed EUD4SH: a model for the smart home implementation related to the ICF assessment of the inhabitants. We have firstly analyzed how the ICF can be used to evaluate the health condition of a smart home inhabitant. Then, the end user (the physician, the caregiver etc.) can match the inhabitants' needs with the current (and available) smart home IoT appliances. Their behavior can be edited or tailored to the user's needs before getting used in the existing smart home context. This results in a last step in which a new assessment of the inhabitants ICF values takes place. At the center of the proposed model we included the EUD techniques, in order to involve the end users in the process for the development of their own solution.

We are currently applying the EUD4SH model in a couple of case studies. One of them is concerned with speech therapy with children, where the role of the smart home is to create a wraparound environment by means of smart lights and a wireless audio speaker. Children are encouraged by such an environment to correctly complete daily speech therapies. Parents and speech therapists play the role of end-user developers by customizing the smart home according to the child's preferences and therapy requirements.

References

1. Yang, R., Newman, M.W.: Learning from a learning thermostat: lessons for intelligent systems for the home. In: Proceedings of the 2013 ACM International Joint Conference on Pervasive and Ubiquitous Computing, pp. 93–102. ACM (2013)
2. Hondori, H.M., Khademi, M., Lopes, C.V.: Monitoring intake gestures using sensor fusion (microsoft kinect and inertial sensors) for smart home tele-rehab setting. In: 2012 1st Annual IEEE Healthcare Innovation Conference (2012)
3. Brandt, Å., Samuelsson, K., Töytäri, O., Salminen, A.-L.: Activity and participation, quality of life and user satisfaction outcomes of environmental control systems and smart home technology: a systematic review. Disabil. Rehabil.: Assist. Technol. **6**(3), 189–206 (2011)
4. Helal, A., Mokhtari, M., Abdulrazak, B.: The Engineering Handbook of Smart Technology for Aging, Disability and Independence. Wiley, Hoboken (2008)
5. Rashidi, P., Cook, D.J.: Keeping the resident in the loop: Adapting the smart home to the user. IEEE Trans. Syst., Man, Cybern. Part A Syst. Hum. **39**(5), 949–959 (2009)
6. Hwang, A.S., Hoey, J.: Smart home, the next generation: closing the gap between users and technology. In: AAAI Fall Symposium: Artificial Intelligence for Gerontechnology (2012)
7. Kolozali, S., Bermudez-Edo, M., Puschmann, D., Ganz, F., Barnaghi, P.: A knowledge-based approach for real-time IoT data stream annotation and processing. In: 2014 IEEE International Conference on, and Green Computing and Communications (GreenCom), IEEE and Cyber, Physical and Social Computing (CPSCom), pp. 215–222. IEEE (2014)
8. Lieberman, H., Paternò, F., Klann, M., Wulf, V.: End-user development: an emerging paradigm. In: End User Development, pp. 1–8. Springer (2006)
9. Fischer, G., Fogli, D., Piccinno, A.: Revisiting and broadening the meta-design framework for end-user development. In: New Perspectives in End-User Development, pp. 61–97. Springer (2017)
10. Cabitza, F., Fogli, D., Lanzilotti, R., Piccinno, A.: Rule-based tools for the configuration of ambient intelligence systems: a comparative user study. Multimed. Tools Appl. **76**(4), 5221–5241 (2017)
11. Caivano, D., Cassano, F., Fogli, D., Lanzilotti, R., Piccinno, A.: We@home: a gamified application for collaboratively managing a smart home. In: International Symposium on Ambient Intelligence, pp. 79–86. Springer, Heidelberg (2017)
12. World Health Organization: International Classification of Functioning. ICF World Health Organization, Disability and Health (2001)
13. Steiner, W.A., Ryser, L., Huber, E., Uebelhart, D., Aeschlimann, A., Stucki, G.: Use of the ICF model as a clinical problem-solving tool in physical therapy and rehabilitation medicine. Phys. Therapy **82**(11), 1098–1107 (2002)
14. Martin, S., Kelly, G., Kernohan, W.G., McCreight, B., Nugent, C.: Smart home technologies for health and social care support. In: The Cochrane Library (2008)
15. Cassano, F.: EUD models and techniques for the smart-home. IS-EUD **2017**, 103 (2017)
16. Atzori, L., Iera, A., Morabito, G., Nitti, M.: The Social Internet of things (SIoT)-when social networks meet the Internet of Things: concept, architecture and network characterization. Comput. Netw. **56**(16), 3594–3608 (2012)

Detecting Activities of Daily Living Using the CONSERT Engine

Mihai Trăscău, Alexandru Sorici[(✉)], and Adina Florea

University Politehnica of Bucharest, Bucharest, Romania
mihai.trascau@cti.pub.ro, {alexandru.sorici,adina.florea}@cs.pub.ro

Abstract. Context-awareness is central to many applications in Ambient Intelligence. The paper presents the functionality of a context reasoning engine, called CONSERT, for developing ambient intelligence applications, which builds on semantic complex event processing. The implementation of core rule processing using the DROOLS Fusion framework and the advantages gained through this approach are discussed. The resulting engine capabilities are presented for the case of recognizing activities of daily living (ADL) contained in the CASAS dataset, together with an evaluation of the obtained results.

Keywords: Context management · Complex event processing
Ambient intelligence · Activity recognition

1 Introduction

Context-awareness is a central topic of investigation in many applications, from domains such as the Internet-of-Things (IoT) [1], Ambient Intelligence (AmI) [2] or Active and Assisted Living (AAL). The semantics-based approach [3] to representing and reasoning about context information is very promising. It allows to easily incorporate existing domain knowledge into applications and offers options for adaptability and explainability which are highly important in user-centric AmI applications.

In previous research we have developed a context modeling and reasoning framework called CONSERT [4], which adopts an approach that combines ontology modeling and complex event processing (CEP). The proposed CONSERT Meta-Model is realized as an ontology and provides a foundation for modeling context information with a high degree of flexibility and expressiveness. The meta-model is exploited by the CONSERT Engine whose reasoning cycle is customized for AmI applications (e.g. addressing temporal reasoning and uncertainty of provided information).

However, the prototype CONSERT Engine implementation [4] suffered from performance limitations and difficulty in interpreting some of the desired reasoning expressions. The purpose of this paper is to present a new version of the

© Springer Nature Switzerland AG 2019
P. Novais et al. (Eds.): ISAmI 2018, AISC 806, pp. 94–102, 2019.
https://doi.org/10.1007/978-3-030-01746-0_11

engine, offering improved reasoning and throughput capabilities for AmI application development, based on the core capabilities of the DROOLS Fusion[1] CEP engine.

Specifically, Sect. 2 presents the functionality principles of the CONSERT Engine, while Sect. 3 details how the reasoning cycle is replicated on top of DROOLS. In Sect. 4 we evaluate engine functionality, showing that it can be successfully used for detecting ADLs based on a dataset from CASAS [6]. We compare to related work in Sect. 5 and conclude in Sect. 6.

2   CONSERT Engine Working Principles

The CONSERT Engine can be succinctly described as a semantic complex event processing engine that can implicitly handle temporal validity extensions of derived context situations. It leverages context information modeled using the CONSERT Ontology [4]. The latter is the realization of the CONSERT Meta-Model expressing context as *predicates* of arbitrary arity defined over *role entities*, which can be further *annotated* by meta-properties including provenance and quality of information data (e.g. $cleaning(Person), personLocated(Person, Room) : \{\lambda_{validity}, \lambda_{certainty}\}$).

Based on the ontology, the engine provides services for: insertion of new events (called *ContextAssertions* in CONSERT), prioritized execution of inference rules, detection and resolution of constraint violations (e.g. value or uniqueness constraints), as well as managing queries and subscriptions to the inferred situations.

Unlike traditional CEP engines, which execute various filtering or aggregation operations on a stream of data, the CONSERT Engine offers a specific information flow that addresses the reasoning challenges commonly found in AmI applications. Using the CONSERT Ontology, a developer can specify the means of acquiring new context information: *static* (it counts as a *fact* and is unchanging), *profiled* (specified directly by the user or the application itself and changing rarely), *sensed* (acquired from physical or virtual - e.g. a new message on a social media platform - sensors) or *derived* (inferred by the engine itself).

For the *sensed* and *derived* acquisition types the developer can instruct the CONSERT Engine to enact the following execution cycle:

1. **continuity analysis:** the engine checks if the content of the new ContextAssertion matches that of the recently inserted ones. Upon a content match, the engine looks at the annotations (meta properties such as timestamp, temporal validity, certainty, provenance) and uses annotation specific operators (cf. Sect. 3) to check if they indicate a *continuity* in the detected situation (i.e. the new ContextAssertion *extends* an existing one). If annotations allow for continuity, the old ContextAssertion is removed from the knowledge base and an updated event is inserted, making use of special operators to compute the new values of the annotations (cf. Sect. 3). Otherwise the new ContextAssertion is passed *as is* to the next step.

---

[1] http://drools.jboss.org/drools-fusion.html.

2. **constraint analysis:** A rule-based approach is used (cf. to details in [4])
to establish whether the new ContextAssertion violates defined constraints.
In case of violation, the resolution mechanism (cf. Fig. 1) decides whether
to keep or discard the new event or a previous one (e.g. in case of uniqueness constraints). *Prefer newer* and *Prefer certain* resolution mechanisms are
provided by default, but developers can provide their own.

In the prototype implementation [4] throughput of the engine was affected
by limitations in the underlying rule execution implementation (based on transactions in Jena TDB[2]). Furthermore, an adequate mechanism to indicate rule
priority was missing, as well as one that keeps count of event expiration (i.e.
lifetime of events in the runtime knowledge base). Lastly, the rule execution
implementation lacked means to *delay* rule consequences when *not* operators
relating to temporal windows were contained in rule conditions. For example, a
rule expressing that no motion event from sensor M1 has been observed for the
past five seconds, needs to explicitly wait for five seconds before executing the
rule consequences.

These limitations are now addressed given the new version of the CONSERT
Engine core implemented on top of DROOLS Fusion.

3   CONSERT Engine on Top of DROOLS Fusion

DROOLS[3] is a business rule management system in wide use today. One of
its modules, Drools Fusion, specifically addresses the domain of complex event
processing.

The DROOLS Fusion module offers an operation mode called *streaming*, in
which the temporal characteristics of inserted events are exploited (e.g. interval
logic operators, time-window based aggregation functions such as count, average,
min, max). It is this operation mode that makes it very suitable for exploitation
in upgrading the CONSERT Engine core functionality. However, the CONSERT
Engine specific reasoning flow described in the previous section (especially step
1 - continuity analysis) cannot be transferred directly on top of DROOLS Fusion.
A custom implementation is required which enables the continuity check using
newly defined operators that manipulate the annotations of ContextAssertions.
In what follows, we present the adaptations that were needed to bring about
the CONSERT Engine functionality using DROOLS Fusion, to achieve both the
initially designed functionality and address the limitations presented previously.

Event Representation. From a representation perspective, we define an object
oriented view[4] of the CONSERT Meta Model, meaning that we define classes
that capture ContextAssertions of different arities (unary, binary and n-ary),
as well as their associated annotations. For annotations, we currently model
temporal validity, timestamp and certainty where the latter two are defined *with*

---

[2] https://jena.apache.org/documentation/tdb/.

[3] https://www.drools.org/.

[4] GitHub link to CONSERT Model implementation: https://goo.gl/jWavX5.

respect to the validity interval. Annotations can be attached individually to a ContextAssertion, but an *annotation bundle* is offered as a convenience class to handle the ensemble of mentioned annotation types.

We use a Java library called RDFBeans[5] to convert to and from the ontology-based representation of a ContextAssertion and its POJO (Plain Old Java Object) equivalent.

Custom Annotation Operators. The validity extension mechanism requires a set of custom methods. *Annotation access and create/update methods* are required for handling meta-properties during inference. For example, DROOLS uses a `start` + `duration` combination, with the interpretation that an event introduced at timestamp t_{start}, with a duration of Δt, ends at timestamp $t_{end} = t_{start} + \Delta t$.

However, when performing situation continuity extension as explained in Sect. 2, the interpretation may change to imply that the insertion timestamp of a new ContextAssertion is its *latest update time*, while the start of the validity interval precedes the update: $t_{start} < t_{update}$. This means that the duration of an event expresses how long that contextual situation *has been valid*, instead of how long it will be.

During the continuity check process, extension of a previous situation is permitted only if the annotations of both existent and new ContextAssertion instances allow it. We implement *annotation operators for validity extension permission* that work on a threshold logic. For example:

- timestamp: allow extension *iff* update timestamps are within Δ_{max_ts}
- certainty: allow extension *iff* $c_{new} > 0.5$ and $|c_{new} - c_{exist}| < \Delta_{max\_cert}$

When the conditions of a context inference rule involve two or more ContextAssertions, the issue is how to compute annotation values of the *derived* ContextAssertion. Two cases are distinguished:

- *extension:* an existing ContextAssertion instance is extended by a new one atomic one (e.g. a *personLocation(Joe, kitchen)* event is extended by a new update from a motion sensor in the kitchen).
- *combination:* a ContextAssertion of a new type is inferred as a consequence of two or more ContextAssertions of different types (e.g. infer *cooking* situation based on *handling_food* and *using_burner*).

In either case, a developer can opt to set the new values manually using the *annotation access and update operators*. However, for convenience, we also implement default *annotation extension and combination operators* to automatically compute the value: timestamp (*extension: max, combination: max*), validity interval (*extension: union, combination: intersection*) and certainty (*extension: avg, combination: avg*).

Situation Validity Extension Mechanism. The CONSERT Engine reasoning cycle on top of DROOLS Fusion is illustrated in Fig. 1.

---

[5] https://rdfbeans.github.io/index.html.

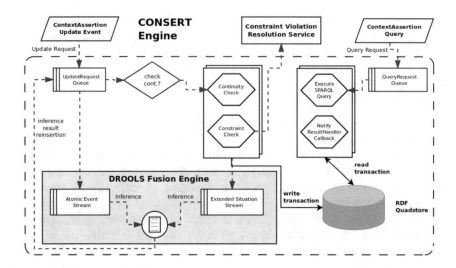

Fig. 1. Information and execution flow in the CONSERT Engine.

When a new ContextAssertion instance is inserted or derived, the engine first checks if it needs to undergo the verification for situation validity extension. The event is inserted as an atomic instance (i.e. as an instant in time) in either case. If it also undergoes the continuity verification, the event will be inserted in a separate stream. During inference, a developer can thus use both the individual atomic instances, as well as the temporally extended versions of a ContextAssertion type. The result of rule inferences in DROOLS Fusion are inserted back in the *Update Request Queue* to undergo the same continuity and constraint check processing.

An additional benefit of DROOLS Fusion is the use of the implicit event expiration policy, which eliminates ContextAssertions which can no longer affect any of the defined rules from the runtime knowledge base.

Each output of the DROOLS core (new ContextAssertions, or old ones that need to be removed) is converted also to its RDF representation using constructs of the CONSERT Ontology and stored/updated in an RDF quadstore (using transactions). This enables SPARQL queries over the resulting context information.

4 Evaluation: ADL Detection

Within the domain of context-aware AmI applications it is natural to consider the problem of recognizing activities of individuals. Of particular interest are activities which may yield some indication regarding a persons usual home behavior.

In order to evaluate the robustness and flexibility of the CONSERT Engine we decided to use CASAS [6], an established ADL dataset. The monitored smart environment is used to record 5 types of activities (*Telephone Use, Hand Washing, Meal Preparation, Eating and Medication Use* and *Cleaning*), performed *in sequence* by 24 different users. It is important to note that performing these activities implicitly activates multiple types of sensors (e.g. motion, cold water use, stove burner use, cooking utensils use), in different rooms of the smart environment, with slight variations in the sensor activations from person to person.

Starting from this setup and by manually analyzing the data, we have implemented several rules within the CONSERT Engine in order to detect the activities recorded in the CASAS dataset. We have defined functional zones in the environment whose spatial significance for certain activities are key in their correct identification. For example, being in the dining table area is a very strong signal that a person is eating. Likewise, *being* in the kitchen is necessary when preparing a meal.

```
rule "Remain at dining table"
    salience 20
    when
        $loc : PersonLocation(loc1 : loc == "DiningTable")
        not( exists PersonLocation(loc2: loc, loc2 != loc1,
            this annOverlappedBy[0s, 5s] $loc || $loc annIncludes this))
        not( exists Motion(status == "ON", this annHappensAfter[0s, 5s] $loc))
    then
        long ts = eventTracker.getCurrentTime();
        DefaultAnnotationData ann = new DefaultAnnotationData(ts);
        PersonLocation sameLoc = new PersonLocation("DiningTable", ann);

        eventTracker.insertAtomicEvent(sameLoc);
end
```

Fig. 2. Localization rule example

The rule described in Fig. 2 is an example of a person localization rule. It exemplifies the extension mechanism discussed in Sect. 3 in the sense that the current location is reconfirmed every 5 s as long as the person remains in place. It is also worthy to note the distinctive use of the *not* operator for CONSERT rules which allows expressing conditions regarding the absence of certain events in a given time window. Upon insertion of the inferred event, the continuity analysis mechanism will extend the validity of the previous *PersonLocation* event, which in turn will again trigger the rule, allowing to implicitly deduce the duration of a situation.

In order to correctly infer all the activities in the dataset using CONSERT's rule engine we have implemented a total of 23 rules[6], 6 of which are localization specific, while the rest refer directly to each of the ADLs (3–4 rules per activity). The approach we have taken is to model and derive the minimal amount of intermediary events needed to infer the correct ADL. For example, to infer the *Meal Preparation* activity we derive intermediate events such as *HandlingFood* and *HeatingFood*.

In the CASAS dataset, the ground-truth start and end timestamps of each activity are explicitly specified. In order to determine if the rules have assigned

---

[6] Rules used in the CASAS ADL Normal dataset evaluation: https://goo.gl/A8zkL2.

Table 1. Results for the CASAS dataset

Activity	Detected	Median start diff. (ms)	Median end diff. (ms)
Cleaning	24/24	−2300	14050
Meal Preparation	24/24	69700	7250
Eating	24/24	38700	43200
Telephone Use	24/24	22100	8050
Hand Washing	22/24	25350	8250

the correct activity labels within their respective intervals, we compute the difference between the ground-truth labels and the validity intervals detected by CONSERT. We report the median values for each activity in Table 1. It is worth mentioning that the larger differences observed for some activities are due to eager activity start and end labeling in the dataset. For example, the *Hand Washing* activity starts while the person is still in the living room, even before moving towards the kitchen. Our approach infers intervals which are most specific to that activity (e.g. *Hand Washing* starts only when the water tap is turned on), so we miss labeling 2 *Hand Washing* activities due to not having the water sensor running in the labeled ground-truth labeled interval. On the other hand, despite the variations (timings or sensor activations) between different trials for the same activities, the rule engine manages to correctly identify and label all the other instances.

Performance wise, the DROOLS core is a significant improvement over the previous prototype implementation. In a controlled load setting like the one in [4], with 689 atomic events and 276 derived ones, the *insertion delay, insertion processing* and *deduction delay* are all reduced to an average of 2 ms. However, the current reimplementation does not handle ontology reasoning, which is left for future work.

5 Related Work

Many of the early works in general context reasoning for AmI applications adopt semantics-based methods. COPAL [7] provides an processor-composition framework, where each processor can perform CEP-style data split, join, filtering and aggregation operations. SOCAM [9] and CoCA [8] rely on ontologies to model context information and on rules written in SWRL to augment the ontology reasoning capabilities. However, these works do not explicitly address temporal or uncertainty based reasoning and there is no apparent mechanism to account for situation expiration (once a situation is derived, how long does it remain valid in the knowledge base). Bikakis and Antoniou [2] address ambiguity and uncertainty of information, as well as context consistency via the Multi-Context Systems paradigm coupled with defeasible logics. While we do not handle non-monotonic reasoning, the explicit temporal validity of situations and the constraint resolution mechanism achieve an equivalent functionality.

With respect to recognizing ADLs, a number of recent works also use semantics-based methods. Riboni et al. [11] employ ontological reasoning to obtain necessary conditions for sensor events to belong to a defined activity. They additionally use rules to capture both knowledge-based and time-based constraints and dependencies between activity instances and sensor events. The work uses a fine-grained ontological modeling of expected place, duration and type of sensor activations for each activity type. In contrast, our approach includes the same of type of background knowledge *directly* in the rules that infer each activity type. This results in a simpler model that uses only strictly required domain and temporal dependencies between sensor activations.

In the iKnow [10] system an ontology model for activity recognition is proposed, which relies on existing ontology standards (DnS, DUL, SPARQL). The authors introduce the notion of *telicity* (endpoint) in recognizing interleaved activities. By modeling partial contexts and *observation neighborhoods* that help connect partial contexts into situations, iKnow manages to infer the start and end of activity instances as well as to link the instances that compose an interrupted activity. However, iKnow is meant to be used to interpret intermediate-level situations, where, for example, sensor observations are already annotated with the interval of validity. By comparison, our implicit temporal validity extension mechanism can operate directly on timestamped sensor activations, requiring no preprocessing steps.

6   Conclusions and Future Work

In this work we presented the reasoning cycle of the CONSERT context inference engine and its implementation on top of the DROOLS Fusion complex event processing framework. The main highlights of the CONSERT Engine are the implicit mechanism for extending the temporal validity of derived situations, as well as the explicit operators for manipulating annotations (including quality of data) of events during inference. Additionally, we exploit the DROOLS Fusion capabilities of formulating rule conditions based on the *non-triggering* of events during defined time-windows, as well as the inherent rule saliency mechanism and implicit expiration of events (once no more time-window based rule needs them).

We have evaluated the usefulness of these features on hand of an activity recognition task based on a dataset from CASAS. With respect to this, it can be argued that managing a growing set of interdependent rules for recognizing each activity type can become unwieldy in time. However, it is to be noted that the CONSERT Engine is intended as a general context reasoning solution, not dedicated solely to activity recognition. Additionally, the engine is meant to be exploited by the CONSERT Middleware [5], which offers a means to logically split the context model of an application, such that the rules relevant for a specific space or activity can be more easily managed.

In future work we plan to incorporate ontology reasoning into the inference cycle of the CONSERT engine. We intend to start by exploiting subsumption reasoning at first, such that a rule for detecting a very specific activity is complemented by inferring that the superclass activity is also valid during the same interval. A side benefit of this approach is that of offering control of granularity in context sharing/access control policies. The resulting improvement will be tested on further datasets stemming from the AAL domain.

Acknowledgement. This research was supported by the national grant PN-III-P2-2.1- 34PED.

References

1. Perera, C., Zaslavsky, A., Christen, P., Georgakopoulos, D.: Context aware computing for the internet of things: a survey. IEEE Commun. Surv. Tutor. **16**(1), 414–454 (2014)
2. Bikakis, A., Patkos, T., Antoniou, G., Plexousakis, D.: A survey of semantics-based approaches for context reasoning in ambient intelligence. In: European Conference on Ambient Intelligence, pp. 14–23. Springer, Heidelberg (2007)
3. Bettini, C., Brdiczka, O., Henricksen, K., Indulska, J., Nicklas, D., Ranganathan, A., Riboni, D.: A survey of context modelling and reasoning techniques. Pervasive Mob. Comput. **6**(2), 161–180 (2010)
4. Sorici, A., Picard, G., Boissier, O., Zimmermann, A., Florea, A.: CONSERT: applying semantic web technologies to context modeling in ambient intelligence. Comput. Electr. Eng. **44**, 280–306 (2015)
5. Sorici, A., Picard, G., Boissier, O., Florea, A.: Multi-agent based flexible deployment of context management in ambient intelligence applications. In: International Conference on Practical Applications of Agents and Multi-Agent Systems, pp. 225–239 (2015)
6. Cook, D.J., Schmitter-Edgecombe, M.: Assessing the quality of activities in a smart environment. Methods Inf. Med. **48**(5), 480 (2009)
7. Li, F., Sehic, S., Dustdar, S.: COPAL: an adaptive approach to context provisioning. In: IEEE 6th International Conference on Wireless and Mobile Computing, Networking and Communications (WiMob), pp. 286–293 (2010)
8. Ejigu, D., Scuturici, M., Brunie, L.: Hybrid approach to collaborative context-aware service platform for pervasive computing. J. Comput. **3**(1), 40–50 (2008)
9. Gu, T., Pung, H.K., Zhang, D.Q.: A service-oriented middleware for building context-aware services. J. Netw. Comput. Appl. **28**(1), 1–18 (2015)
10. Meditskos, G., Kompatsiaris, I.: iKnow: ontology-driven situational awareness for the recognition of activities of daily living. Pervasive Mob. Comput. **40**, 17–41 (2017)
11. Riboni, D., Sztyler, T., Civitarese, G., Stuckenschmidt, H.: Unsupervised recognition of interleaved activities of daily living through ontological and probabilistic reasoning. In: Proceedings of the 2016 ACM International Joint Conference on Pervasive and Ubiquitous Computing, pp. 1–12 (2016)

Comparative Results with Unsupervised Techniques in Cyber Attack Novelty Detection

Jorge Meira[✉], Rui Andrade, Isabel Praça, João Carneiro, and Goreti Marreiros

GECAD – Research Group on Intelligent Engineering and Computing for Advanced Innovation and Development, Institute of Engineering, Polytechnic of Porto (ISEP/IPP), Porto, Portugal
{janme,rfaar,icp,jomrc,mgt}@isep.ipp.pt

Abstract. Intrusion detection is a major necessity in current times. Computer systems are constantly being victims of malicious attacks. These attacks keep on exploring new technics that are undetected by current Intrusion Detection Systems (IDS), because most IDS focus on detecting signatures of previously known attacks. This paper explores some unsupervised learning algorithms that have the potential of identifying previously unknown attacks, by performing outlier detection. The algorithms explored are one class based: the Autoencoder Neural Network, K-Means, Nearest Neighbor and Isolation Forest. We use these algorithms to analyze two publicly available datasets, the NSL-KDD and ISCX, and compare the results obtained from each algorithm to perceive their performance in novelty detection. Our results show that the applied technics are indeed capable of identifying a considerable amount of the unknown attacks present in both datasets. Furthermore, our work has also addressed some pre-processing techniques and how they can be combined with the unsupervised algorithms to achieve the best combination in anomaly detection.

Keywords: Unsupervised learning · Anomaly detection · Outlier detection
Novelty detection

1 Introduction

Cyber-Security is a field that is constantly evolving, the rate by which new threats and attacks appear is enormous and this requires a constant research for vulnerabilities and ways of solving them by the people responsible for the Security Systems [1]. Intrusion Detection Systems (IDS) are tools based on attack detection techniques to finding out new vulnerabilities. IDS tend to follow one of two different approaches: signature based, or anomaly based. Signature based detection requires previous knowledge of an attack before being able to identify it, on the other hand anomaly base detection only requires knowledge of regular data, and any potential deviation from that norm can correspond to an attack, even if the attack has not been discovered yet [2]. This is an arduous task, and classification algorithms can be used to aid in this scenario. Some algorithms, called supervised learning algorithms are well suited for problems where exiting classified examples can be used as training data for the algorithm. However, with new vulnerabilities there are no classified examples for supervised algorithms to learn from. One

© Springer Nature Switzerland AG 2019
P. Novais et al. (Eds.): ISAmI 2018, AISC 806, pp. 103–112, 2019.
https://doi.org/10.1007/978-3-030-01746-0_12

possibility to help with this problem is the usage of unsupervised learning algorithms. Unsupervised learning algorithms can learn what is *normal* data and find deviations from that, which in this case would indicate a possible attack previously unknown (anomaly based).

The motivation for this work comes from the project SASSI [3]. The goal of the project is to create an Intelligent Decision Support System for detection, prediction and prevention of cyber-attacks in computer systems. In this paper, we focus on the detection of unknown attacks studying the performance of unsupervised algorithms in the task of detecting attacks in public datasets. These datasets are the NSL-KDD and the ISCX. The NSL-KDD contains examples of normal activity and several attacks divided in four categories: Denial of Service (DoS), Remote to Local (R2L), User to Root (U2R) and Probe. The ISCX contains samples of normal activity and attacks from: HTTP, SMTP, SSH, IMAP, POP3 and FTP protocols.

Our results show that the applied algorithms present good results in the detection of anomalies, especially when combined with the appropriate pre-processing technics. We also conclude that further research might be beneficial to lower the amount of normal activity classified as an anomaly.

This paper has the following structure: Sect. 2 presents some related work on this topic, Sect. 3 describes the datasets and algorithms used in our approach, Sect. 4 presents and discusses the results obtained in our exploration, and Sect. 5 draws some final considerations about this work and some comments and ideas of future work.

2   Related Works

Intrusion Detection Systems (IDS) is a frequent study topic in the literature, some authors have proposed and studied interesting techniques to deal with the problem of unknown attacks, this section presents some of those works.

Goldstein and Uchida [4] presented a comparative evaluation of unsupervised algorithms used in the context of Anomaly Detection. The algorithms were applied to a group of different datasets, one of each is the KDD 99, described in Sect. 3.1, however the analyses only used part of the dataset regarding HTTP traffic. It's important to note that an improved version of this dataset called NSL-KDD is presented and used in this paper.

A two-tier architecture to analyze network activity was proposed by Zanero and Serazzi [1] as a base for an IDS. The first tier consists in a data capturing algorithm that tries to separate the captured data by its similarities such as protocol type, the authors refer to this as "a pattern recognition problem applied to packet payloads" [1]. The second tier is an algorithm responsible for detecting outliers in the data, these outliers representing possible attacks to the network.

Aleroud and Karabatis [5] explored the detection of zero-day attacks, with an approach that combines already existing methods with linear data transformation techniques such as discriminant functions that separate the data in normal patterns from attack patterns, and anomaly detection techniques using the One Class Nearest Neighbor algorithm to identify the zero-day attacks. This approach was evaluated with the NSL-KDD dataset.

Casas and Mazel [2] presented the concept of an Unsupervised Network Intrusion Detection System (UNIDS), using Sub-Space Clustering and Multiple Evidence Accumulation techniques for outlier detection. Their evaluation of this system includes its application to the KDD 99 dataset.

Noto et al. [6] studied anomaly detection using an approach called FRaC, feature regression and classification. The FRaC technique builds a model of normal data and the distances of its features and uses the learnt model to detect when an anomaly occurs. Noto et al. [6] also compare their approach with other commonly used techniques, such as, Local Outlier Factor (LOF), One class support vector machines (One class SVM) and Cross-feature analysis (CFA).

3   Experimental Work

In our approach, we study the behavior of some unsupervised algorithms based in one class classification, to verify if these techniques are a viable solution to discover and detect unknown attacks. In this section, we describe the entire workflow of the network anomaly detection as shown in Fig. 1. We start with the description of the datasets used and the pre-processing techniques applied. Then the anomaly detection algorithms behavior is explained, such as the parameters applied in each of them.

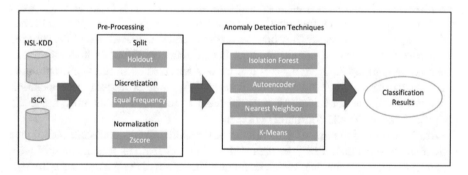

Fig. 1.   Anomaly detection workflow

3.1   Datasets and Preprocessing

In our exploration, we analyzed the NSL-KDD [6, 7] and the ISCX datasets [9]. These datasets contain samples from normal activity and from simulated attacks in computer systems and are commonly used in the literature. Before using the learning algorithms, we've used some pre-processing methods to prepare the data.

NSL-KDD

In 1999, in the third international competition in the conference Knowledge Discovery and Data Mining (KDD), the KDD 99 dataset [7] was presented to the scientific community. This dataset is highly used in the literature of IDS evaluation, and it contains simulated network activity samples, corresponding to normal and abnormal activity divided in five categories:

- **Denial of Service (DoS):** An intruder tries to make a service unavailable (contains 9 types of DoS attacks);
- **Remote to Local (R2L):** An intruder tries to obtain remote access to the victim's machine (contains 15 types of R2L attacks);
- **User to Root (U2R):** An intruder with physical access to the victim's machine tries to gain super-user privileges (contains 8 types of U2R attacks);
- **Probe:** An intruder tries to get information about a victim's machine (contains 6 types of Probe attacks);
- **Normal:** It constitutes the normal operations or activities in the network.

Tavallaee et al. [8] improved on KDD 99 and the result was the NSL-KDD dataset. This dataset is already organized in two subsets: one to train the algorithms, and another one to test them. Each data sample contains 43 features where four of them are nominal type, six are binary and the rest of them are numerical type. As we are testing one class classification algorithms, it was selected a portion of normal data from the training set and a portion of both normal and attack data from the test set, where the attack data contains all four attack categories and represents 10% of the test set.

Some pre-treatment techniques were applied to the dataset before preforming the discretization and normalization operations, as shown in Fig. 1. Some features were removed namely: 'Wrong_fragment', 'Num_outbound_cmds', 'Is_hot_login', 'Land' and 'level_difficulty' because they have redundant values in at least one of the subsets. In the case of the 'level_difficulty' feature, it represents the difficulty level of attacks detection by learning algorithms. This feature was removed because its information is not relevant in a real-world anomaly detection problem. Another pre-treatment operation to the data, was the conversion of nominal features to numerical features, since the algorithms cannot handle non-numerical data.

After performing the cleaning of the subsets, two different pre-processing techniques were applied to the data. First, the data with continuous features was discretized with the equal frequency technique. In this technique, the values of the features were divided in k bins in the way that each bin contains approximately the same number of samples. Thus, each bin has $\frac{n}{k}$ adjacent values. The value of k is a user defined parameter, and to obtain this value we used the heuristic $\sqrt{n}$ where n is the number of samples. The discretization technique can provide better accuracy and fast learning in certain anomaly detection algorithms, since the range of values is smaller [9].

The second pre-processing technique was the data normalization, to have all the features within the same scale. This way, prevents some classification algorithms to give more importance at features with large numeric values. Once the features are all on the same scale, the classifiers assign the same weight to each attribute. Z-Score was the normalization technique applied to the data. This technique transforms the input, so the mean is zero and the standard deviation is one. We tested the algorithms with each pre-processing technique and both combined to evaluate which techniques improve the performance of the algorithms.

ISCX

ISCX is a dataset developed by Shiravi et al. [10] at the Canadian cybersecurity institute. This dataset is based on the concept of profiles that contain detailed descriptions of abstract intrusions and distributions for applications, protocols, services, and low-level network entities. To create this dataset, real network communications were analyzed to create profiles for agents that generate real traffic for HTTP, SMTP, SSH, IMAP, POP3 and FTP protocols. In this regard, a set of guidelines have been established to delineate a valid dataset that establishes the basis for profiling. These guidelines are vital to the effectiveness of the dataset in terms of realism, total capture, integrity, and malicious activity [10]. Each data sample in the ISCX dataset contains 21 attributes. There is a total of 7 days of network traffic captured with 4 different attack types shown in Table 1.

Table 1. ISCX captured activity

Capturing date	Network activity
11/06/2010	Normal
12/06/2010	Normal
13/06/2010	Normal + Internal infiltration into the network
14/06/2010	Normal + HTTP Denial of Service
15/06/2010	Normal + Distributed Denial of Service using a Botnet IRC
16/06/2010	Normal
17/06/2010	Normal + Brute Force SSH

For this dataset, we did the following changes before applying the pre-processing techniques shown in Fig. 1 and described in the NSL-KDD dataset:

- All nominal features where converted to numerical – the algorithms used cannot handle non-numeric features;
- All "Payload" features were removed – These are string features, so it's not possible to train and test the algorithms with these features;
- The source and destination IP address features were removed – There is no interest in training the algorithms with these features, since the IP addresses are constantly changing;
- A new feature was created to represent the time interval of an operation on the network, defined as the difference between the features "stop date time" and "start date time".

3.2   Unsupervised Learning Algorithms

Unsupervised learning algorithms are suitable for scenarios where the objective is to perform outlier detection to a dataset. Some of these algorithms follow the basic idea of learning from a training dataset that only contains normal samples, and in the classification the output is either "normal" if it resembles the learnt set or "outlier" if it does not. These algorithms are denominated by one-class classification and appear to be good candidates for the problems of discovering unknown attacks, since every attack can be considered an outlier. In this work, we applied four one-class algorithms, namely

Autoencoder, Nearest Neighbor, K-Means, and Isolation Forest and evaluate is performance in the NSL-KDD and ISCX datasets.

Autoencoder
Autoencoder [11] is a type of neural network designed around the idea of data compression and decompression, where the number of neurons gets smaller in each layer to a certain minimum and after that it increases again to the original number of neurons. Its output is the reconstructed data and a reconstruction error. The autoencoder network can be used in anomaly detection by learning only with normal data. After the learning process, when a new sample is processed the reconstruction error is expected to be high if the sample represents an attack, so an error threshold is defined and when the reconstruction error is above the error threshold that sample is classified as an attack, represented in red dots, as shown in Fig. 2. In our exploration, autoencoder obtained the best results with a configuration of 50-5-50 neurons on the hidden layers and applying a hyperbolic tangent activation function in both datasets.

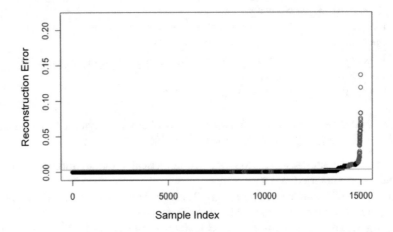

Fig. 2. Autoencoder reconstruction threshold example

One-Class K-Means
K-Means [12] is one of the most used clustering algorithms. One-Class K-Means works by first learning with a training dataset. The algorithm calculates the distances in between the features of all samples in the training dataset and makes clusters (groups of data) with the sample that has small distances between their features. To apply this algorithm in outlier detection the learning process should only use normal data examples, after that in the classification process the algorithm calculates the distance to the closest cluster. If that distance is higher that a defined threshold that sample is classified as an attack. In both the NSL-KDD and ISCX datasets the ideal number of clusters was 4.

One-Class Nearest Neighbor
The One-Class Nearest Neighbor is an adaptation of the original K-Nearest Neighbor supervised algorithm [13], it's used in outlier detection in a way similar to the One-Class

K-Means algorithm but instead of using clusters, this algorithm uses the distances between neighbors. The One-Class Nearest Neighbor algorithm starts by learning the distances in a training set, containing only normal data. Those distances are used to identify a maximum distance and to define a threshold. Then, in the classification process the algorithm starts by calculating the distance from the sample to its first nearest neighbor from the training set. If that distance is higher than the maximum distance threshold then the sample in question is classified as an outlier.

Isolation Forest

Isolation Forest [14] is a method for outlier detection that uses data structures called trees, such as binary trees. Each tree is created by partitioning the instances recursively, by randomly selecting an attribute and a split value between the maximum and minimum values of the selected attribute [14]. Each splitting phase represents a node in a tree, and the number of partitions required to isolate a point is equivalent to the path length from the root node to a terminating node. Lui and Ting [14] observed that the path length to isolate a normal point is greater than the path length of the anomaly point. So, this algorithm calculates the average path length of a set of trees and uses this value to determinate the anomaly score. In our tests, we used the default algorithm parameter of 100 trees in both datasets, since the variation of this parameter did not show any substantial impact in performance.

4   Comparative Evaluation

To compare the anomaly detection performance of the unsupervised algorithms we used four evaluation metrics [15]:

- The accuracy which corresponds to the number of correct predictions made as a ratio of all predictions made;
- Recall or sensitivity. It is the number instances from the anomaly class that actually predicted correctly;
- Precision. Is the proportion of anomaly class that were correctly identified;
- F1 score, representing the harmonic mean which combines recall and precision metrics with an equal weight.

We tested all combinations of pre-processing techniques with the unsupervised learning algorithms and graphically presented the results of the best techniques applied to each algorithm for NSL-KDD and ISCX datasets.

As shown in Figs. 3 and 4, and as expected, the algorithms One-class K-means and Nearest Neighbor had the best performance applying the Z-Score technique since they work with Euclidian distances. The Isolation Forest algorithm, had the best results without any kind of data transformation as it uses binary trees in the process of data recursive partitioning [14]. In the case of the Autoencoder, it got better performance in detecting anomalies applying Z-score and Equal Frequency (EF) techniques in the pre-processing phase.

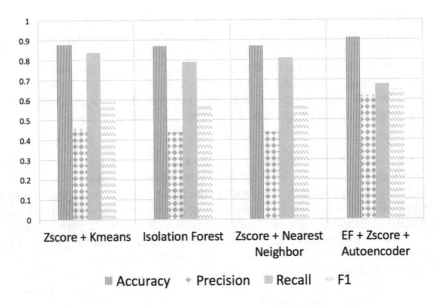

Fig. 3. Anomaly detection results in NSL-KDD dataset

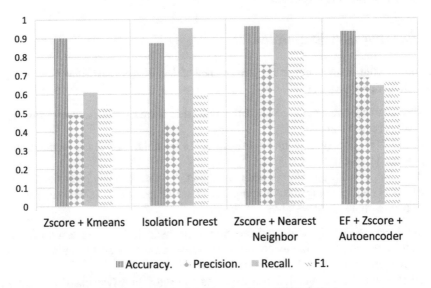

Fig. 4. Anomaly detection results in ISCX dataset

In both datasets all the algorithms had a high accuracy. That was expected as most of the samples are from normal activity. To achieve better conclusions about the algorithms efficacy the F1 metric will also be analyzed, since it combines the metrics recall (true positive rate) and precision (positive predicted values), where anomalies are represented by the positive class. Starting by the NSL-KDD dataset shown in Fig. 3, we can see that all of the algorithms had a similar result close to 60% of F1. It is not a high-performance score, however the recall was much higher than precision in K-Means, Nearest Neighbor and Isolation forest algorithms, around 80%, which means that the false negatives were much less than the false positives. In cybersecurity it is important to have a low false negative rate, since it represents data predicted as normal, while in fact it represents malicious or abnormal activity.

In the ISCX dataset (Fig. 4) all algorithms showed a slightly better performance, except the Nearest Neighbor algorithm which had a much higher score compared to the NSL-KDD. These results were expected in the ISCX, whereas this dataset only has 4 different types of attacks compared to the 38 different types in the NSL-KDD dataset. So, it can be said that these 4 types of anomalies could be isolated by the distance length between normal and abnormal instances.

5 Conclusion and Future Work

Threats in information systems have become increasingly intelligent and they can deceive the basic security solutions such as firewalls and antivirus. Anomaly based IDS, allows monitored traffic classification or computer system calls classification in normal activity or malicious activity. The efficiency of intrusion detection depends on the techniques used in these systems. In this paper, we presented a comparative evaluation of some one-class unsupervised based techniques, paired with pre-processing techniques, in the ambit of novelty detection using the NSL-KDD and ISCX datasets. The results showed that these unsupervised techniques combined with the best pre-processing techniques could detect most of the anomaly instances but also generate a lot of false positives. These occurs due to the similarity between normal and abnormal instances. Therefore, as future work, different approaches will be studied with the intent of reducing the problem of the high rate of false positives. Also, the techniques used in this paper will be applied to a new dataset, where the sensors of the SASSI system will collect the data. Then novelty detection techniques will be compared to this work. After creating a consistent predictive model, the aim will be the study of action rules to prevent the detected attacks.

Acknowledgements. This work was supported by SASSI Project (ANI‖P2020 17775), and has received funding from FEDER Funds through P2020 program and from National Funds through FCT - Fundação para a Ciência e a Tecnologia (Portuguese Foundation for Science and Technology) under the project UID/EEA/00760/2013.

References

1. Zanero, S., Serazzi, G.: Unsupervised learning algorithms for intrusion detection. In: IEEE Network Operations and Management Symposium, NOMS 2008, pp. 1043–1048. IEEE (2008). https://doi.org/10.1109/noms.2008.4575276

2. Casas, P., Mazel, J., Owezarski, P.: Unsupervised network intrusion detection systems: detecting the unknown without knowledge. Comput. Commun. **35**, 772–783 (2012). https://doi.org/10.1016/j.comcom.2012.01.016

3. SASSI Project. http://sassi.visiontechlab.com/

4. Goldstein, M., Goldstein, M., Uchida, S.: A comparative evaluation of unsupervised anomaly detection algorithms for multivariate data. PLoS One 1–31 (2016). https://doi.org/10.7910/dvn/opqmvf

5. Aleroud, A., Karabatis, G.: Toward zero-day attack identification using linear data transformation techniques. In: Proceedings of the 7th International Conference on Software Security and Reliability, SERE 2013, pp. 159–168 (2013). https://doi.org/10.1109/sere.2013.16

6. Noto, K., Brodley, C., Slonim, D.: FRaC: a feature-modeling approach for semi-supervised and unsupervised anomaly detection. Data Min. Knowl. Discov. **25**, 109–133 (2012). https://doi.org/10.1007/s10618-011-0234-x

7. UCI Machine Learning Repository: KDD Cup 1999 Data. 92697 (2015)

8. Tavallaee, M., Bagheri, E., Lu, W., Ghorbani, A.A.: A detailed analysis of the KDD CUP 99 data set. In: IEEE Symposium on Computational Intelligence in Security and Defense Applications, CISDA 2009, pp. 1–6 (2009). https://doi.org/10.1109/cisda.2009.5356528

9. Liu, H., Hussain, F., Tan, C.L.I.M., Dash, M.: Discretization: An Enabling Technique, pp. 393–423 (2002)

10. Shiravi, A., Shiravi, H., Tavallaee, M., Ghorbani, A.A.: Toward developing a systematic approach to generate benchmark datasets for intrusion detection. Comput. Secur. **31**, 357–374 (2012). https://doi.org/10.1016/j.cose.2011.12.012

11. Chen, J., Sathe, S., Aggarwal, C., Turaga, D.: Outlier detection with autoencoder ensembles. In: Proceedings of the 2017 SIAM International Conference on Data Mining, pp. 90–98 (2017)

12. Yoon, K.A., Kwon, O.S., Bae, D.H.: An approach to outlier detection of software measurement data using the K-means clustering method. In: First International Symposium on Empirical Software Engineering and Measurement (ESEM 2007), pp. 443–445 (2007)

13. Angiulli, F., Pizzuti, C.: Fast outlier detection in high dimensional spaces. Princ. Data Min. Knowl. Discov. 15–27 (2002). https://doi.org/10.1007/3-540-45681-3_2

14. Liu, F.T., Ting, K.M., Zhou, Z.-H.: Isolation-based anomaly detection. ACM Trans. Knowl. Discov. Data **6**, 3:1–3:39 (2012). https://doi.org/10.1145/2133360.2133363

15. Mitchell, T.M., Tom, M., Hansson, K., Yella, S., Dougherty, M., Fleyeh, H., Pham, D.T., Afify, A.A., Wuest, T., Weimer, D., Irgens, C., Thoben, K.-D.: Machine learning in manufacturing: advantages, challenges, and applications. Proc. Inst. Mech. Eng. Part B J. Eng. Manuf. **4**, 1–13 (2016). https://doi.org/10.5923/j.ajis.20160601.01

Extracting Clinical Information from Electronic Medical Records

Manuel Lamy[1], Rúben Pereira[1], João C. Ferreira[1(✉)], José Braga Vasconcelos[2,3], Fernando Melo[4], and Iria Velez[4]

[1] Instituto Universitário de Lisboa (ISCTE-IUL), ISTAR-IUL, Lisbon, Portugal
{manuel_lamy,rfspa,jcafa}@iscte-iul.pt
[2] Universidade Europeia, Laureate International Universities, Lisbon, Portugal
jose.vasconcelos@universidadeeuropeia.pt
[3] Centro de Administração e Politicas Publicas (CAPP) da Universidade de Lisboa, Lisbon, Portugal
[4] Hospital Garcia de Orta, Almada, Portugal
{fmelo,iria.velez}@hgo.min-saude.pt

Abstract. As the adoption of Electronic Medical Records (EMRs) rises in the healthcare institutions, these resources' importance increases because of the clinical information they contain about patients. However, the unstructured information in the form of the narrative present in those records makes it hard to extract and structure useful clinical information. This limits the potential of the EMRs, because the clinical information these records contain, can be used to perform important operations inside healthcare institutions such as searching, summarization, decision support and statistical analysis, as well as be used to support management decisions or serve for research. These operations can only be done if the clinical information from the narratives is properly extracted and structured. Usually, this extraction is made manually by healthcare practitioners, what is not efficient and is error-prone. This research uses Natural Language Processing (NLP) and Information Extraction (IE) techniques in order to develop a pipeline system that can extract and structure clinical information directly from the clinical narratives present in Portuguese EMRs, in an automated way, in order to help EMRs to fulfil their potential.

Keywords: Electronic Medical Records · Information Extraction
Machine Translation · Natural Language Processing · Text mining

1 Introduction

Electronic Medical Records (EMRs) are computerized medical systems that collect, store and display a specific patient clinical information [1]. These records are used "by healthcare practitioners to document, monitor, and manage health care delivery within a care delivery organization (CDO). The data in the EMR is the legal record of what happened to the patient during their encounter at the CDO and is owned by the CDO" [2]. Many types of clinical information are stored in EMRs, such as x-rays, prescriptions,

© Springer Nature Switzerland AG 2019
P. Novais et al. (Eds.): ISAmI 2018, AISC 806, pp. 113–120, 2019.
https://doi.org/10.1007/978-3-030-01746-0_13

physician's notes, diagnostic images and other types of medical documentation [3]. EMRs became one of the most important new technologies in healthcare [4].

Hospitals play a central role in the healthcare domain and in any society. These health care institutions produce large amounts of digital information, mainly with the broad utilization of EMRs. In the United States, a study [5] from 2012 showed that 72% of office-based physicians used an EMR system. In Portugal, statistics from 2014 [6] show that the number of hospitals using EMRs rose from 42% in 2004 to 83% in 2014.

EMRs usually contain unstructured information in the form of narrative [7] written by the healthcare practitioners, concerning the patients. However, the amount of unstructured information that is contained in the EMR presents a barrier to realizing the potential of EMRs [8]. This free-text form used by healthcare practitioners is advantageous to "demonstrate concepts and events but is difficult for searching, summarization, decision support or statistical analyses" [9].

Healthcare practitioners extract clinical information from the EMRs' narratives "by employing of domain experts to manually curate such narratives" [8]. This practice is not efficient, is error-prone and consumes human resources that could be used for other tasks [10]. The desirable scenario is to be able to extract clinical information from the EMRs' narratives in an automated and reliable way. The proposed system in this research aims to provide a solution to extract clinical information from the clinical narratives of Portuguese EMRs in an automated and structured way, in order to facilitate the day-to-day activities of healthcare practitioners and help the healthcare institutions to have faster and reliable access to the patients' clinical information.

2   Review of Literature and Positioning of This Work

There are already several NLP systems capable of extracting clinical information from EMRs, with the significant part of them only working for the English language. A review of some of these systems is depicted in this study [9]. However, a review of the literature in the NLP domain revealed that there are not many studies and systems that focus on extracting clinical information from EMRs written in the Portuguese language.

An NLP system capable of extracting clinical terms from discharge records written in European Portuguese is MedInX, developed in the Institute of Electronics and Telematics Engineering of Aveiro in 2011 [10]. Despite having a good performance, this is still a proprietary system, and it was only tested concerning the extraction of clinical terms related to hypertension.

There are also studies made for the Brazilian variant of the Portuguese language, concerning IE from clinical data. A study conducted by the Faculty of Medicine, University of São Paulo, in 2007, proposed a pipeline system capable of extracting clinical terms from clinical reports, by coupling Machine Translation (MT) and an NLP system together [11]. However, despite having a good performance, this study only aimed to extract twenty-two different types of clinical terms and was limited to chest x-rays reports. To add to that, this research is from 2007, and since then the MT and NLP systems were improved. Nonetheless, this study showed that is indeed possible to achieve success, by coupling MT and NLP together to extract clinical terms correctly.

In this research, the authors pretend to use a similar approach to the problem by using Machine Translation(MT) first, in order to translate the clinical texts from Portuguese to English and only after performing the clinical information extraction from EMRs using NLP and IE techniques. This approach is justified by the fact of almost all dictionaries and ontologies being much more mature for the English language, allied to the predominance of the English language in the biomedical field. To add to that, the significant part of MT and NLP systems are built and optimized to work with the English language and are more improved than ever. The authors also pretend to use the only open-source software during the whole process and analyse the findings, in order to build a solution immediately available to everyone.

3 Information Extraction of Electronic Medical Records

Natural Language Processing (NLP) is a "theoretically motivated range of computational techniques for analysing and representing naturally occurring texts at one or more levels of linguistic analysis for the purpose of achieving human-like language processing for a range of tasks or applications" [12]. Information Extraction (IE), considered a subfield of NLP, is the task of retrieving certain types of information from unstructured natural language texts by processing them in an automatic way. IE differs from Information Retrieval (IR) since the goal of IR is retrieving a subset of documents relevant to a query and the goal of IE is to extract information from the documents themselves. NLP, expressed by means of IE, can then be used to extract clinical information from the unstructured clinical narrative texts present in EMRs.

4 Importance and Motivation of Extracting Clinical Information

The narrative parts present in the EMRs have many purposes, such as give a description of the patient clinical situation, be used for management decisions, support research and serve legal issues. The clinical information present in these narratives, if extracted and structured, can be used to perform operations like search, summarize or to execute statistical analyses. The significant part of the time, manual extraction of clinical information is done, typically by healthcare practitioners [13]. However, when the extraction, standardization and structuring of clinical information are performed manually by the healthcare practitioners, text ambiguity and personal differences can lead to inconsistencies [14]. Replacing the narratives with structured information would also not be ideal since significant information is lost because of limiting the expressive power of narratives [15].

The motivation of this research is to show that is possible to extract clinical information in an automated way from EMRs written in the Portuguese language, with precision and neither losing the expressive power of the narrative nor clinical information. The clinical information that the authors propose to extract from EMRs is concerned only with the patients' diseases, symptoms, medications, anatomical regions and clinical procedures.

This automated extraction helps to reduce the amount of time and resources used by the hospital for performing manual analyses of the EMRs, by automating the process of extracting clinical information, using an NLP system and IE techniques. Therefore, by extracting clinical information in an automated way, operations like searching, summarizing and doing statistical analyses of the clinical information extracted, can be done faster and efficiently by CDOs, since once the clinical information is extracted it can efficiently be structured in a database and worked on in order to extract valuable clinical knowledge from it.

5   Methodology Applied

In order to extract clinical information from the Portuguese EMRs, a whole system based on open-source modules is coupled together. Firstly, the authors need a translator able to translate the EMRs' clinical narratives from Portuguese to the English language, with reasonable performance. Secondly, the authors need an open-source NLP system responsible for performing the clinical information extraction from the EMRs' narratives. To explain the methodology applied, a top-down approach is used. First, a high-level overview of the whole system followed then by a more refined explanation of each sub-system. A high-level depiction of the pipeline system the authors aimed to build on this research is shown in Fig. 1.

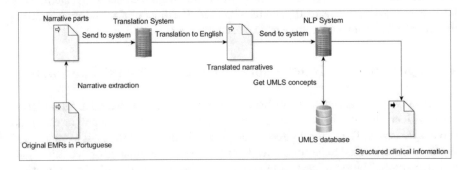

Fig. 1. A high-level view of the whole pipeline system

As can be seen in Fig. 1, the whole process initiates by extracting the narratives from the EMRs. After that, the authors send the narratives to the open-source translation system, in order to translate each one of them to the English language. After all the narratives being translated, the authors send each one of them to the NLP system, in order to extract diseases, symptoms, medications, anatomical regions and clinical procedures. The NLP system uses a database filled with clinical terms and concepts from the Unified Medical Language System (UMLS) in order to be able to identify and extract the clinical entities found in the narratives. UMLS is a repository of biomedical vocabularies developed by the US National Library of Medicine, containing more than 2.2 million concepts and 8.2 million concept names, some of them in different languages than English [16]. Finally, the NLP system extracts diseases, symptoms, medications, anatomical regions and clinical procedures, from the EMRs' narratives, outputting a file with all of the information extracted, plus the correspondent EMR identifier that gave origin to those extractions and the correspondent clinical speciality.

6 Machine Translation and Natural Language Processing

Machine Translation is considered a sub-field of computational linguistics that consists in using software to translate text or speech from one language to a different one. The authors are still studying which open-source translator fits best this research's needs and ambitions. From the translators already considered, an open-source translator called OpenNMT [17], developed at Harvard University in 2017 and with major source contributions by a proprietary translator seems a strong possibility. OpenNMT uses neural machine translation, a recently proposed approach to MT, that unlike statistical and rule-based MT, aims to build a single neural network that can be tuned and trained to achieve maximum translation performance. The authors are working on this process to obtain the best performance possible in the translation from Portuguese to the English language.

The research was already conducted by the authors to verify which open-source NLP system should be used. From several options, an open-source NLP system developed in the Mayo Clinic College of Medicine in Rochester called cTAKES [8], is our final decision to use in this research. This system is currently maintained by the Apache Software Foundation. This system was already used with success to identify the patients smoking status from clinical texts [18], confirm cases of hepatic decompensation in radiology reports [19] and extract clinical information concerning Crohn's disease and ulcerative colitis from EMRs [7].

7 Data Description

The hospital made available 34295 EMRs in an Excel file. The EMRs were collected with permission and previously de-identified by the hospital itself. Of these EMRs, 11358 are from ambulatory care, while the others 22937 are from inpatient care. The EMRs from ambulatory care are from different specialities of the hospital, such as gastroenterology, haematology, nephrology, neurology, medical oncology, paediatrics, pulmonology, rheumatology, pain unit, urology and oncology. The three clinical specialities more represented by the EMRs can be seen in Table 1.

Table 1. Most represented specialities

Speciality	Number of EMRs
Medical oncology	3199
Pain unit	2058
Haematology	1810

Each EMR is composed of different fields, such as a sequence number, number of the clinical episodes, speciality description, diagnosis description, date and clinical narrative text. Table 2 presents the top 5 of the most frequent diagnoses. Table 3 presents the obtained statistical results regarding the clinical narratives.

Table 2. Diagnosis count

Diagnosis	Occurrences
Tumors(neoplasms)	3972
Blood and hematopoietic organs diseases	2188
Osteoarticular system diseases	2023
Infectious and parasitic diseases	1494
Nervous system diseases	455

Table 3. Statistical analysis of EMRs' clinical narratives

Criteria/Type of Care	Ambulatory	Inpatient
Total number of words	465862	3627576
Mean number of words per narrative	41	158
Most used the clinical word	Transfusion	Pain

8   Results

This section presents the results of some initial experimental tests performed, using part of the EMRs and processing each one individually. No metric-based evaluations were made yet. For these tests, the translator system used was OpenNMT, and it was not pre-trained at all. The NLP system used was cTAKES. The authors only wished to verify the behaviour and performance of the systems coupled together, without making any kind of tuning and configurations yet. An example of a translated clinical narrative already processed by cTAKES is shown in Fig. 2. Figure 2 has some handmade annotations in order to facilitate the explanation of the figure. In the right side of Fig. 2, one can see the clinical narrative translated into English that is processed by cTAKES while the left side presents the analysis results shown by the system.

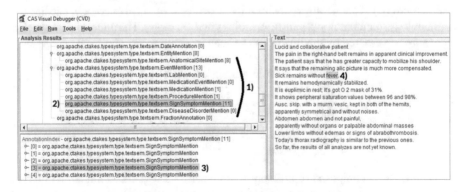

Fig. 2.   - Example of cTAKES' processing results

Considering Fig. 2 and concerning clinical entities, by default cTAKES is able to identify several clinical mentions in the text, as can be seen in the annotation (1) in Fig. 2. In the annotation (2) it can be seen that is possible to select a group of mentions and iterate by each one of them, as is being done in annotation (3) right below. By selecting each one of the mentions, its position appears immediately highlighted in the clinical narrative, as shown by the symptom annotation (4) of "fever", in Fig. 2.

The cTAKES system can also output the processing results in different formats, such as XMI, XML, HTML, plain text or directly to a database. This allows a simple structuration of the data automatically extracted, preparing this data to be easily used as clinical information, what opens a wide range of possibilities that can hugely benefit CDOs and their healthcare practitioners in their day-to-day activities.

9   Conclusions

The authors were already capable of doing some experimental tests of the pipeline system aimed to build with open-source software and with real EMRs from the hospital. Those tests already extracted clinical information from the EMRs' narratives with reasonable success. A precious factor in this research is having thousands of real EMRs, from a hospital itself, working with in order to tune the systems used.

By translating the narrative parts of EMRs given by the Portuguese hospital, not all terms and concepts were correctly translated into English. Not an unexpected finding since the MT system was not trained yet. This is an issue to improve, by training the translating system and adding some rules to it as needed. However, grounded on the performed tests, the authors conclude that the significant part of each narrative was well translated. Improving the MT process and its performance is the authors' immediate objective in the near future. A pipeline system like this one could be of great use by any healthcare institution that works with Portuguese clinical data, as a tool to extract clinical information in a fast and automated fashion, that could be used to improve the hospital operations that rely on that said clinical information.

References

1. Boonstra, A., Broekhuis, M.: Barriers to the acceptance of electronic medical records by physicians from systematic review to taxonomy and interventions. BMC Health Serv. Res. **10**(1), 231 (2010)
2. Garets, D., Mike, D.: Electronic Medical Records vs. Electronic Health Records: Yes, There Is a Difference. HIMSS Analytics, EMR vs. EHR: Definitions The marke. Heal, pp. 1–14 (2006)
3. Meinert, D., Peterson, D.: Anticipated use of EMR functions and physician characteristics. IGI Glob. **4**(June), 1–16 (2009)
4. Murphy, E.C., Ferris, F.L., O'Donnell, W.R., O'Donnell, W.R.: An electronic medical records system for clinical research and the EMR EDC interface. Invest. Ophthalmol. Vis. Sci. **48**(10), 4383–4389 (2007)
5. Hsiao, C.-J., Hing, E.: Use and characteristics of electronic health record systems among office-based physician practices: United States, 2001–2012. NCHS Data Brief. 1–8 (2012)

6. Instituto Nacional de Estatistica, Statistics Portugal (2014). https://www.ine.pt/xportal/xmain?xpgid=ine_main&xpid=INE. Accessed 03 Feb 2018
7. Ananthakrishnan, A.N., et al.: Improving case definition of crohn's disease and ulcerative colitis in electronic medical records using natural language processing. Inflamm. Bowel Dis. **19**(7), 1411–1420 (2013)
8. Savova, G.K., et al.: Mayo clinical Text Analysis and Knowledge Extraction System (cTAKES): architecture, component evaluation and applications. J. Am. Med. Informatics Assoc. **17**(5), 507–513 (2010)
9. Meystre, S.M., Savova, G.K., Kipper-Schuler, K.C., Hurdle, J.F.: Extracting information from textual documents in the electronic health record: a review of recent research. IMIA Yearb. Med. Informatics Methods Inf. Med. **47**(1), 128–144 (2008)
10. Ferreira, L.: Medical Information Extraction in European Portuguese, p. 262 (2011)
11. Castilla, C.: Instrumento de Investigação Clínico-Epidemiológica em Cardiologia Fundamentado no Processamento de Linguagem Natural, p. 112 (2007)
12. Liddy, E.D.: Natural language processing. Encycl. Libr. Inf. Sci. (2001)
13. Thomas, A.A., et al.: Extracting data from electronic medical records: validation of a natural language processing program to assess prostate biopsy results. World J. Urol. **32**(1), 99–103 (2014)
14. Suominen, H.: Machine Learning and Clinical Text. Supporting Health Information flow, no. 125. TUCS Dissertations No 125 (2009)
15. Lovis, C., Baud, R.H., Planche, P.: Power of expression in the electronic patient record: structured data or narrative text? Int. J. Med. Inf. **58–59**, 101–110 (2000)
16. Kleinsorge, R., Tilley, C., Willis, J.: Unified medical language system (UMLS). Encycl. Libr. Inf. Sci. 369–378 (2002)
17. Klein, G., Kim, Y., Deng, Y., Crego, J., Senellart, J., Rush, A.M.: OpenNMT: Open-Source Toolkit for Neural Machine Translation (2017)
18. Savova, G.K., Ogren, P.V., Duffy, P.H., Buntrock, J.D., Chute, C.G.: Mayo Clinic NLP system for patient smoking status identification. J. Am. Med. Informatics Assoc. **15**(1), 25–28 (2008)
19. Garla, V., et al.: The Yale cTAKES extensions for document classification: architecture and application. J. Am. Med. Informatics Assoc. **18**(5), 614–620 (2011)

Time Orientation Training in AAL

M. A. Guillomía, J. I. Artigas, and J. L. Falcó[✉]

University of Zaragoza, Zaragoza, Spain
jfalco@unizar.es

Abstract. A time display and time training device has been developed in collaboration between University and Special Education Schools, with which an empirical intervention is to begin shortly. Former trials in children provided evidences of time managing improvement, reduction of stress in changes of activity and usefulness in managing behavior contention.

This communication describes current version of the prototype which has been adapted to the experimental setup to be started this year which integrates feedback obtained in previous trials, in accordance with previewed experimental intervention.

Time passing is represented by turning off sequentially and gradually a row of luminous elements initially on, which provides for association with time perception. Besides luminous elements, tasks display provides for association of agenda-information. Tasks management support and anticipation to changes is supported by setting different colors to the luminous elements and by providing luminous or sound messages. Disruptive behavior control training is also supported by added flexibility in configuring different times and feedbacks.

Training tools are integrated in an AAL platform that includes home control and AAL specialized interface, keeping in mind its potential application to other cognitive impairments as Alzheimer disease.

Keywords: Time orientation · AAL interface · Human Machine Interface
Training in AAL · Special education schools

1 Introduction

1.1 Aims and Scope

Temporal Orientation has been prioritized as a basic capacity to improve personal autonomy by education professionals of Alborada Special Education School in Zaragoza [1]. There has been a maintained collaboration between the University of Zaragoza and Alborada Special Education School in societal awareness of disability and in looking for solutions to children difficulties for more than 10 years. One of the collaboration areas is time orientation.

In such frame, we seek deeper understanding of training time-cognitive functions as well as a practical tool to improve this area in children with special needs. In order to assess and formalize intervention results, this action plans to perform a classical empirical research methodology with pre-intervention time-capacities assessment,

© Springer Nature Switzerland AG 2019
P. Novais et al. (Eds.): ISAmI 2018, AISC 806, pp. 121–128, 2019.
https://doi.org/10.1007/978-3-030-01746-0_14

intervention and post-intervention assessment. Goal population is children in special education schools.

Goals of such training with children include (1) time perception capacity, time and agenda management; (2) make task-changes emotionally softer by supporting awareness of their vicinity; and (3) altered-behavior control support.

Additionally, since its integration in an AAL platform, the aims are boosted to a global Human Machine Interface (HMI) as far as time is concerned.

1.2   Previous Works and State of the Art

Time Orientation Concept. Classical human functions classification by the World Health Organization is a key document in cognitive functions [2]. It sets human time orientation as body functions and participation functions. As mental body functions it is included in items b1140 "awareness of day, date, month and year"; as part of superior cognitive functions it is included in item b1642 "time management by ordering events in a time sequence" and b1802 "subjective experience of duration and passing of time". As participation functions it is considered in the item 2306 "adapting to time demands". Another key document in time orientation concept is the PhD thesis of Gunnel Janeslätt [3, 4] which further details those items by gradual cognitive evolution, from time perception to time orientation, to time management and to global time processing ability. In that work, the assessment tool KaTid is developed, which we have found very well suited for our intervention with children.

Time Orientation Intervention. The original idea from which our intervention originates comes from Arne Svensk's report about the use of a quarter of an hour as an understandable time unit and about sequential turning off of luminous elements in a row as an comprehensible representation of time passing [5–7]. He reported the use of this approach for phone assistance to cognitive disabled adults for their time management, as shown in [6]. The quarter hour clock [8] was derived from his experience, which is a reference in assistance devices and is being extensively used in AAL field. Three cases studies about the use of "Quarter Hour Watch" have been written by two occupational therapists (Rose-Marie Remvall and Karin Mansson). One of the studies is based on his extensive experience in the introduction of this special watch in the early nineties, where it was used to perform the various daily tasks [7].

In our previous studies [9, 10], tasks description images were also used for the various activities (in school, day center and at home) together with the row of luminous elements. The joint use of the foregoing caused the perception of time to show a substantial improvement.

Other few studies report good results by using time aids: in [11] authors showed that independence and autonomy should be considered as two separate concepts and report the importance of having frequent communication with the user to understand the usability of the aids. A recent study [12] concludes the effectiveness of time aids to improve time-processing ability and managing one's time in children with intellectual and developmental disabilities.

2 Description of the Time Orientation Device

2.1 Forework and Specifications

From [7–10] some aspects have been considered to drive the time aid design:

- User's attention must be gained and advance notice must be given and use of sound and lighting/images information must be designed to improve understanding and the ability to gain user's attention. Prototype follows such suggestions from teachers.
- Provide enough flexibility by facilitating teaching professionals the programming of time indicators adapted to each user. Prototype considers this flexibility.
- Provide information about the environment through the programming of specific tasks. As part of a platform, software may use environmental information.

Following these premises, specific objectives were set to be included in the Time Orientation Device (DOT following Spanish initials). In particular, it is essential to try and obtain a tool that supports people with disabilities in:

- Perceiving the passage of time.
- Improving time orientation: awareness of time of day, week and time of year.
- Training for time management or task planning, so increasing independence.
- Anticipation of task changes to reduce associated stress.

Prototype described here is the fourth version of the DOT and part of a model developed in 2006 [9], 2011 [10], and later redesigned in 2015. Currently there is a prototype running at school for teacher's acquaintance and testing.

DOT is integrated into an AAL platform designed for schools, an extension included within the TICO-home-control system presented in [13]. TICO-home-control was installed during the last school year in the special education school to study adaptations and training strategies. Case studies and structured and semi-structured observations were applied with five children.

2.2 Description of the Time Panel

Hardware device consists of a translucent plastic panel of 81x19x6 cm that integrates a series of 40 luminous elements (LEDs) which are configurable in color and brightness. Each element symbolizes a configurable time unit (quarter of an hour by default). Each LED either gets off or changes color with the passing of time. Next events are announced in a visual and acoustic way by a control device located in the upper part of the DOT. So stimuli to draw attention are given with sufficient time to avoid anxiety and to motivate the user. The device is ready to be hung on any surface (Fig. 1).

Fig. 1. (A) Initial aspect of the DOT, (B) Current aspect

2.3 Electronics

DOT electronics consists of two parts, the first one is formed by five LED boards, each one including eight high-brightness RGB LEDs together with its control chips (MAX7315). Every board has three control chips, one for each basic color, and is managed by an I2C interface. All LED boards are connected in series to be managed by a single I2C port. The second part is currently based on a Raspberry Pi (RPI) which is a low cost single board computer specifically developed for educational purposes. Previous versions used a Zigbee module to operate the LED boards from a remote computer, or an autonomous microcontroller board to operate the LEDs which provided autonomy from the classroom computer however limiting functionality about multimedia feedback and connectivity (Fig. 2A). Current version is equipped with a RPI boosting its connectivity, multimedia feedback and database management capacities (Fig. 2B).

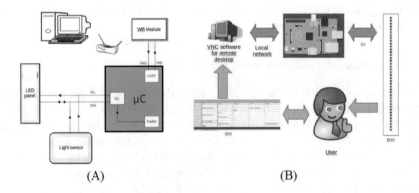

(A)   (B)

Fig. 2. (A) DOT version 2011, (B) DOT version 2015

Suggested by teachers, RPI has been equipped with a 3.5" 320x480 TFT screen to display the pictogram associated with each of the scheduled tasks, a set of speakers to reproduce the sound associated with the task, a Wi-Fi dongle for wireless communication and a webcam. LED boards are controlled with the GPIO connector of RPI.

2.4   Software

DOT uses Raspbian Jessie operating system, which is free and open source operating system optimized for the RPI hardware. It includes an open source HTTP web server (apache) as well as a general programming language of server-side code for dynamic web content development (php), a database data management system (MySQL) and a tool to manage MySQL databases (phpMyAdmin).

Multiple web applications have been developed to handle several databases (users and permissions, DOT configuration, logs…) besides some Python applications for the management of RF433 devices, webcam, mentioned I2C communication, video and audio playback applications.

2.5   Communications

Communication with the system can be done through the connection, as a client, through an SSH terminal, remote desktop, either locally (connecting directly to the RPI via Ethernet or Wi-Fi) or remotely since the system is provided with a DDNS service. This allows modifying the DOT tasks by programming them remotely. This communication can be established from any computer, smartphone, tablet, regardless of the operating system used (Windows, Linux, Android, Mac, etc.) in order to improve the usability and flexibility of the system.

3   User Interface Description

User interface integrates feedback obtained by workshops with teachers and previous trials, trying to maintain it simple for basic and standard configurations and providing enough flexibility for all intervention goals.

When connecting to the RPI website, one can see the control environment shown in Fig. 3. Main menu screen is divided into two windows: scheduled task information and task introduction or editing.

First one shows current date and time, list of scheduled tasks for the whole day and the whole week, selection tool to delete or edit tasks one by one, task status in time (past, current and pending). Task status is shown with different color patterns: background color regards days: gray for days gone by, yellow for tasks for the current day and cyan/blue for tasks for future days. Font color displays task time regarding present moment: finished tasks have green font, current tasks purple and pending tasks blue.

The bottom window in Fig. 3 shows the editing window, where you can: (1) create new tasks; (2) edit any scheduled task to change the day of the week, start time, end time, name of the task, as well as the possibility to include some notes; (3) send picto-grams and sounds to a folder for later use; (4) associate a pictogram and a sound to a task; (5) load/save tasks configured to an external file; (6) initialize the list of tasks; (7) display start time of daily activity; (8) turn on/off a webcam to assist user by visualization of the elapsed time or the environment.

A pop-up screen for each task shown in Fig. 4 allows for further editing: (9) enable and disable touch screen; (10) configure preferences in color codes and brightness of

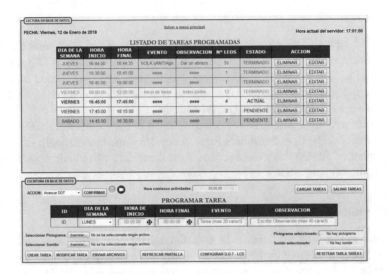

Fig. 3. Spanish DOT HMI main menu

pending, current, finished, unscheduled tasks and set the tasks that turn on and off LEDs. It gives an option to perform LED color tests to directly check color outcome; (11) configure time base (hours, minutes, seconds) and multiplicative factor for each luminous dot (x1, x5, x10, and x15), having minutes and 15 as default (each dot means 15 min);

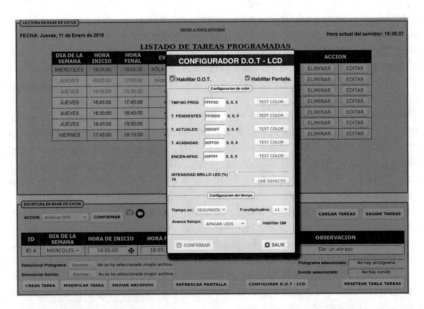

Fig. 4. DOT color and brightness - with test option - and time configuration window

Configuration is saved in a MySQL database. The way to access it is by username and password through the application phpMyAdmin or by saving the contents of the table in a file.

4   Future Work

Current prototype integrates collected needs for next intervention in training children. Intervention results will be assessed by evaluating time cognitive capacities of participants previous and after intervention with KaTid tool [3, 4]. A period of 2 to 3 months is estimated for intervention, in which class agenda is to be introduced in described tool and multimedia feedback configured for participants. We are confident that good results in several areas of time orientation will be measured, from perception to daily time management.

Parameters to vary in intervention are currently in draft design and will be the focus of a future communication. Study of results will hopefully highlight which keys in training methodology are best adequate for each cognitive situation, from which training protocols and methods will be derived and written in user manuals. Special education teachers already have working communities to share education, ideas and materials, so they will take the baton to integrate and maintain a living AAL training tool.

Emotions recognition systems will be included in future set-ups to better know which time functions are being most stressful for participants to focus support and training on them. Works have started to include physiological parameters to measure anxiety level among others.

Future target population is expected to widen to elderly with dementia and cognitive impaired adults. New target groups considered are people with frequent memory failures, who periodically forget daily tasks and people in early stages of dementia, for whom support tool would show the task to be performed and also tutor the user through its realization with different configurable levels of detail.

References

1. CPEE Alborada Home Page. https://cpeealborada.wordpress.com. Accessed 21 Jan 2018
2. International Classification of Functioning, Disability and Health (ICF), Spanish Edition by Spanish Ministry of Work and Social Affairs, Imserso, ISBN 84-8446-034-7, World Health Organization (WHO) (2001)
3. Janeslätt, G.: Time for time: assessment of time processing ability and daily time management in children with and without disabilities, thesis for doctoral degree (Ph.D.). Karolinska Institutet (2009)
4. Janeslätt, G., Granlund, M., Kottorp, A.: Measurement of time processing ability and daily time management in children with disabilities. Disabil Health J. 2(1), 15–19 (2009). https://doi.org/10.1016/j.dhjo.2008.09.002
5. Svensk, A.: Design for cognitive assistance, Licentiate, Certec - Rehabilitation Engineering and Design, Lund Sweden (2001)

6. Magnusson, C., Svensk, A.: A research method using technology as a language for describing the needs both of people with intellectual disabilities and people with brain injuries. In: Proceedings of the 2nd European Conference on the Advancement of Rehabilitation Technology (ECART 2), Stockholm, Sweden (1993)
7. CERTEC Home Page. http://www.certec.lth.se/english. Center for Rehabilitation Engineering Research Lund Institute of Technology. Accessed 21 Jan 2018
8. Abilia Home Page. http://www.abilia.com/en. Web site of Abilia (developing assistive technology for people with cognitive disabilities based on their cognitive abilities). Accessed Jan 2018
9. Falcó, J.L., Muro, C., Plaza, I., Roy, A.: Temporal orientation panel for special education. In: Miesenberger K., Klaus J., Zagler W.L., Karshmer A.I. (eds.) Computers Helping People with Special Needs. ICCHP. LNSC, vol 4061. Springer, Heidelberg (2006)
10. Blanco, T., Asensio, A., Cirujano, D., Marco, A., Falcó, J.L., Casas, R.: Time Orientation Device for people with disabilities: do you want to assess it? In: CSUN Assistive Technology Conference (2011)
11. Janeslatt, G., Kotorp, A., Granlund, M.: Evaluating intervention using time aids in children with disabilities. Scandinavian Journal of Occupational Therapy (2013), (1–10)
12. Arvidsson, G., Jonsson, H., The impact of time aids on independencie and autonomy in adults with developmental disabilities. Occup. Ther. Int. **13**(3), 160–175 (2006). Published online in Wiley Interscience. https://doi.org/10.1002/oti215
13. Guillomía, M.A., Falcó, J.L., Artigas, J.I., Sánchez, A.: AAL platform with a 'de facto' standard communication interface (TICO): Training in home control in special education. Sensors **17**(10) (2017)

Real Time Object Detection and Tracking

Dária Baikova[1], Rui Maia[1], Pedro Santos[1], João Ferreira[2(✉)],
and João Oliveira[2]

[1] INOV, Lisbon, Portugal
{daria.baikova,rui.maia,pedro.santos}@inov.pt
[2] Information Sciences, Technologies and Architecture Research Center
(ISTAR-IUL), Instituto Universitário de Lisboa (ISCTE-IUL), Lisbon, Portugal
{jcafa,joao.p.oliveira}@iscte-iul.pt

Abstract. The present work proposes a real-time multi-object detection and tracking system to be implemented in commercial areas. The purpose is to gather and make sense of costumer behavior data extracted from surveillance footage (available from ceiling cameras) in order supply retailers with a set of analytics, management and planning tools to help them perform tasks such as planning demand and supply chains and organizing product placement on shelfs. To achieve this goal, deep learning techniques are used, which have been yielding outstanding results in computer vision problems in recent years.

Keywords: Deep Learning · Computer Vision · Object detection
Object tracking

1 Introduction

Today, big grocery retail chains account for the great majority of grocery shopping made globally, generating about \$4 trillion annually, which makes this industry one of the biggest worldwide. To keep competitive, retail chains like WalMart (U.S.), Carrefour (France), TESCO (United Kingdom) adopt a wide range of strategies in order to collect information about its customers, to build profiles and to analyze shopping habits, such as targeting clients with personalized vouchers or using loyalty cards. However these strategies don't provide the retailer with sufficient information to have a good understanding about the shopping patterns of its costumers, which leads to inefficiencies felt throughout the current model adopted by most retailers. This includes performing tasks such as planning demand and supply chains or placement on shelfs, monitoring costumer satisfaction and overseeing the payment process in the supermarket checkouts. To oversee and fix these problems in real-time, retailers need to access to information that is currently impossible to obtain, for instance: knowing how many customers are inside the store, determining the distribution of costumers inside the stores, keeping track of the path taken by the costumers throughout the store or learning the costumers shopping patterns under different conditions.

© Springer Nature Switzerland AG 2019
P. Novais et al. (Eds.): ISAmI 2018, AISC 806, pp. 129–137, 2019.
https://doi.org/10.1007/978-3-030-01746-0_15

The present work is focused on: (1) studying and implementing computer vision algorithms based on Deep Learning, which are the state-of-the art for computer vision tasks; and (2) Apply these techniques to a domain-specific application (retail), which will be focused on obtaining the information retailers need.

2   Related Work

The main purpose of object tracking it to detect multiple objects in a video frame and maintain these identities in the subsequent frames (over time) in order to identify the trajectory of each object. Classical approaches to object tracking include Multiple Hypothesis Tracking (MHT) [20], recursive methods which make use of Kalman filtering to predict locations [5], Joint Probabilistic Data Association (JPDA) [15] or Particle filtering techniques [11]. However, with the recent success of Deep Learning techniques in tasks such as image classification [12,14,16,22,24] and object detection [9,10,17,19,21], and more access to data, deep learning models also started being applied to object tracking. Namely, tracking by detection has become a popular paradigm to solve the tracking problem [2,7]. This type of framework consists of two steps: applying a detector, and subsequently, a post-processing step witch involves a tracker to propagate detection scores over time. The main challenge is grouping the detected targets in contiguous frames in order to represent the targets by their trajectory over all frames. The detection step can be solved with high performing deep learning detection model [9,10,17,19,21], which in may use deep learning classification models as its base - models such as the ones in [12,23,24,24]. The detection association step has many solutions: some approaches include associating different tracks by calculating the Intersection over Union (IOU) of the bounding boxes in consecutive frames [6]; other approaches leverage Kalman filters or Hungarian algorithm such as the SORT method (Simple Online and Realtime Tracking) [4]. Other approaches skip tracking by detection and use fully Deep Learning based methods, for instance, methods based on Recurrent Neural Networks (RNNs) [8,18] or siamese convolutional networks architectures [3]. The siamese CNN (Convolutional Neural Network) proposed in [13] is the base for the present work.

3   Proposed Architecture

At a high level the project consists of three modules represented in Fig. 1: a frame reader module, responsible for reading frames from a real-time RTSP stream and writing these to a multi-threading queue which also serves as a buffer, a detection module, responsible for outputting object bounding boxes, and the tracking module which uses the previous detection outputs in order to compute the final verdicts about the tracks of each object. Both detection and tracking modules use recent deep learning techniques: the detection module was tested with different CNN detection architectures and the tracking module uses the architecture introduced in [13], codenamed GOTURN (Generic Object Tracking Using Regression Networks). The following sections describe these modules.

3.1 Object Detection Module

The tasks in this module consist of reading frames from the queue, applying a multi-object detector to these frames and finally sending the detections to the tracking module. The results sent to the tracker include the found bounding boxes and detection scores (the extent to which the model is certain of that detection). The first problem is to select the detection model, which is a trade-off between accuracy and speed (inference time). In order to find the optimal model for this application two types popular CNN detection architectures were tested: Faster-RCNN and SSD models. The R-CNN family models [9,10,21] have demonstrated excellent performance in recent years. The most advanced of these models, Faster-RCNN [21] tackles object detection as a three step problem: firstly, it uses category-independent region proposal algorithms, e.g. [25], in order to generate possible detection candidates, which are then fed to a CNN that serves as a feature extractor, and lastly, a set of class-specific classifiers are applied on top of the features in order to find the final verdict about the classes of the subjects present in the image. However, due to the region proposal step, the R-CNN models have a slower inference time than single-shot CNN alternatives such as SSD [17] architectures. Speed is a significant feature in the context of the present work due to the real-time requirement of the project. The key idea of an SSD architecture is that each of the last few layers of the network is responsible for the prediction of progressively smaller bounding boxes, and final prediction is the union of all these predictions. In this way, these models are able to simultaneously predict the bounding box and the object class in one iteration, eliminating the region proposal step altogether and are able to achieve an improvement in speed. In contrast 7 frames per second (FPS) speed with 73.2% mAP on VOC2007 test with a Faster-RCNN, SSD operates at 59 FPS with 74.3% mAP on the same dataset. The object detection models tested for this module are: a MobileNet SSD, an Inception-v2 SSD and a ResNet101 Faster-RCNN.

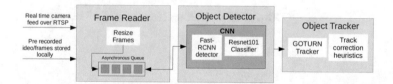

Fig. 1. Architecture overview.

3.2 Tracking Module

The tracking module leverages the single object tracking architecture proposed in [13] codenamed GOTURN (Generic Object Tracking Using Regression Networks). This architecture (Fig. 2) has the structure of a Siamese CNN and is able to perform generic object tracking (which is not trained for a specific class of objects) at very high speeds (100 FPS) making it appropriate for real time applications. The network is trained entirely offline and learns the generic relationship between appearance and motion in order to be able to recognize novel objects online. In order to learn this relationship, at each training iteration, the network takes two inputs: a crop of a frame at time $t-1$ and a crop of the next frame at time t. The crop of the frame at time $t-1$ contains the object the network will be looking for -trying to track-, *i.e.*, the target. The second input is a crop of the frame at time t, which represents the area the network will be searching for the target object in (search area), which is centered on the same point that the previous crop was, however, it is scaled by a factor of k. If the objects don't move too quickly, this scaling ensures the target object will still be present in the search area. The network directly regresses the bounding box coordinates of the tracked object in the search area of the next frame.

Since the GOTURN architecture is only able to track one object at a time, in order to perform multi-object tracking in real-time the tracker runs multiple times per frame (one time for each object detected when it first enters the field of view). Furthermore, in order to initialize, maintain and end a track additional rules are needed. These rules are described below.

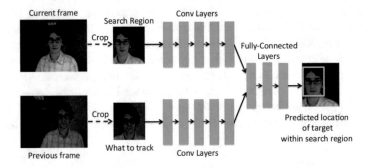

Fig. 2. GOTURN architecture.

Tracker Initialization and Maintenance and Ending. The tracker is initialized when it receives the first detection from the detection module. This detection constitutes the first input fed to the tracker (target) and defines what object the tracker will be looking for during it's runtime. At each iteration, the tracker uses the previously predicted frame crop (at time $t-1$) and the current frame search area (scaled area of the frame at time t, centered on the previous frame target's center) as an input. In order to make sure the tracker does not lose it's target, from N to N frames the detection re-runs on the current frame and the result is compared with the tracker's prediction via IOU (Intersection Over Union) measure; The tracker bounding box prediction is then switched by the bounding box predicted by the detector since the detection bounding boxes tend to be more accurate. When an object leaves the camera field of view it's track should be removed. In order to achieve this, a parameter is used: the track of an object is considered finished when no detections of that object occurred in M frames.

4   Dataset and Network Training

In order to create a new training dataset and to simulate a commercial area, a controlled test environment was created at the INOV-INESC building entrance area where hundreds of different people pass through daily. The setup consists of one top camera which captures one frame per second in the period from 07 h 00 am to 20 h 00 pm. These frames are afterwards stored in an SMB storage system remotely accessible for future use. The live video feed can be accessed in real time via RTSP protocol. The training for the detection models consists of labeled images annotated with object bounding boxes. The tracking dataset has a similar structure with the addition of the object IDs in each frame. Since the tracking module takes a pair of frames as inputs, an additional pre-processing step was to create pairs of target/search area images. The initial labels were created by hand, and afterwards, the ensable learning technique was used to generate new data. Three different detectors were trained on the initial dataset, and when shown a new frame, each of the models "voted" on where it though a bounding box should exist and the final annotation was a conjugation of all the predictions. In this way, the final training dataset had about 10000 labeled images. All the models (detection and tracking) were trained offline with the previously described datasets using the Tensorflow [1] framework. In order to reduce training time and data overfitting, transfer learning was used in the detection models, which were all pre-trained on the COCO dataset.

5   Results

In order to determine which of the object detection models is more suitable for this project, mean average precision (mAP) and inference time of the predictions for each model is measured and shown in Table 1. Between the three tested detection models, the Resnet101/Faster R-CNN model yields the best results Fig. 3(a), however, inference time is the highest, which could cause a delay in reading and processing frames from the queue and consequently cause the queue to overflow and discard frames. This would be a problem since the tracking module relies on the assumption that the objects move slowly through time, and can lose track of an object if there are gaps in the frame sequences. The Mobilenet/SSD model is more appropriate for a real time applications since it is very fast (inference takes 5 ms), however, the detections are faulty, and in some cases, it even fails to detect completely for numerous frames (Fig. 3(b)). The Inception-v2/SSD model seems a good compromise between inference time and precision, which is able to have a good mAP value while maintaining a fast inference time. In conclusion, if the stream reading rate is set between 5 FPS and 10 FPS, Resnet101-Faster R-CNN is still a better choice since no frames are lost in the queue (the queue writing rate is lower or equal to the frame processing rate). Since this range of frame rates does not damage tracking results, Faster-RCNN/Resnet101 was the chosen architecture for the detection module.

(a) Resnet classifier with Faster R-CNN detector   (b) Mobilenet base network with SSD detector   (c) Inception-v2 base network with SSD detector

Fig. 3. Samples of the test set inference outputs for the different detection models.

Table 1. Detection results.

Model	mAP	Inference time (s)
Resnet101/Faster R-CNN	0,9198	0.1
MobileNet/SSD	0,8015	0,005
Inception-v2/SSD	0,9	0.04

The tracking module is able to use detection bounding boxes and produce and maintain the ID of each object through the object's path (Fig. 4). It works well even in situations where multiple people go through the camera field of view. The results are the best when the M parameter (number of frames where no detection exists before the track is considered finished) is set to two frames, *i.e.*, when no detection of an object is found in two consecutive frames, the track is considered finished. If this number is greater than two, there is the risk of a new object being associated with another object's track. Despite the results being seemingly good, at this time there is still no way of measuring the exact tracking accuracy however the tracker seems to perform well. This approach is suitable for real time tracking due to the fast nature of the performed computations in contrast to more elaborate trackers that use image information (features) to predict the next position of the tracked objects.

6   Conclusion and Future Work

The present work proposes a real time detection and tracking system aimed at implementation in commercial areas, in which the tracking by detection framework is used. Based on the results, the models (both tracking and detection)

Fig. 4. Tracking sequence. The tracker attributes IDs and draws colored bounding boxes around the objects in every frame which it considers belong to the same object-each object has it's own ID and color. The tracker is able to maintain these IDs even with multiple people going through the field of view. Here, four different IDs (from 2 to 5) are represented in different colors.

are already yielding good results and with more training data should achieve higher levels of precision. The next steps for the tracking module are to annotate new training and test data with paths of each individual object in order to properly measure tracking accuracy. This task can now be easily achieved since the information returned by the tracking module can be recycled (the tracker does most of the work and human input is only needed for minor adjustments). When these tasks are finished, the next steps will be to train deep learning models (for instance R-NNs) to predict object positions since they are the current state-of-the-art and a really promising research field.

References

1. Abadi, M., Agarwal, A., Barham, P., Brevdo, E., Chen, Z., Citro, C., Corrado, G.S., Davis, A., Dean, J., Devin, M., Ghemawat, S., Goodfellow, I., Harp, A., Irving, G., Isard, M., Jia, Y., Jozefowicz, R., Kaiser, L., Kudlur, M., Levenberg, J., Mane, D., Monga, R., Moore, S., Murray, D., Olah, C., Schuster, M., Shlens, J., Steiner, B., Sutskever, I., Talwar, K., Tucker, P., Vanhoucke, V., Vasudevan, V., Viegas, F., Vinyals, O., Warden, P., Wattenberg, M., Wicke, M., Yu, Y., Zheng, X.: TensorFlow: large-scale machine learning on heterogeneous distributed systems (2016)
2. Andriyenko, A., Schindler, K., Group, R.S.: Multi-target tracking by continuous energy minimization (2014)
3. Bertinetto, L., Valmadre, J., Henriques, J.F., Vedaldi, A., Torr, P.H.: Fully-convolutional siamese networks for object tracking. In: Lecture Notes in Computer Science (Including Subseries Lecture Notes in Artificial Intelligence and Lecture Notes in Bioinformatics). LNCS, vol. 9914, pp. 850–865 (2016)
4. Bewley, A., Ge, Z., Ott, L., Ramos, F., Upcroft, B.: Simple online and realtime tracking. In: Proceedings - International Conference on Image Processing, ICIP 2016, pp. 3464–3468, August 2016
5. Black, J., Ellis, T., Rosin, P.: Multi view image surveillance and tracking. In: Proceedings - Workshop on Motion and Video Computing, MOTION 2002, pp. 169–174 (2002)

6. Bochinski, E., Eiselein, V., Sikora, T.: High-speed tracking-by-detection without using image information. In: Proceedings of the IEEE International Conference on Advanced Video and Signal-Based Surveillance (AVSS), August 2017

7. Choi, W.: Near-online multi-target tracking with aggregated local flow descriptor. In: Proceedings of the IEEE International Conference on Computer Vision 2015 Inter, pp. 3029–3037 (2015)

8. Gan, Q., Guo, Q., Zhang, Z., Cho, K.: First step toward model-free, anonymous object tracking with recurrent neural networks, pp. 1–13 (2015)

9. Girshick, R.: Fast R-CNN. In: Proceedings of the IEEE International Conference on Computer Vision 2015 Inter, pp. 1440–1448 (2015)

10. Girshick, R., Donahue, J., Darrell, T., Malik, J.: Rich feature hierarchies for accurate object detection and semantic segmentation. In: Proceedings of the IEEE Computer Society Conference on Computer Vision and Pattern Recognition, pp. 580–587 (2014)

11. Green, P.: Reversible jump Markov chain Monte Carlo computation and Bayesian model determination. Biometrika, **82**(4), 711–732 (1995). http://biomet.oxfordjournals.org/content/82/4/711.short

12. He, K., Zhang, X., Ren, S., Sun, J.: Deep residual learning for image recognition, December 2015

13. Held, D., Thrun, S., Savarese, S.: Learning to track at 100 FPS with deep regression networks. Lecture Notes in Computer Science (Including Subseries Lecture Notes in Artificial Intelligence and Lecture Notes in Bioinformatics). LNCS, vol. 9905, pp. 749–765 (2016)

14. Krizhevsky, A., Sutskever, I., Hinton, G.E.: ImageNet classification with deep convolutional neural networks. In: Proceedings of the 25th International Conference on Neural Information Processing Systems, NIPS 2012, vol. 1, pp. 1097–1105. Curran Associates Inc., USA (2012)

15. Kuhn, H.: The Hungarian method for the assignment problem. Nav. Res. Logist. **52**(1), 7–21 (2005)

16. LeCun, Y., Boser, B., Denker, J.S., Henderson, D., Howard, R.E., Hubbard, W., Jackel, L.D.: Backpropagation applied to handwritten zip code recognition (1989)

17. Liu, W., Anguelov, D., Erhan, D., Szegedy, C., Reed, S., Fu, C.y., Berg, A.C.: SSD: single shot multibox detector (2015)

18. Milan, A., Leal-Taixe, L., Reid, I., Roth, S., Schindler, K.: MOT16: a benchmark for multi-object tracking, pp. 1–12 (2016). http://arxiv.org/abs/1603.00831

19. Redmon, J., Divvala, S., Girshick, R., Farhadi, A.: You only look once: unified, real-time object detection. In: The IEEE Conference on Computer Vision and Pattern Recognition (CVPR), pp. 779–788 (2016)

20. Reid, D.: An algorithm for tracking multiple targets. IEEE Trans. Autom. Control **24**(6), 843–854 (1979). http://ieeexplore.ieee.org/lpdocs/epic03/wrapper.htm?arnumber=4046312, http://ieeexplore.ieee.org/lpdocs/epic03/wrapper.htm?arnumber=1102177

21. Ren, S., He, K., Sun, J., Girshick, R.: Faster R-CNN: towards real-time object detection with region proposal networks. IEEE Trans. Pattern Anal. Mach. Intell. **39**(6), 1137–1149 (2017)

22. Simonyan, K., Zisserman, A.: Very deep convolutional networks for large-scale image recognition, pp. 1–14 (2014)

23. Szegedy, C., Ioffe, S., Vanhoucke, V., Alemi, A.: Inception-v4, Inception-ResNet and the Impact of Residual Connections on Learning (2016)

24. Szegedy, C., Liu, W., Jia, Y., Sermanet, P., Reed, S., Anguelov, D., Erhan, D., Vanhoucke, V., Rabinovich, A.: Going deeper with convolutions. In: Proceedings of the IEEE Computer Society Conference on Computer Vision and Pattern Recognition, 7–12 June, pp. 1–9 (2015)
25. Uijlings, J.R.R., Van De Sande, K.E.A., Gevers, T., Smeulders, A.W.M.: Selective search for object recognition (2012)

Predictive Analysis in Healthcare: Emergency Wait Time Prediction

Filipe Gonçalves[1], Ruben Pereira[1], João Ferreira[1(✉)], José Braga Vasconcelos[2,3],
Fernando Melo[4], and Iria Velez[4]

[1] Instituto Universitário de Lisboa (ISCTE-IUL), ISTAR-IUL, Lisbon, Portugal
{fsgsa,rfspa,jcafa}@iscte-iul.pt
[2] Universidade Europeia, Laureate International Universities, Lisbon, Portugal
jose.vasconcelos@universidadeeuropeia.pt
[3] Centro de Administração e Politicas Publicas (CAPP) da Universidade de Lisboa,
Lisbon, Portugal
[4] Hospital Garcia da Orta, Almada, Portugal
{fmelo,iria.velez}@hgo.min-saude.pt

Abstract. Emergency departments are an important area of a hospital, being the major entry point to the healthcare system. One of the most important issues regarding patient experience are the emergency department waiting times. In order to help hospitals improving their patient experience, the authors will perform a study where the Random Forest algorithm will be applied to predict emergency department waiting times. Using data from a Portuguese hospital from 2013 to 2017, the authors discretized the emergency waiting time in 5 different categories: "Really Low", "Low", "Average", "High", "Really High". Plus, the authors considered as waiting time, the time from triage to observation. The authors expect to correctly evaluate the proposed classification algorithm efficiency and accuracy in order to be able to conclude if it is valuable when trying to predict ED waiting times.

Keywords: Healthcare · Big data · Predictive analytics
Emergency department

1 Introduction

Emergency departments (ED) are an important and complex area of a hospital and are the major entry point to the healthcare system [1]. With the increase of life expectancy, population aging and bigger amount of health issues, ED tend to have greater demand [2]. If hospitals and more specifically, ED, are not ready, this will increase emergencies crowding, creating a big problem to authorities and hospital management since resources are limited. According to the American College of Emergency Physicians (ACEP) "Crowding occurs when the identified need for emergency services exceeds resources for patient care in the emergency department, hospital or both" [3]. Lack of beds, patients in hallways, greater amount of people in the waiting rooms, longer waiting times, greater patient length of stay and general patient dissatisfaction are some of the consequences

© Springer Nature Switzerland AG 2019
P. Novais et al. (Eds.): ISAmI 2018, AISC 806, pp. 138–145, 2019.
https://doi.org/10.1007/978-3-030-01746-0_16

of this phenomenon. It is an international problem and it is vital for hospitals [3] to solve it due to the life-threatening context of the area.

ED wait times are the second most referred theme regarding patient experience [4] which indicates that this area requires intervention to increase care quality and resource efficiency to achieve greater patient satisfaction. That can be achieved using Predictive Analytics (PA) which has the potential to improve the operational flexibility and throughput quality of ED services [5]. Waiting time prediction would help clinicians prioritize patients and adjust work flow to minimize time spent [6]. Predictive Analytics allows to predict future events or trends using retrospective and current data [7]. It could be applied in several healthcare areas, taking advantage of the big data in healthcare. According to [8], predictive analytics is a tactic that healthcare organizations should adopt, allowing the stratification of risk to predict outcomes, that in healthcare can be harmful to the patients. Other advantages would be the adoption of more sensor based technologies that would help patients to be more aware about their health, provide life-style suggestions by determining some diseases that he could suffer if he kept the same lifestyle [9], help the management of high risk and high cost patients during hospital care and after discharge follow-up care [8], etc.

In this research, the authors applied Predictive analytics grounded on data collected from a real ED and studied the performance of the Random Forest algorithm to predict patient waiting time.

2 Related Work

The authors started by searching about big data in healthcare and predictive analytics applications in healthcare industry. Later, the authors focused on ED waiting time, to understand the variables that can have the most influence on the patient waiting time, plus the used algorithms.

Regarding predictive analytics in the healthcare industry, some authors studied its advantages and possible applications, like Malik et al. [10] that reviewed and analyzed applications of predictive analytics and data mining in the healthcare industry or Chauhan and Jangade [9] that claim that predictive analytics in healthcare can be beneficial as it would allow for patient disease prediction, fraud detection and cost management initiatives.

Another author that defends predictive analytics importance in the healthcare industry is Palem [11], defending that predictive analytics can be helpful on various areas of the healthcare industry like "life-sciences, healthcare providers, insurance providers, public health, individuals". Janke et al. [5] also defend that predictive systems can be beneficial to the ED. They studied big data and predictive analytics implementation challenges and opportunities and how it could improve the ED patient flow.

The aforementioned models are defined by Kaul et al. [12], that defined predictive models as models that "concentrate upon analyzing a set of relevant data and predict a future implication or a meaningful pattern", analyzing how they can be applied in healthcare, for example, providing alerts about disease outbreaks. They have studied healthcare data stating that 80% of it is unstructured and difficult to analyze.

Some of the authors also analyzed the advantages of predictive analytics but focusing on some specific areas. One of those cases is Bates [8] that provided some use cases of predictive analytics application on high risk and cost patient management, defining predictive systems as "software tools that allow the stratification of risk to predict an outcome", defending that, in the future, healthcare organizations will use predictive analytics.

Focusing on ED waiting times, Barad et al. [3] studied the ED of an Israeli hospital in order to find the reasons for ED crowding. Started by conducting interviews with the clinicians and analyzed the communication between departments. In the research, they used the American College of Emergency Physicians definition of ED crowding, "Crowding occurs when the identified need for emergency services exceeds resources for patient care in the emergency department, hospital or both".

Some authors focused on predicting the ED waiting times, like Bruballa et al. [2] that created an agent based simulation to study the patient length of stay, considering it as a major problem for the healthcare system worldwide. They also defend that the existence of information or a recommendation system showing emergency department state information would help avoiding long waiting times in the services.

Chong et al. [13] developed a system dynamic model to study the patient flow in the emergency department of a hospital in Hong Kong. They concluded that by increasing staff and the amount of beds, the time spent by patients in the ED could be reduced.

Others studied the ED waiting times, using machine learning techniques, like quantile regression, Q-Lasso or expectation maximization.

Sun et al. [6] were some of the authors that used quantile regression to develop a model to predict emergency department waiting time, based on triage information. Did not use the predicted mean waiting time since it is affected by possible outliers, instead, predicted "a range of the 50th percentile to the 95th percentile". They defined waiting time as the "interval from triage end time to the physician's consultation time", and considered that the patient flow rates of other acuity levels could impact on other levels since clinicians could move between queues. This developed model ignored patient characteristics which could be a limitation. Other authors that used quantile regression were Ding et al. [14] that created a system to predict length of stay in ED. They claim that "providing patients with an expected LoS at triage may result in increased patient satisfaction". They considered three phases for the length of stay: waiting time, treatment time and boarding time, and used "acuity level, arrival day and time, arrival mode, chief complaint and patient characteristics." as variables.

Q-Lasso was used by Ang et al. [15] to predict ED waiting time, using data from four different hospitals from the United states of America. They defined Q-Lasso as an algorithm that is a combination of the "queueing theory and the lasso method, that uses a penalty to correct estimation errors".

3   Work Methodology

The work methodology can be divided in three different processes: Data Collection, where the dataset and the correspondent ED is described and analyzed, Data Pre-Processing, which is the process the data is manipulated towards the waiting time prediction and finally, Data-Mining, where the Random Forest algorithm is applied to the previously processed data in order to predict the ED waiting time.

3.1   Data Collection

In this process, the authors describe the data used in this research, that was provided by an ED of a Portuguese hospital and includes registers from January 1 of 2013 to December 31 of 2017. Before the data was provided to the authors of this research, information that could identify the patient, doctor or nurse of each record was anonymized due to privacy regulations.

This hospital's ED flow has five main processes: Admission, Triage, Observation, Discharge and Administrative Discharge (Fig. 1). The first step occurs when the patient is admitted to the ED, then, in the second process is when the patient is submitted to testing, being categorized according to the Manchester Triage Protocol (MTP) (Table 1). The third step corresponds to the observation, where the patient will be observed by a doctor and treated. The last two steps of this ED flow are correlated, the discharge occurs when a doctor, after evaluating the patient, considers that he is ready to be moved to another hospital, department or even to go home, and finally, the administrative discharge, occurs when all the documentation necessary for the patient to leave is approved and the patient leaves the ED.

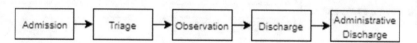

Fig. 1.   ED flow, in a five strae process

Table 1.   Manchester Triage Protocol standards

Class	Treatment	Target time (minutes)
Red	Immediate	0
Orange	Very urgent	10
Yellow	Urgent	60
Green	Standard	120
Blue	Non-urgent	240

As aforementioned, this ED is compliant with the MTP, a protocol for hospital triage system, that defines the advisable time limit patients must wait to be treated. Following this, all patients are categorized on five different triage colors: "red", "orange", "yellow", "green" and "blue", from the most urgent to the least urgent respectively. In this ED, there is also another category called "others" for the other cases, where the patient doesn't follow the usual triage system. In this ED, most of the patients are either in green, yellow or "others" categories and on 50% of the cases, the advisable time was exceeded.

The ED is divided in three departments: general department (GD), pediatric department (PD) and obstetrics department (OD).

In those four years of records, 672720 patients attended the ED on that hospital. Each record on this dataset contains 20 attributes represented on Fig. 2 with the respective acronyms. The first 10 are the dates and times for each step of the previously mentioned ED flow, so the date (dd/mm/yyyy) and time (hh:mm:ss) for the admission (D1 and T1 respectively), the triage (D2 and T2 respectively), observation (D3 and T3 respectively), discharge (D4 and T4 respectively) and administrative discharge (D5 and T5 respectively). Other attributes are the patient triage color according to the MTP (defined as TC in Fig. 2), ED sub-department (DEP in Fig. 2), discharge status, discharge destination, readmission flag and an anonymized patient id, doctor id and nurse id.

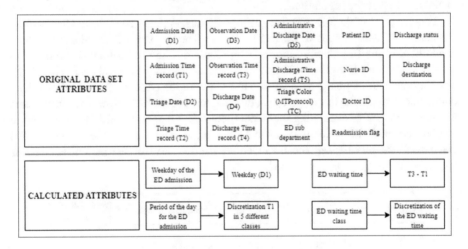

Fig. 2. Attributes used in this research

3.2   Data Pre-processing Process

In this process, the authors manipulated the data to analyze it and apply the Random Forest algorithm in order to predict the ED waiting time (Sect. 3.3).

The authors started by removing the attributes that would not be needed according to the goals of the project, like patient, doctor and nurse ids and readmission flag. Records with null values were also removed, resulting in a cut of 5.1% of the original data.

Then, the authors calculated the weekday where the event took place, based on the admission date (D1 as described in Fig. 2), using the weekday function from excel.

Another important step was to discretize the hour of the day where the event took place, based on the admission time (T1). The authors created six different day periods: 21–3 h, 3–7 h, 7–12 h, 12–15 h, 15–19 h and 19–21 h and then aggregated the events based on those intervals. This allowed to ease the analytical process.

Another calculated attribute was the waiting time. The authors defined patient waiting time as the time from admission to observation (T3 – T1 from Fig. 2). As expected, the minimum time is 00:00:00h, since the MTP defines that patients classified with red color category should be attended immediately. The maximum waiting time was 9:50:55h and it corresponds to a patient that was has with the green color, while the average waiting time was 00:56:02h.

Then, this calculated time was discretized on five different custom waiting time categories: "really low", "low", "average", "high", "really high". This process of discretization started with the calculation of the total average waiting time, that was used as reference in order to categorize all the time events. The events that fit the first 20% of the average waiting time (00:00:00 to 00:11:12) were categorized as "really low". Then, for the second category ("low") the author categorized the events between 20% and 70% of the average (from 00:11:12 to 00:50:26). For the "average" category, the author used the 20% around the average waiting time, 10% under and 10% above (00:50:26 to 1:01:38). For the next category, "high", added 70% of the average waiting time to the average waiting time (01:01:38 to 1:40:51). Finally, the last category, "really high", was based on the maximum waiting time, since all the events had to be covered. This discretization process is described on Fig. 1. These discretization processes were made using Microsoft Excel 2016 (Fig. 3).

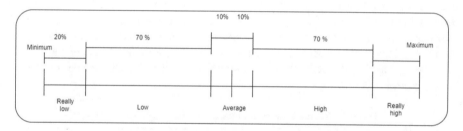

Fig. 3. ED waiting time classification

The "Low" category is the category where there are more occurrences on every department. As expected, on the pediatric department, most of the patients have short waiting times, around 81% of the entries are either "low" or "really low" category. On the other hand, the obstetric department also has a lot of entries with the "low" category (41%), but it is the department where the patients have to wait the most since the "high", and the "really high" categories represent 46% of the entries.

In terms of triage color, as expected, the most urgent colors have smaller waiting times, for example, on red, which is the most urgent color, 94% of the entries fit into the "low" or "really low" categories. For the blue triage color, the non-urgent cases, 61% of the entries fit on the higher waiting times categories, "high" and "really high" categories.

About the time periods, the "Low" category is the category with more occurrences of every period. During the night periods, 21–3 h and 3–7 h, there is an increase of the "really high" category events, reaching 25% of the occurrences on the 3–7 h period.

The "low" category is also the category with more occurrences on all of the days of the week. During the whole week, the "high" and "really high" categories have more events than the "average" and "really low" categories. On Sundays, there is a clear increase of "really high" category events, covering 28% of all the occurrences.

3.3 The Data Mining Process

The goal was to analyze the waiting times in the ED and to do so, the authors applied Random Forest (RF), using R in R-Studio. The library "rminer" was used to compute evaluation metrics like accuracy, precision, true positive rates and F1. The varimp function from R was used to analyze the input variables importance.

The Random Forest algorithm builds several decision trees and calculates the mean or majority class for all decision trees. Random Forests are better than decision trees because it will help avoiding overfit by creating smaller subsets of trees, while the decision tree algorithm has a unique decision tree, making it denser and deeper, which might cause overfitting.

This algorithm achieved an accuracy of 50.09% (see Table 2 for full results). The classes that had more events, "Really low" and "low", were the ones with better precision. In general, all the classes had low true positive rate except for the "low" category. Since the category "really low" has a good precision, but low true positive rate, we can assume that this low true positive rate is being caused by a big number of false negatives, because true positive rate or recall is the number of true positives divided by the sum of true positives and false negatives. The "average" category was expected to perform the worst since it is the category with the smallest number of events (6.08%). Analyzing the input variables importance, triage color stands as the most important for the desired prediction, reaching an overall of 317, followed by weekday (176.8), day period (161.4) and department (151.8).

Table 2. Random forest results with total average waiting time

Class	Precision	True positive rate	F1-score
Really Low	66.67	1.34	2.62
Low	51.97	88.72	65.55
Average	50	0.34	0.68
High	28.16	4.23	7.36
Really high	43.60	33.29	37.75

4   Conclusion

The authors conclude that the success of the RF on predicting ED wait times is highly dependent on the amount of available data and how it is discretized. This can be proven by the fact that the categories with the highest number of events have better precision, while the ones with less events have lower precision. Discretizing some of the fields with different methods, like equal areas, should also be explored, since all the possible classes of a certain input, would have the same amount of events, which would avoid having some classes with a low number of events, that as mentioned before can cause low precision results.

Complementing the data with other variables could also improve the predictive capability. For example, it could be useful to add weather information like temperature, precipitation rate or humidity, since those factors can have impact on the ED adherence.

Other algorithms like Naïve Bayes or Neural Networks could also be applied, allowing to compare the algorithms efficiency and possibly getting better results.

References

1. Liu, Z., Rexachs, D., Luque, E., Epelde, F., Cabrera, E.: Simulating the micro-level behavior of emergency department for macro-level features prediction. In: 2015 Winter Simulation Conference (WSC), vol. 2016, Febru, pp. 171–182 (2015)
2. Bruballa, E., Wong, A., Epelde, F., Rexachs, D., Luque, E.: A model to predict length of stay in a hospital emergency department and enable planning for non-critical patients admission. Int. J. Integr. Care 16(6), 1–2 (2016)
3. Barad, M., Hadas, T., Yarom, R.A., Weisman, H.: Emergency department crowding. In: 19th IEEE International Conference on Emerging Technologies and Factory Automation, ETFA 2014 (2014)
4. Sonis, J.D., Aaronson, E.L., Lee, R.Y., Philpotts, L.L., White, B.A.: Emergency department patient experience. J. Patient Exp. 5(2), 101–106 (2017). https://doi.org/10.1177/2374373517731359
5. Janke, A.T., Overbeek, D.L., Kocher, K.E., Levy, P.D.: Exploring the potential of predictive analytics and big data in emergency care. Ann. Emerg. Med. 67(2), 227–236 (2016)
6. Sun, Y., Teow, K.L., Heng, B.H., Ooi, C.K., Tay, S.Y.: Real-time prediction of waiting time in the emergency department, using quantile regression. Ann. Emerg. Med. 60(3), 299–308 (2012)
7. Dinov, I.D.: Methodological challenges and analytic opportunities for modeling and interpreting big healthcare data. Gigascience 5(1), 12 (2016)
8. Bates, D.W., Saria, S., Ohno-Machado, L., Shah, A., Escobar, G.: Big data in health care: Using analytics to identify and manage high-risk and high-cost patients. Health Aff. 33(7), 1123–1131 (2014)
9. Chauhan, R., Jangade, R.: A robust model for big healthcare data analytics. In: 2016 6th International Conference - Cloud System and Big Data Engineering (Confluence), pp. 221–225 (2016)
10. Malik, M.M., Abdallah, S., Ala'raj, M.: Data mining and predictive analytics applications for the delivery of healthcare services: a systematic literature review. Ann. Oper. Res. 63(2), 357–366 (2016). https://doi.org/10.1016/j.pcl.2015.12.007
11. Palem, G.: The practice of predictive analytics in healthcare, July 2013
12. Kaul, C., Kaul, A., Verma, S.: Comparitive study on healthcare prediction systems using big data. In: 2015 International Conference on Innovations in Information, Embedded and Communication Systems (ICIIECS), pp. 1–7 (2015)
13. Chong, M., et al.: Patient flow evaluation with system dynamic model in an emergency department: data analytics on daily hospital records. In: Proceedings of the 2015 IEEE International Congress on Big Data, BigData Congress 2015, pp. 320–323 (2015)
14. Ding, R., McCarthy, M.L., Lee, J., Desmond, J.S., Zeger, S.L., Aronsky, D.: Predicting emergency department length of stay using quantile regression. In: 2009 International Conference on Management and Service Science, vol. 45(2), pp. 1–4 (2009)
15. Ang, E., Kwasnick, S., Bayati, M., Plambeck, E.L., Aratow, M.: Accurate emergency department wait time prediction. Manuf. Serv. Oper. Manag. 18(1), 141–156 (2016)

A Robotic Haptic Feedback Device
for Immersive Virtual Reality Applications

Keren Jiang[1], Xinlei Piao[1], Mohammed Al-Sada[1,2],
Thomas Höglund[3], Shubhankar Ranade[1], and Tatsuo Nakajima[1(✉)]

[1] Department of Computer Science and Engineering,
Waseda University, Tokyo, Japan
{jiangkeren,xinlei1020,alsada,
shubhi,tatsuo}@dcl.cs.waseda.ac.jp
[2] Department of Computer Science and Engineering,
Qatar University, Doha, Qatar
[3] Department of Electrical Engineering and Automation,
University of Vaasa, Vaasa, Finland
thomas.hoglund@uva.fi

Abstract. The use of virtual reality technology has become popular in modern theme parks across the world. Attractions using virtual reality technologies offer highly immersive experiences as if people are really staying in fictional worlds that the theme parks like to create. Haptic feedback is an important piece in virtual reality, where it can increase the immersion and enjoyment, but most virtual reality attractions in theme parks do not offer enough haptic feedback. In this paper, we present HapticDaijya, which is a waist-worn robot capable of giving various haptic and tactile feedback on the torso, neck, face, arms and hands. We present the design and implementation of HapticDaijya. Then, we show a user study and analyze the results for investigating its feasibility. Also, we present preliminary evaluation results that gauged the user's accuracy in distinguishing the locations of taps applied on the chest, as well as general usability and user acceptance.

Keywords: Wearable robot · Haptic feedback · Tactile feedback

1 Introduction

Theme parks is a popular destination for enjoying numerous various types of entertainment attractions, rides, and other events in a single location. Disneyland is one of the most famous theme parks in the world, where it has different fictional characters that visitors can interact with. Each attraction in the park is an immersive experience based on a Disney story. Because these stories are very popular, when people visit Disneyland, they feel that these fictional characters exist in the real world, that they can meet these characters and that they can enjoy being with them during their visit. In particular, incorporating fictionality through virtuality is an essential approach to enhance user experiences in theme parks [8].

© Springer Nature Switzerland AG 2019
P. Novais et al. (Eds.): ISAmI 2018, AISC 806, pp. 146–154, 2019.
https://doi.org/10.1007/978-3-030-01746-0_17

Recently, the progress of virtual reality (VR) technologies offer highly immersive experiences so that the technologies will significantly change user experience in theme parks by incorporating their fictional stories through virtuality. Attractions using virtual reality technologies offer highly immersive experiences as if people are in fictional worlds that the theme parks like to create. Actually, some theme parks already adopted such technologies and successfully offer impressive immersive experiences: for example, in Japan, VR Zone Shinjuku's Ghost in the Shell: Arise Stealth Hounds[1], and Dragon Ball VR Master the Kamehameha[2], and globally, Disneyland's Star Wars: Secrets Of The Empire[3], and Universal Studio's The Repository[4]. Also, Triotech sells some VR attractions like Virtual Rabbids: The Big Maze[5]. Creative Works has developed Hologate which is a VR arcade game.[6] In these VR attractions, a head-mounted display (HMD) is a key technology to offer visual immersion, but the current VR attractions do not offer enough haptic feedback to users, where haptic feedback is essential for enabling truly immersive user experiences within VR. As most virtual reality attractions within theme parks lack haptic feedback, the existence of haptic feedback is essential in increasing the enjoyment and immersion within such attractions.

In this paper, we present HapticDaijya, which is a waist-mounted six degrees of freedom (DoFs) serpentine robot that is capable of providing various haptic experiences. HapticDaijya is able to enhance immersion through haptic feedback that can be delivered to different parts of the body. Our approach attempts to fulfil two design targets. First, HapticDaijya can be used in a variety of haptic and tactile feedback methods, such as producing normal or shear forces, as well as gestural output [1, 7], such as poking or stretching the skin. Second, HapticDaijya is capable of haptic and tactile feedback in multiple locations on the body. We present our prototype system, our preliminary evaluation and the future direction.

The structure of the paper is as follows. In Sect. 2, we introduce haptic feedback in VR and related work. Section 3 present an overview of HapticDaijya. Then, Sect. 4 shows results of the preliminary user study, and Sect. 5 describes its current potential pitfalls in this study. Section 6 concludes the paper.

2 Haptic Feedback in Virtual Reality and a Robotics Haptic Feedback Device

2.1 Motivation and Scenario

Haptic feedback has long been investigated as a method to increase the immersion or enhance the interaction within VR. Many modern VR HMDs, like HTC Vive[7] and

---

[1] https://vrzone-pic.com/en/activity/koukaku.html.

[2] https://vrzone-pic.com/en/activity/dragon.html.

[3] https://disneyworld.disney.go.com/attractions/disney-springs/star-wars-secrets-empire/.

[4] https://www.facebook.com/pages/VR-the-Repository-Universal-Orlando/1264916250217177.

[5] http://trio-tech.com/products/vr-attraction/.

[6] https://thewoweffect.com/products/hologate/.

[7] https://www.vive.com/.

Oculus Rift[8], allow players to move around physically in a tracked space while being engaged in VR. Accordingly, numerous consumer products and research literature investigated wearable haptic feedback methods for areas like the arms, hands and torso. Yet, other body areas, like the neck, face, head or others, have largely been unexplored for their validity for haptic or tactile feedback, especially within the context of VR.

Figure 1 shows how a flexible robotic haptic device offers a more immersive user experiences to users. In the future, when a user is being punched on the chest and chin in the VR world, a haptic force with similar magnitude and location is applied to his chest and chin through the robotics haptic feedback device. The approach offers various opportunities to enhance the current VR attractions in theme parks.

Zero Latency is an emerging VR attraction that is globally available in various locations[9]. It is a free-roam multiplayer game. Each player wears an HTC Vive VR HMD, headphones, a backpack containing a high-performance computer. Here, we would like to explain how a robotics haptic feedback device can be used to enhance the game experience through a scenario. We explain how our approach may significantly improve a player's overall VR user experience.

Each player holds a controller which is a simulated weapon. Eight players are divided to two teams to fight and they will also meet zombies in the game which exist in the VR game world. While the game is played, a player is attacked by an enemy who is near him. The bullet shoots the player's chest, and at the same time, the device applies a similar haptic stimulus by hitting player's chest with matching location and power magnitude. The player runs away and hides in the back of a box. At this time, there is another enemy trying to approach the player. Since the player holds a radar property which is able to detect live person's activity in five meters, the device gently pats the player's shoulder twice, to notify him about the approaching player. The player turns back quickly and finds the enemy. He shoots the enemy and avoids being attacked. Suddenly the device holds his hand, the player turns around and finds that there is a zombie that holds his hand in the VR game world. He casts off the device and gives the zombie a shot.

2.2 Related Work

Previous researches have investigated a variety of feedback methods that can enhance VR experiences. Several researches explored vibrotactile feedback at various locations on the body, especially the chest [4, 5]. Other researches attempted to simulate impacts and pressure using solenoids a vest [1]. Yet, such feedback remains confined to pre-determined points and is limited to a single type.

While there exists a large body of works around vests for vibrotactile feedback around the torso [4, 10], such works remain limited in terms of the diversity of haptic or tactile feedback as well as their capability to deliver feedback to other locations on the body. Also, surveyed literature and products were mainly confined to delivering feedback to fixed stimulation points (as in [5]) and were mostly capable of vibrotactile feedback.

---

[8] https://www.oculus.com/rift/.

[9] http://tokyo-joypolis.com/attraction/1st/zerolatency/.

Fig. 1. Robotics haptic feedback device in a VR environment

Alternate reality experiences [3] refer to a human experience felt by refining the meaning of real space by incorporating virtuality [9]. An alternate reality experience is typically achieved by modifying our eyesight or replacing our five senses to others, and offers a promising possibility for implicitly influencing our attitudes and behaviors in everyday life. For example, human eyesight can be altered using head-mounted displays (HMDs), which modify real-world views captured by video cameras attached to the HMDs with virtual reality technologies. Similarly, visual color images can be transformed to sounds for color-blind people, or visual stimuli can be converted into vibrations to reduce the visual cognitive load.

3   HapticDaijya

3.1   Basic Design

In this section, we present the basic design issues in HapticDaijya, where Daijya (大蛇) means a big snake in Japanese. Our device looks like a snake that is capable of providing various of haptic feedback through its snake-like behavior. The main design objective of our approach is rich haptic feedback in a wearable form. To further extend previous works by diversifying haptic feedback, we designed a waist mounted serpentine-shaped robot with an end effector as shown in Fig. 2. We have chosen the serpentine morphology as it's high DoFs allow the attached end effector to deliver a variety of haptic feedback. Moreover, such flexibility allows the robot to also reach the user's face, neck, shoulders and arms.

Using the robot end effector as described in Sect. 3.2, HapticDaijya can apply various types of normal and shear forces with varied durations and magnitudes. Furthermore, by varying and combining forces, HapticDaijya can provide a variety of feedback, such as pushing, pulling, hitting, scratching and pinching. Gestural feedback [7] can also be created by applying directional and tangential forces on the user's body.

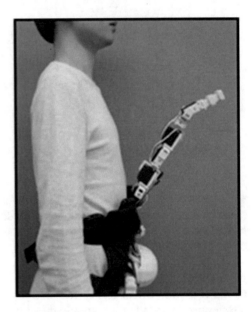

Fig. 2. Overview of HapticDaijya

3.2   Device Structure and Functionality

HapticDaijya consists of the following components.

Robot: Our implementation uses six hobby servomotors (EZ Robot [2], Stall torque = 19 kg/cm) connected serially in a serpentine formation (Fig. 2). The total length of the robot is 51 cm and weighs 742 g. The robot is mounted on a base, which holds an EZ-B robot microcontroller [2].

Vest: The base of the robot is strapped to a vest, weighing 300 g. The vest makes the robot comfortable and easy to wear or take off.

Control: The EZ-B microcontroller is remotely controlled by a PC through Wifi. The control software was developed under the EZ-Builder framework and integrated with the Unity3D game engine using a client-server architecture.

With exchangeable end effectors, HapticDaijya can deliver a variety of haptic feedback in a variety of locations as shown in Fig. 3. Such capability is not limited to expanding the range of haptic feedback types, but to also match distinct user preferences or ergonomic requirements. For instance, taller users may use bigger or longer end effectors so the robot arm may reach all their torso areas, and sensitive individuals may use softer or padded end effectors.

The haptic and tactile feedback can be used for purposes beyond VR games. For example, drawing the user's attention to hazards and emergencies, like earthquakes, or to deliver smartphone notifications. In combination with exchangeable end effectors, additional applications such as feeding users during VR gaming are possible.

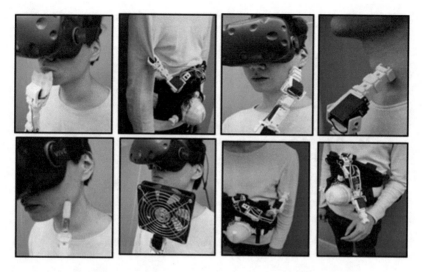

Fig. 3. Various feedback effects by HapticDaijya

4 Preliminary User Study

4.1 Research Method

To perform a preliminary test and evaluate our robot, we designed an experiment that gauges users' accuracy in determining the location of taps that are applied to various locations of the torso. We accordingly followed the experiment with questionnaires and interviews to evaluate general usability aspects.

We hired ten college students (Age m = 22.80, SD = 2.94, 6 Females). Each participant was first introduced to the robot and took a profiling questionnaire. Next, we carried out the *calibration* process, followed by the *tutorial*, which comprised a single dry run for each of the 16 calibrated points. Such process familiarized participants with the feedback on all 16 locations. As shown in the left part in Fig. 4, the torso is segregated into 16 cells. Cells 1 through 4 are aligned horizontally to four points on the collar bone and shoulders of each participant. The remaining 12 cells are aligned with 5-8 cm vertical spacing. The robot was calibrated to tap the center of each cell from an approximate distance of 10 cm using the maximum servo speed and full torque. The test took approximately 20 min per participant.

The *trials phase* started by first blindfolding the participants to simulate a VR experience. Each trial included a single tap on one of the 16 points, then each participant verbally indicated the point at which they believed they received the feedback.

We repeated the trials three times for each of the 16 points, thereby subjecting each participant to 48 taps. The trials were randomized to avoid possible learning effects. In total, we carried out 480 successful trials.

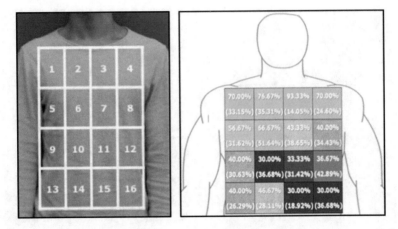

Fig. 4. User study: procedure and results

4.2   Results

Figure 4-Left illustrates the haptic feedback point-matrix on the user's torso. Each cell represents a region where the robot applied haptic feedback. Overall, participants achieved the highest accuracy levels on the first row and the sides, after which their accuracy gradually drops as shown Fig. 4-Right (SD values in brackets).

Our questionnaire used a 5-point Likert scale (1 is Disagree/Bad, 5 is Agree/Good). Participants rated "*I can easily distinguish the feedback among different points*" with 3.40 (SD = 1.07) and "*I can distinguish feedback among contiguous points*" with 2.70 (SD = 0.95). Most participants also indicated that identifying feedback on the edges of the torso is easier than the center, asserting that feedback on cells 1 through 4 is easier to identify as it is near the collarbones and shoulders.

Participants rated the comfort of our device with 3.80 (SD = 0.79) and the weight with 2.3 (SD = 0.95), thus we conclude that the wearability of the device was generally acceptable. Lastly, they rated their overall satisfaction with 3.80 (SD = 0.92).

Overall, we concluded that other factors, such as the intensity of the taps as well as our chosen cell locations and dimensions, may have contributed to these results. Nevertheless, we believe such results are intriguing to validate further.

5   Some Pitfalls Found in the Current Experiences

From the experiences with the current prototype, we have extracted the following potential pitfalls. In particular, analyzing our prototype within the context of the scenario described in Sect. 2 allowed us to extract a number of usability and implementation shortcomings. For example. although our prototype allows for manually changeable end-effectors to deliver various feedback types, future iterations must embed an automated end-effector changing mechanism to deliver a variety of haptic feedbacks without delay. Such limitations may restrict the possibilities of games significantly, but it may not be a good idea just simply to increase a number of robotic components.

In the following, we also present some pitfalls found during the design, implementation, and evaluation.

Visuo-Haptic Synchronization: Despite the variety of possible haptic feedback, the serpentine morphology imposes several limitations. Since the robot has to move to different points to apply feedback, there is an unavoidable delay in orienting and moving the robot. This is especially prevalent if the visual feedback in VR is much faster or very frequent, such that it outpaces the robot's ability to synchronously deliver haptic feedback in accordance with visual stimuli.

Simultaneous Haptic Feedback: Another shortcoming of the serpentine morphology is its incapability to deliver multiple haptic feedback impulses in parallel. For example, as described in the scenario shown in Sect. 2, a player may be attacked at two different positions on his/her body by two different enemies. The current hardware cannot provide the corresponding haptic feedbacks simultaneously at two different positions.

Movement Obstruction: The workspace of HapticDaijya may easily collide with hand movements of the user, especially since most modern VR platforms utilize hand controllers or gestures for interaction. This problem may be solved by context awareness to avoid collisions, or a smaller design that stays closer to the user's chest.

Calibration: An easy and precise calibration method ensures replicable and high quality user experiences. A quick calibration method is important for instantly fulfilling a user's ergonomic differences. Moreover, thick clothes, like jackets, could absorb feedback, thus, delivered feedback should be adapted to variance in users' clothing. Lastly, delicate Areas like the neck or the shoulders present calibration and safety challenges for delivering haptic feedback.

6   Conclusion

In this paper, we present HapticDaijya, a wearable haptic and tactile feedback robot. We presented our initial design and implementation, followed by an analysis of advantages and limitations. Our initial evaluations are overall encouraging to pursue further development, and the survey results are intriguing for further investigation.

In the future, we would like to realize a complete a multi-user attraction using HapticDaijya. In the future theme park, such advanced attractions require various haptic feedback and more immersion. We believe our future work contribute with various useful insights that will make future theme parks more enjoyable.

References

1. Corley, A.M.: Tactile Gaming Vest Punches and Slices (2010). spectrum.ieee.org/automaton/robotics/robotics-software/tactile-gaming-vest-punches-and-slices. Accessed 12 Jan 2018
2. EZ-Robot Inc. https://www.ez-robot.com/. Accessed 7 Jan 2018

3. Ishizawa, F., Sakamoto, M., Nakajima T.: Extracting intermediate-level design knowledge for speculating digital–physical hybrid alternate reality experiences. Multimed. Tools Appl., 42 (2018). https://doi.org/10.1007/s11042-017-5595-8
4. Jones, L.A., Nakamura, M., Lockyer, B.: Development of a tactile vest. In: Proceedings of the 12th International Symposium on Haptic Interfaces for Virtual Environment and Teleoperator Systems, Haptic 2004 (2004)
5. Konishi, Y., Hanamitsu, N., Minamizawa, K., Outram, B., Mizuguchi, T., Sato, A.: Synesthesia suit: the full body immersive experience. In: Proceedings of the ACM SIGGRAPH 2016 VR Village (SIGGRAPH 2016), Article 20, 1 p. (2016)
6. Maiman, A.A., Roudaut, A.: Frozen suit: designing a changeable stiffness suit and its application to haptic games. In: Proceedings of the 2017 CHI Conference on Human Factors in Computing Systems, pp. 2440–2448 (2017)
7. Roudaut, A., Rau, A., Sterz, C., Plauth, M., Lopes, P., Baudisch, P.: Gesture output: eyes-free output using a force ARAIG - multi-sensory VR feedback suit. In: Proceedings of the SIGCHI Conference on Human Factors in Computing Systems (2013)
8. Sakamoto, M., Nakajima, T.: Gamifying intelligent daily environments through introducing fictionality. Int. J. Hybrid Inf. Technol. 7(4), 259–276 (2014)
9. Sakamoto, M., Nakajima, T., Akioka, S.: Gamifying collective human behavior with gameful digital rhetoric. Multimed. Tools Appl. 76(10), 12539–12581 (2017)
10. Wu, S.W., Fan, R.E., Wottowa, C.R., Fowler, E.G., Bisley, J.W., Grundfest, W.S., Culjat, M.O.: Torso-based tactile feedback system for patients with balance disorders. In: Proceedings of the International Symposium on Haptic 2010 (2010)

Personalized Hybrid Recommendations for Daily Activities in a Tourist Destination

Tahir Majeed[✉], Aline E. Stämpfli, Andreas Liebrich, and René Meier

Lucerne University of Applied Sciences and Arts,
Suurstoffi 41b, 6343 Rotkreuz, Switzerland
tahir.majeed@hslu.ch
https://www.hslu.ch

Abstract. Valuable recommendations are not effortless to receive in a tourist destination [18]. Considering the daily routine of a person on vacation in a tourist destination, a hybrid Recommender System for a mobile app is proposed. A hybrid system helps in unifying the best aspects of different recommendation algorithms while simultaneously minimizing the drawbacks of the individual algorithms. It is capable of providing personalized, diverse and serendipitous recommendations for the stay in a tourist destination and suggests places to dine, to relax and possibilities for sports activities. As input for the algorithm, the information needs of tourists were examined conducting qualitative studies in an Alpine tourist destination. The proposed Recommender System, the results of the qualitative studies and the basic testing performed using initial data are presented.

Keywords: Recommender system · Personalized
Collaborative filtering · Content-based · Tourism

1 Introduction

In our daily lives, we all have to make decisions. We are faced with questions such as which book should I read? What kind of music should I listen to now? Where could I travel next? To which restaurant could I go for dinner? Which bar could be nice for a drink? In the cases in which we make informed decisions, we use our knowledge or previous experiences, ask a friend, search the internet, ask an expert and so on. However, good advice is difficult to find, expensive, or time consuming [9]. Recommender System fill this gap by providing personalized, high quality, and affordable advice [9].

Within a tourist destination, smartphones and apps have become important tools for tourists navigating in an unknown place. Smartphones are used for searching for information what to do, for facilitating the stay by checking train schedules or the weather, to communicate with home and on-site, and

© Springer Nature Switzerland AG 2019
P. Novais et al. (Eds.): ISAmI 2018, AISC 806, pp. 155–165, 2019.
https://doi.org/10.1007/978-3-030-01746-0_18

for entertainment purposes [28]. Also because of the smartphone driven shorter-term nature of organizing a stay's activities [27], mobile online recommendations in tourism are in demand.

Recommender Systems in diverse domains have been applied over the last 25 years [1,3,16]; however, they have only recently started to be developed in the tourism domain [6,7,10,12,13,19,21]. Tourism is a multi-trillion-dollar industry (the total contribution of tourism to the global economy in 2016 was 7.61tr USD[1]), where the task of providing information was traditionally done by tour operators and travel agents. Nowadays, people are increasingly using Internet-based platforms[2] to collect information and plan their trips, which can be very time-consuming. There is no provision of easy access to the relevant information for users. Thus, Recommender System fill the gap by providing valuable, personalized and accurate recommendations to them. For a short-term planning of trips on-site, specialized Recommender Systems for mobile devices are needed, which can provide valuable and personalized recommendations. This will result in saving time and effort at users' end.

The dynamics of a Recommender System in the tourism domain are generally different from the RSs in other domains such as songs, movies, books or news. A person might need recommendations in quick succession for books, movies, and songs, perhaps every week or every day. In contrast, a user of a tourism Recommender System might need recommendations only when he is on holidays which could be once a year or once in six months. When he goes back to the normal routine life after holidays, he will be inactive until his next holidays. However, a user who is on holidays in the target destination is considered an active user needing recommendations many times a day during his holidays.

In this paper, we propose a hybrid Recommender System to provide personalized recommendations on restaurants, bars, hotels and sports activities in an Alpine tourist destination. It is also intended that the recommendations are diverse and serendipitous for the user. The state of the art regarding recommendation algorithms is presented in Sect. 2, then Sect. 3 provides the details of the proposed hybrid Recommender System, Sect. 4 outlines the studies conducted to identify tags for the content based recommendations, then evaluation findings are presented in Sect. 5, and finally, Sect. 6 concludes the paper.

2   State of the Art

It has been known over the years that all recommendation approaches have their pros and cons. User-User Collaborative Filtering (UUCF) [5,8] can recommend items that are diverse and serendipitous [8]. On the other hand, the approach does not scale well to a large dataset of users and items [23]. To compensate for the scaling issue Item-Item Collaborative Filtering (IICF) was proposed by Sarwar et al. [23]. Although IICF has been shown to be robust to large datasets,

---

[1] Accessed on August 8, 2017: https://www.statista.com/topics/962/global-tourism/.
[2] www.booking.com, www.tripadvisor.com, and www.expedia.com
   accessed on August 8, 2017.

they do not have the potential to recommend serendipitous items. Both UUCF and IICF suffer from the cold-start problem [5]. The cold-start problem is associated with the new user and new item. When a new user/item has not provided/received any or enough ratings then it is not possible to reliably compute recommendation for the new user or to recommend new item using collaborative filtering approaches (UUCF or IICF). The cold-start problem can be addressed with Content-Based Recommendation (CBR) which can still be used even in the absence of user feedback data.

The CBR [11,15] approach associates a profile to every item. The item profile is defined in terms of descriptive keywords or tags that defines the item specific characteristics. A user profile is then created in terms of the same tags and keywords used to define item characteristics. To recommend items to a user a similarity score between the user profile and the item profile is computed and the top ranking items are then recommended to the user. The user profile can be created in a number of ways; using explicit user feedback (ratings), or simply asking the user to provide his preferences to a super-set of tags. Pessemier et al. [18] have used the latter approach to address the cold-start problem by asking the user to provide his preferences to a set of categories and then computing the recommendations. However, the CBR approach requires huge effort to create items profile and in certain domains, it might not be possible to automatically create the profile [2]. This approach is also susceptible to over-specialization [18,29]. Over-specialization occurs when the set of recommended items is homogenous because the items are identical to each other.

Although both IICF and CBR are item based recommendation techniques, they utilize completely different properties of the items. While CBR techniques look at the tags associated with the items, the IICF techniques look at how two items have received ratings to identify similar items and then use user ratings from the identified similar items to compute the recommendations. Generally, a non-homogenous and diverse set of recommendations is preferred that will allow the user more choices to choose from the recommendation list. This explains why many of the new Recommender System tend to be hybrid systems [4,14,17,18].

3   Approach

In this paper, the focus is on a hybrid Recommender System architecture and parameter optimization and due to space constraints, we have chosen not to address the cold-start problem here. First, the system architecture is outlined in Sect. 3.1 and then the mathematical formulation is presented in Sect. 3.2.

3.1   Architecture

Figure 1 shows the architecture diagram of the proposed Recommender System. The source of the data is a mobile app (appCommunicator) specifically built to provide personalized recommendations and collect user feedback on his/her likings and preferences to items. The app provides the interface through which the users can provide explicit feedback in the form of ratings (on a scale 0.5–5.0), while the app collects the implicit user feedback by monitoring users' viewing duration (in seconds).

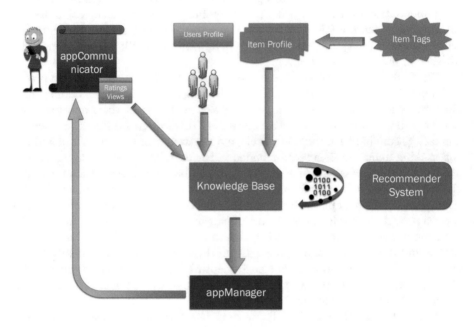

Fig. 1. Architecture diagram of the Recommender System

The feedback data obtained from the app is then stored in the knowledge base which is shared with the Recommender System. The knowledge base also contains users and items information which is stored externally to the app as shown in Fig. 1. Item information is stored in terms of keywords or tags which have been obtained through qualitative studies analyzing tourists' information needs (see Sect. 4). The Recommender System utilizes the data in the knowledge base to provide personalized recommendations. The computed recommendations are then written back to the knowledge base. The appManager queries the knowledge base and asks for recommendations for a particular user. It then sends the computed personalized recommendations to the user. The user sees the recommended items and decides to either view or not to view the item. If the user views the item, then the viewing duration is recorded. If the user decides to consume the item and physically visits the item recommended to him, then he can also rate the item or provide no feedback at all. This completes the Recommender System cycle.

3.2 Mathematical Formulation

The proposed hybrid Recommender System is a linear combination of UUCF, IICF, and CBR which extends the idea of Balabanovic and Shoham [2] and

Pessemier et al. [18]. The recommendations from the three approaches are computed using the user ratings data as depicted in Fig. 2.

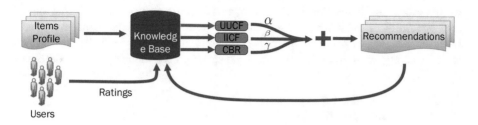

Fig. 2. Proposed hybrid recommendation system

Let $\mathbf{D} = \{(u^{(j)}, i^{(j)}, r^{(j)}_{u,i})\} : j = 0, \ldots, k - 1$ be the set of test dataset consisting of k-samples and the superscript j represents the jth-sample, where $u \in \mathbf{U}|\mathbf{U} : u \in \{0, \ldots, m - 1\}$ is the set of m-users, $i \in \mathbf{I}|\mathbf{I} : i \in \{0, \ldots, n - 1\}$ is the set of n-items in the system, and $r \in \mathbf{R}|\mathbf{R} = \{0.5, 1.0, 1.5, \ldots, 5.0\}$ is the set of ratings. The user u selects a rating from $\mathbf{R}$ for an item i donated by $r_{u,i}$. All unrated items by the user get a value 0.0. The predicted rating of the hybrid Recommender System for user-u about item-i is donated by $\hat{r}_{u,i}$. The predicted rating $\hat{r}_{u,i}$ is a linear combination (see Eq. 1) of $\hat{r}_{uu_{u,i}}, \hat{r}_{ii_{u,i}}, \hat{r}_{cb_{u,i}}$ which are computed using the three approaches UUCF, IICF and CBR respectively utilizing the training data as depicted in Fig. 2.

$$\hat{r}_{u,i} := \alpha \times \hat{r}_{uu_{u,i}} + \beta \times \hat{r}_{ii_{u,i}} + \gamma \times \hat{r}_{cb_{u,i}} \tag{1}$$

$$\alpha + \beta + \gamma = 1 \tag{2}$$

The sum of the parameter α, β, γ is equal to one as given by Eq. 2. Although in literature other types of hybrid Recommender System have been suggested, we have chosen to implement the one based on the linear combination of recommendations because it is simple to combine and optimize the parameters (α, β, γ). A loss function $L(\theta) \mid \theta := (\alpha, \beta, \gamma)$ can be defined over the test dataset $\mathbf{D}$ that computes the error between the predicted item rating $\hat{r}_{u,i}$ and the ground truth rating actually provided by the user $r_{u,i}$. The loss function can be computed using the Eq. 3. Equation 4 defines the objective function which is then minimized to obtain the optimized values $\theta^* = [\alpha^* \ \beta^* \ \gamma^*]$ for which the error between the predicted rating $\hat{r}_{u,i}$ and the actual rating $r_{u,i}$ over the entire test set is minimized.

$$L(\theta \mid \mathbf{D}) = \sum_{j=0}^{k-1} \left(\sqrt{(\hat{r}^{(j)}_{u,i} - r^{(j)}_{u,i})^2} \right) \tag{3}$$

$$\underset{\theta}{\arg\min} \, L(\theta \mid \mathbf{D}) \tag{4}$$

The parameters can be optimized using the optimization algorithm given by Algorithm 1. A brute force strategy has been used to compute the optimal value of the parameters. Predicted ratings of all the items for all the users are computed just once and subsequently used in every iteration. Other optimization strategies could be used for the search of optimal parameters, for example, non-linear optimization methods [25], but the global minimum solution is not guaranteed and generally, a local minimum solution is found. Once the parameters have been learned, the top-N (where N is a predefined constant in the system which is fixed at 10) recommendations for a given user can then easily be computed using Eq. 1.

Result: Optimized θ^*
initialization $\theta_0 = (0, 0, 0)$;
for ($\alpha = 0$; $\alpha \leq 1$; $\alpha = \alpha + 0.05$) {
    for ($\beta = 0$; $\beta \leq (1 - \alpha)$; $\beta = \beta + 0.05$) {
        for ($\gamma = 0$; $\gamma \leq (1 - \alpha - \beta)$; $\gamma = \gamma + 0.05$) {
            for ($j = 0$; $j < k$; $j = j + 1$) {
                $\hat{r}_{u,i}^{(j)} = \alpha \times \hat{r}_{uu_{u,i}}^{(j)} + \beta \times \hat{r}_{ii_{u,i}}^{(j)} + \gamma \times \hat{r}_{cb_{u,i}}^{(j)}$
            }

$$L(\theta \mid \mathbf{D}) = \sum_{j=0}^{k-1} \left(\sqrt{(\hat{r}_{u,i}^{(j)} - r_{u,i}^{(j)})^2} \right)$$

        }
    }
}

Algorithm 1: Parameters optimization

The proposed approach builds upon the idea of Pessemier et al. [18], however, in their approach, the weights of the linear combination of methods are set heuristically, whereas, in the proposed approach the weights are optimized over a test dataset. Another difference is that the user profile representing the user tag affinity is built by asking the user to manually set his preference to the set of tags in the system using sliders which is then used as a descriptor of the user profile whereas, in the proposed approach the user profile is build using the ratings data and finally, no knowledge-based recommendations are used in the proposed approach.

The weights of the proposed approach are learned offline only once and subsequently can be used in online recommendations. It is not required to learn the parameters every time recommendation is required for a user. This strategy of learning weights over a test dataset ensures that if the test dataset is a good representative of the underlying distribution of the user interests and preferences then the optimally learned weights will provide optimal recommendations. We are not aware if someone has used this approach to learning parameters of the target function in the Recommender System community, however, optimizing parameters in such a manner is a standard approach in machine learning domain.

4   Category Definition

To use the content based filtering, the identification of the correct tags is of utmost importance. The tags capture all the aspects of the items, and simultaneously they should be expressive enough to capture user preferences. For identifying tags, tourists' information needs in an Alpine tourist destination were examined by applying the qualitative mobile ethnography approach and conducting qualitative interviews.

4.1   Study 1

The aim of the first study was to explore the information needs of the tourists by capturing the information the tourists consider relevant themselves.

Participants. Participants were recruited on site, with the help of the project partners; hotels and apartments where the tourists were living during their stay. Seven tourists participated in the study. Five of them travelled with their family, one with the partner, and one was an individual tourist. Five participants have visited the destination before five or more times, one between one and five times and one never. Most of the participants had an age between 40 and 49 years (3), two were older than 50 years and two had an age between 18 and 29. Three Swiss participated in the study, two from the United Kingdom, one from Germany and one from the United States. As an incentive, participants received a free one-day ski pass.

Method. The mobile ethnography approach was applied to collect the data. This approach is based on self-report data using smartphones. It allows participants to capture the information they consider relevant regarding the research question and thus facilitates to receive an extensive but relevant feedback [20,24,26].

Materials and Measures and Procedure. In an initial personal briefing, participants filled in a form, giving information about themselves such as the country of origin, sex, age, relation to their fellow travelers (for example, family), and how often they have visited the Alpine tourist destination before. In the briefing, participants were informed about the aim of the study, to capture their experiences with moments of information needs. Further, they were given instructions about downloading and using the smartphone app for capturing the self-report data. The app "ExperienceFellow" was applied. ExperienceFellow works like a diary on the smartphone. Participants were also informed about the voluntary individual interviews which were planned to be conducted after the self-report phase of the study, to precise the self-captured information. Participants captured data during 2–6 days. They were instructed to evaluate each captured moment on a scale from very negative (-2) to very positive ($+2$). Photos, videos, and texts could be added to captured moments.

4.2   Study 2

To complement Study 1, qualitative interviews were conducted in the Alpine tourist destination.

Participants and Method. Eleven tourists participated in the in-depth qualitative interviews in the Alpine tourist destination. Two of them were participants in Study 1. Seven of the interviewees traveled with their family, two with their partner and one with friends (information missing for 1 interviewee; in the following, only the available values are reported). Three of the participants have never visited the destination before, two have visited the destination one to five times, two more than five times and two more than fifty times. The interviewed parents' and their partners' age was five times (counted per person not interview) between 40 and 49 years and three times over 50 years. The children's age was for three families between 10 and 19 years and for one family between 25 and 29 years. Six interviews were conducted with Swiss, five with international tourists (mainly from the United States). As an incentive for their participation, a beverage was offered to the participants.

Materials and Measures. After giving some information about themselves such as age, the interviewees answered questions regarding their information needs before and during their journey to the destination as well as during their stay.

4.3   Definition of Information Need Drivers

The data of both studies were analyzed to identify information need determinants of tourists, taking into account the existing literature and data of the local destination management organizations on their guest structure. To structure the information, the concept of the customer journey was applied, the guests' experience from getting inspired and searching for information, to booking and traveling, up to the experience after traveling.

4.4   Results

The fellow travelers were identified as an important determinant of tourists' information needs. For example, families have different information needs in an Alpine tourist destination than single travelers. Another important determinant identified was, whether tourists were first-time visitors or regulars.

In addition to the determinants of information needs identified, recommendations for providing information were made per subgroups of determinants. One point considered, when providing inputs for the algorithm, was the flexibility in recommendations. For example, if the Recommender System recognizes that a tourist loves eating meat, it should not recommend restaurants offering meat every day of a tourist's stay but recommend the tourist to go for having traditional Swiss fondue, one day.

5   Evaluation

To evaluate the accuracy of the recommendations provided to the users, different sets of evaluation metrics were used. To evaluate the accuracy of the predicted rating $\hat{r}_{ui}$ of the recommendations, Mean Absolute Error (MAE) and Root Mean Squared Error (RMSE) were used [22]. To evaluate the relevance of the recommendations to the user with respect to the Top-5, Top-10 and Top-20 recommendations, additional measures were identified. In Sect. 3 it was mentioned that the Top-N recommendations were fixed at 10, however, for evaluation purpose, the parameter was varied to compute Top-5, Top-10 and Top-20 recommendations and the same number of Top-N recommendations were extracted from the ground truth data. To perform the evaluation based on the Top-N recommendations, Precision, Recall, and F-measure were used [22]. To evaluate the overall hybrid Recommender System performance with respect to the coverage, catalog coverage and user coverage measures were used [22].

As this is a work in progress initial testing was performed using synthetic data and the identified measures were found to provide a good evaluation of the proposed approach. Synthetic data was generated using a random number generator. The random number generator was used to generate user profile which included categories such as age (3 possible values), gender (2 possible values), country (8 possible values), guest-type (2 possible values), visitor-type (3 possible values), and language (4 possible values). This user profile reflects the data obtained through the qualitative studies conducted as explained in Sect. 4. The items data was similarly generated using the random number generator. The item profile consisted of different tags identified through the study conducted. The third part of the synthetic data was to generate user ratings data. For each user first the item IDs were randomly generated and then another random number generator subsequently assigned randomly generated rating to those item IDs. While generating the synthetic ratings data, care was taken to generate data that closely resembled real users.

The identified evaluation metrics presented in the previous paragraph were employed on the randomly generated data. The purpose of the testing at this stage was to evaluate how well the metrics can evaluate the proposed system performance. The results of the evaluation using the synthetic data are not presented because a low or high error will not provide helpful insight into the recommendation accuracy. The collection of real data from an actual Alpine destination is in progress that will enable extensive testing of the proposed approach. The data collected during the studies performed (see Sect. 4), only deals with collecting and identifying relevant item tags and were not meant to collect user ratings.

6   Conclusion

This paper presented a hybrid Recommender System in the tourism domain which is a linear combination of UUCF, IICF, and CBR. The parameters of the hybrid system were learned in an energy minimization framework by iterating over the parameter space and minimizing the predicted loss. A hybrid recommendation system is used because it combines the strong aspects of the different recommendation approaches while simultaneously minimizing their drawbacks. Explicit user feedback data was used for computing the recommendations. Users' information needs were examined conducting qualitative studies in an Alpine tourist destination to capture expressive tags for the algorithm. Initial tests were performed and the Recommender System performance was evaluated using different evaluation metrics. In future work, we plan to perform extensive evaluation using real tourism datasets. We also intend to address the cold-start problem using view duration data. Further tests using non-linear optimization methods will be carried out to compare them with the employed brute force strategy with respect to speed and accuracy.

Acknowledgement. This work was funded in part by Innosuisse - the Swiss Innovation Agency. The authors would also like to thank ipeak Infosystems for their support and for providing the data that made this work possible.

References

1. Adomavicius, G., Tuzhilin, A.: Toward the next generation of recommender systems: a survey of the state-of-the-art and possible extensions. IEEE Trans. Knowl. Data Eng. **17**(6), 734–749 (2005)
2. Balabanovic, M., Shoham, Y.: Fab: content-based, collaborative recommendation. In: Communications of the ACM, pp. 66–72 (1997)
3. Bobadilla, J., Ortega, F., Hernando, A., Gutiérrez, A.: Recommender systems survey. Knowl. Based Syst. **46**(7), 109–132 (2013)
4. Burke, R.: Hybrid recommender systems: survey and experiments. User Model. User-Adapt. Interact. **12**(4), 331–370 (2002)
5. Ekstrand, M.D., Riedl, J.T., Konstan, J.A.: Collaborative filtering recommender systems. Found. Trends Hum.-Comput. Interact. **4**(2), 81–173 (2010)
6. Garcia, I., Sebastia, L., Onaindia, E.: On the design of individual and group recommender systems for tourism. Expert Syst. Appl. **38**(6), 7683–7692 (2011)
7. Gavalas, D., Kasapakis, V., Konstantopoulos, C., Mastakas, K., Pantziou, G.: A survey on mobile tourism recommender systems. In: International Conference on Communications and Information Technology (ICCIT), pp. 131–135 (2013)
8. Herlocker, J.L., Konstan, J.A., Borchers, A., Riedl, J.: An algorithmic framework for performing collaborative filtering. In: International Conference on Research and Development in Information Retrieval, pp. 230–237 (1999)
9. Jannach, D., Zanker, M., Felfernig, A., Friedrich, G.: Recommender Systems: An Introduction. Cambridge University Press, Cambridge (2011)
10. Kenteris, M., Gavalas, D., Mpitziopoulos, A.: A mobile tourism recommender system. In: Computers and Communications, pp. 840–845 (2010)
11. Lang, K.: NewsWeeder: learning to filter netnews. In: Machine Learning Proceedings, pp. 331–339. Elsevier (1995)
12. Liu, Q., Chen, E., Xiong, H., Ge, Y., Li, Z., Wu, X.: A cocktail approach for travel package recommendation. IEEE Trans. Knowl. Data Eng. **26**(2), 278–293 (2014)

13. Liu, Q., Ge, Y., Li, Z., Chen, E., Xiong, H.: Personalized travel package recommendation. In: International Conference on Data Mining, pp. 407–416 (2011)
14. Lucas, J.P., Luz, N., Moreno, M.N., Anacleto, R., Figueiredo, A.A., Martins, C.: A hybrid recommendation approach for a tourism system. Expert Syst. Appl. **40**(9), 3532–3550 (2013)
15. Mooney, R.J., Roy, L.: Content-based book recommending using learning for text categorization. In: Conference on Digital libraries, pp. 195–204 (2000)
16. Park, D.H., Kim, H.K., Choi, Y., Kim, J.K.: A literature review and classification of recommender systems research. Expert Syst. Appl. **39**(11), 10059–10072 (2012)
17. Pazzani, M.J.: A framework for collaborative, content-based and demographic filtering. Artif. Intell. Rev. **13**(5), 393–408 (1999)
18. Pessemier, T.D., Dhondt, J., Martens, L.: Hybrid group recommendations for a travel service. Multimedia Tools Appl. **76**(2), 2787–2811 (2016)
19. Petrevska, B., Koceski, S.: Tourism recommendation system: empirical investigation. J. Tour. **14**(4), 11–18 (2012)
20. Plesner, A., Clatworthy, S.: Lessons learned. In: Stickdorn, M., Frischhut, B. (eds.): Service Design and Tourism, pp. 110–117. Books on Demand, Norderstedt (2012)
21. Ricci, F.: Mobile recommender systems. Inf. Technol. Tour. **12**(3), 205–231 (2010)
22. Ricci, F., Rokach, L., Shapira, B., Kantor, P.B. (eds.): Recommender Systems Handbook. Springer, Heidelberg (2010)
23. Sarwar, B., Karypis, G., Konstan, J., Riedl, J.: Item-based collaborative filtering recommendation algorithms. In: International Conference on World Wide Web, pp. 285–295 (2001)
24. Stickdorn, M., Frischhut, B., Schmid, J.S.: Mobile ethnography: a pioneering research approach for customer-centered destination management. Tour. Anal. **19**(4), 491–503 (2014)
25. Swann, W.: A survey of non-linear optimization techniques. FEBS Lett. **2**(S1), 39–55 (1969)
26. Verhelä, P., Stickdorn, M.: In search for authentic user insights. In: Stickdorn, M., Frischhut, B. (eds.) Service Design and Tourism, pp. 52–63. Books on Demand, Norderstedt (2012)
27. Wang, D., Park, S., Fesenmaier, D.R.: The role of smartphones in mediating the touristic experience. J. Travel Res. **51**(4), 371–387 (2011)
28. Wang, D., Xiang, Z., Fesenmaier, D.R.: Adapting to the mobile world: a model of smartphone use. Ann. Tour. Res. **48**, 11–26 (2014)
29. Yu, C., Lakshmanan, L.V.S., Amer-Yahia, S.: Recommendation diversification using explanations. In: IEEE International Conference on Data Engineering, pp. 1299–1302 (2009)

Monitoring Rehabilitation Process
Using Microsoft Kinect

Miguel A. Laguna[(✉)] and Irene Lavín

GIRO Research Group, University of Valladolid, Campus M. Delibes, 47011 Valladolid, Spain
{mlaguna,irene}@infor.uva.es

Abstract. One of the most remarkable applications of the pervasive computing paradigm is the development of smart assistants to increase the quality of life of people with diverse physical difficulties. In the context of limb rehabilitation of injured or stroke affected people, this article presents a solution for exercise recognition and monitorization using a camera based sensor, the Microsoft Kinect. The system uses the data obtained from Kinect as spatial positions of the joints. Applying the algorithms described in the article, the system quantitatively measures the improvement of limbs mobility during a rehabilitation process.

Keywords: Rehabilitation · Monitoring · Kinect

1 Introduction

The concept of pervasive health has recently emerged with the aim of enhancing the quality of life of patients (including elderly, injured people during rehabilitation process, etc.) using Information and Communication Technologies for health care and wellness [1, 2]. A straightforward application of this technology is the development of smart assistants to help patients alone at home or to allow their remote monitoring. Many applications for social assistance have been developed and usually subsidized by governments or non-profit organizations. Alert panic buttons are common, but different sensors can provide data about heart rate, oxygen saturation, temperature, etc. [2]. In this context, the GIRO research group has developed several projects of monitoring applications [3] deployed for diverse stationary and mobile platforms and connected to a central system for continuous surveillance. In the case of physical activity monitoring, accelerometers, gyroscopes or cameras can be used. In our case, we decided to use the Microsoft Kinect device [4], an inexpensive camera based sensor, in several projects in this field.

The aim of this paper is to report the experience in the development of a physical exercise monitoring system to help therapists to control limbs rehabilitation processes, initially focused on their hemiplegic patients. The system is based on Kinect, connected via USB to a PC running the local application. Kinect automatically detects the spatial position of a person and differentiates the parts of her body, more exactly the skeletal joints. The system controls that the patients do correctly the proposed exercises and automatically measures the degree of improvement in the movement capability of the

P. Novais et al. (Eds.): ISAmI 2018, AISC 806, pp. 166–173, 2019.
https://doi.org/10.1007/978-3-030-01746-0_19

extremities in the rehabilitation process. In addition, the temporal sequence of the data associated with the exercise (flexion or extension angles, etc.) are recorded automatically for subsequent analysis by the therapist.

The rest of the article is organized as follows: Sect. 2 establishes the motivation for the work and a summary of the requirements of the system. Section 3 discusses the technological possibilities of the various sensors and refers some of the existing solutions and related work. Section 4 describes the design and implementation of the monitoring system for rehabilitation processes. Finally, Sect. 6 concludes the article and summarizes future work.

2 Physical Activity and Monitoring

The main goal of our work on monitoring systems in GIRO is to assist dependents in getting increased autonomy and a higher quality of life. In this context, the GIRO group has developed several fall detection projects [3] for diverse mobile and fixed platforms, including a version based on Kinect. The prototype we developed with the first version of the Kinect sensor aroused the interest of some rehabilitation professionals as a possibility to improve the recovery processes of their patients. The main problem they face is the intensive manual required work that ideally could be replace by automatic control. The collaboration with the occupational therapists allowed us to focus on the appropriate exercises. In addition to controlling the execution of the exercises, the system should record the detailed parameters (positions and angles reached or duration of the exercise) that could serve the assigned therapist to verify the achieved improvements. The goal was to use the application in rehabilitation processes after a fall, stroke, accident, etc. In summary, the main objectives of the proposal were:

- To manage an agenda with the physical exercises that a person must perform
- To automatically detect, using the Kinect sensor, the correct execution of these exercises and update the agenda
- In the case of applying a specific rehabilitation program, to record the improvements using the measures of reached angles of shoulder or elbow movements.

3 Sensor Technology for Activity Monitoring

The requirements gathered in the previous Section can be structured in two sets:

- Information and communication requirements (user management, exercises definition, information logging about their execution)
- Monitoring requirements (specifically to analyze the user movements to measure the shoulder or elbow angles of flexion, extension, etc.)

In monitoring systems, the acquisition and analysis of sensor data is the most critical part. There are profuse literature about the use of sensors for monitoring, including commercial solutions [1, 2]. We surveyed the literature specifically dedicated to activity monitoring, searching for affordable approaches to keep the costs as low as possible.

For the detection of physical activity, including fall detection, there are several algorithms that use 2D and 3D image analysis, based on motion and shape changes [1, 5, 6]. Most of these works are ad hoc solutions, valid for concrete platforms and sensors, for example in [1] a solution based on floor vibration sensors is described, being impracticable as universal solution. The approaches described in [5] and [6] need complex arrays of cameras to follow the displacement of a person at home, even in a unique room. Presence detectors is an alternative for detecting continuous activity but although these detectors are cheaper, the provided information is less reliable. We discarded these possibilities considering that we need a much more focused system. The intention was not to monitor a complete house but only a reduced space where the patient can perform her exercises.

For that limited activity monitoring we decided to use the Microsoft Kinect sensor. This sensor detects the relative spatial position of the users and this position can be evaluated by an application running on a PC connected via USB with the sensor. Microsoft provides the required libraries for software development [4]. The sensor basically consists of an infrared (IR) camera and a projector of the same frequency spectrum, which allows collecting information about the situation of the objects in front of the sensor in a three-dimensional space. A secondary standard camera is also provided, together with four microphones for speech recognition and a motor that controls the camera angle. This camera is connected to a 1600 × 1200 CMOS IR and visible sensor, although only uses it partially, being the practical resolutions supported by the camera 640 × 280 and 1280 × 1024.

The IR projector transmits an infrared point cloud. The pattern consists of nine rectangles and the approximate detection range is 0.8 to 4 m from the sensor. However, it depends also of the size of the object [7] and also it is possible to detect an object at less than 0.8 m from the sensor, using the Kinect software. The IR camera detects the infrared information whit variable resolution. To provide a 3D image using the IR camera and the projection pattern, Kinect uses disparity (the lateral displacement between two retinal images corresponding to a scene in the real world). This technique allows calculating the relative distance between two objects. In Fig. 1, an imaginary pyramid is defined by the observation point and the plane that contains the image observed on the viewer. The first discernible plane further delimits the figure, resulting in the truncated pyramid that defines the Kinect range (from 0.8 to 4 m).

Microsoft provides a SDK [4] with set of basic functions to use in a Kinect based applications. The most interesting are:

- Video Stream (standard VGA camera stream)
- Depth And Player Index Stream (IR disparity stream with player recognition)
- Skeleton Tracking (the user code can activate and control this advanced possibility)

The third data source provides the spatial position of the user "bones" (more exactly, points that the Kinect internal algorithm identifies as head, hands, hips, knees, ankles, etc.). This information is very useful for activity monitoring, as an exercise can be defined as a sequence of the expected relative positions of the "bones" as described in next Section.

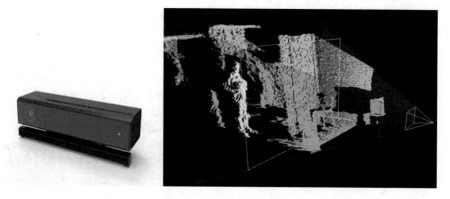

Fig. 1. The Kinect camera and the truncated pyramid that defines the Kinect range (0.8 to 4 m, red lines), and a human figure being recognized (in yellow)

4   Design and Implementation of the System

Following the suggestions of occupational therapy professionals, the prototype focused on monitoring standardized rehabilitation exercises, well documented in the literature. Using the relative positions of three joints, the angle formed when making a movement can be measured. To achieve this, the standardized vectors formed by the points of the skeleton are used, that is, the segments that connect the joints used for the measurement. For example, in the case of the angle of the elbow during flexion or extension movements, the shoulder, elbow and wrist are used. The practical result is that, by capturing these three positions, the angle of flexion or extension can be obtained at any position in space.

The objective of the rehabilitation exercises of the prototype in its current state of implementation is the progressive recovery of the movement of the upper limbs taking as reference the shoulder and elbow. The exercises are usually composed of two complementary movements: flexion and extension or adduction and abduction (although in future versions rotational movements will be incorporated). Each type of movement is associated with a limit that marks the range of values a person can reach. The list of movements to be performed is based on the Manual of Articular Physiology proposed by Kapandji [8], and are the following:

- Elbow flexion and extension movements
- Flexion and extension movements of the shoulder
- Adduction and abduction movements of the shoulder
- Horizontal flexion and extension movements of the shoulder

Although it is easy to measure the angles of flexion or extension while performing elbow exercises, the rest of the exercises are more complex. For the flexion-extension of the shoulder it is necessary to take into account the simultaneous movement of the trunk and introduce adjustments, which guarantees at all times that the angle measured is the real one. In addition, once a certain angle is reached the shoulder inevitably moves upwards and this displacement must be compensated. In the case of adduction-abduction of the shoulder, when the arm is in front of the body there is a displacement of the shoulder which must be corrected. In both cases, these observed displacements are consistent with those described in [8]. Finally, the horizontal flexion-extension movement of the shoulder can present the problem of temporary occlusion (the sensor cannot distinguish the shoulder from the elbow or the wrist). In this case, a reference plane was introduced as an aid to calculate the angle. All these corrections have been introduced as a result of the contrast tests performed with people of different body typology. The objective was to verify that the error regarding the measurements obtained manually by the therapist following a well-established protocol was always under the 10% limit.

Regarding the details of the measurement algorithms, two examples of exercise are shown in Fig. 2 and succinctly analyzed. For the shoulder flexion-extension of the left of Fig. 2 the procedure is as follows:

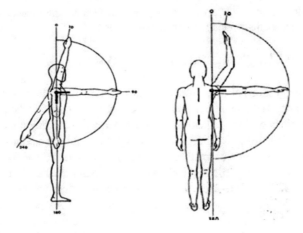

Fig. 2. Some rehabilitation exercises: Shoulder flexion-extension and abduction

1. Calculate the normal of the sagittal plane from the vector that goes from the right hip to the left hip (this vector is perpendicular to the sagittal plane)
2. Calculate the vectors that represent the arm and the spine. To obtain the vector of the arm, the position of the elbow and the position of the shoulder have been used. For the spine, the positions of the "central shoulder" and the spine are used
3. Project these two vectors in the sagittal plane calculated during step 1
4. Calculate the angle between the two projected vectors

For the abduction exercise (right of Fig. 2) the following steps have been followed:

1. Calculate the normal of the frontal plane. For this calculation, the plane was obtained through 3 points: shoulder (left or right according to the calculated side), "central shoulder", and spine
2. Calculate the vectors that represent the arm and the spine in a similar way to the previous exercise
3. Project both vectors in the previously calculated plane. If not projected, the angles obtained would not always be exactly those corresponding to abduction
4. Finally, calculate the angle between these two vectors

The movement improvement can be evaluated by comparing the evolution over time of the values of the angles achieved in each exercise. This will not only show the patient the results that indicate the improvement of the affected limb, but she will be also motivated to continue the rehabilitation work.

During the exercise, different information can be observed in the side menu, and different options can be chosen, such as the type of exercise or the affected side of the body in treatment (Fig. 3). The detection of the exercise begins when the corresponding button is pressed by the therapist (or a voice command is used, detected by Kinect microphones), showing the formed angle in real time. When the results need to be recorded, pressing a button will store the angles and timestamp data (see Fig. 4).

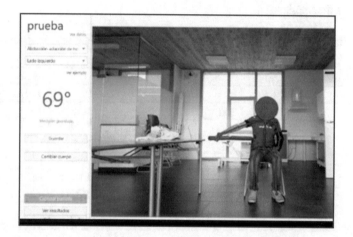

Fig. 3. User interface of the rehabilitation application

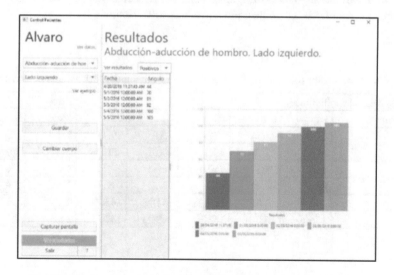

Fig. 4. User interface of the rehabilitation application: results

5   Results and Discussion

For each type of exercise, a first set of tests have been carried out by an occupational therapist using a conventional goniometer. Several angles were chosen, repeating the test 5 times for each angle. The process was as follows: The joint and limb of the subject is placed at a certain angle using the goniometer, and the angle obtained by the application is recorded and both are compared. As representatives examples, Tables 1 and 2 present a summary of the validations for the mentioned exercises, clearly under the aimed 10% limit. For the elbow flexion-extension the results are comparable. However, in the case of horizontal flexion-extension considerable errors are obtained in angles around 30° or less, due to the occlusion problem described above and it is certainly a limitation of the application.

Table 1. Tests results (compared with a conventional goniometer) for flexion-extension movement of shoulder

Kinect data (abs. difference)						Error	
Goniometer	P1	P2	P3	P4	P5	Mean	Relative
30	30(0)	30(0)	29(1)	29(1)	30(0)	0,4	1%
45	43(2)	44(1)	44(1)	46(1)	45(0)	1	2%
75	77(2)	77(2)	74(1)	75(0)	75(0)	1	2%
90	88(2)	91(1)	91(1)	90(0)	89(1)	1	1%
135	138(3)	135(0)	134(1)	136(1)	132(3)	1,6	1%
150	150(0)	149(1)	150(0)	150(0)	151(1)	0.4	<1%

The application is currently in her validation phase with real patients by Occupational Therapy professionals at the Orhu rehabilitation cabinet (http://orhu.es). The requirements to install the application are, apart from the Kinect sensor, a personal computer with an approved USB 3.0 connection and Microsoft Windows 10.

Table 2. Tests results (compared with a conventional goniometer) for abduction movement

Kinect data (abs. difference)						Error	
Goniometer	P1	P2	P3	P4	P5	Mean	Relative
30	27(3)	28(2)	27(3)	30(0)	29(1)	1,8	6%
45	44(1)	42(3)	46(1)	45(0)	45(0)	1	2%
75	73(2)	78(3)	75(0)	76(1)	76(1)	1,4	1%
90	93(3)	94(4)	90(0)	92(2)	91(1)	2	2%
135	140(5)	137(2)	134(1)	141(6)	142(7)	4	2%
150	156(6)	153(3)	153(3)	149(1)	152(2)	3	2%

6   Conclusions

A system has been developed to monitor the rehabilitation processes based on a low cost and easily available sensor. The initial goal of the system was the rehabilitation of hemiplegic patients, for whom a set of specialized exercises has been selected. Specifically, the system measures not only the actual number of performed repetitions of the exercises prescribed by the therapist but also the improvement in the reached angles of flexion, extension, etc. The application is currently in her validation phase with hemiplegic patients at the Orhu rehabilitation cabinet. As future work, more exercises can be incorporated to extend the functionality of the system, following other protocols including rehabilitation of wrist, knee, etc. Once the usefulness of the system has been verified by the occupational therapists, the usability of the system can be improved to make it accessible directly to patients, although always under professional supervision.

Acknowledgments. This work has been funded by the Junta de Castilla y León (FUESCYL) and the General Foundation of the University of Valladolid. The authors thank Hilario Ortiz (Orhu cabinet) for his advice, and Alavaro Garzo for his work developing parts of the application.

References

1. Alemdar, H., Cem, E.: Wireless sensor networks for healthcare: a survey. Comput. Netw. **54**(15), 2688–2710 (2010)
2. Tocino, A.V.: Personal health monitor. In: New Directions in Intelligent Interactive Multimedia Systems and Services 2, pp. 465–475 (2009)
3. Laguna, M.A., Finat, J.: Remote monitoring and fall detection: multiplatform Java based mobile applications. In: III International Workshop of Ambient Assisted Living, pp. 1–8 (2011)
4. Microsoft: Kinect for Windows. https://developer.microsoft.com/en-us/windows/kinect. Accessed 12 Feb 2018
5. Rougier, C., Meunier, J., St-Arnaud, A., Rousseau, J.: Fall detection from human shape and motion history using video surveillance. In: Proceedings of AINAW, pp. 875–880 (2007)
6. Sixsmith, A., Johnson, N.: A smart sensor to detect the falls of the elderly. IEEE Pervasive Comput. **3**(4), 42–47 (2004)
7. Smisek, J., Jancosek M., Padjla T.: 3D with Kinect; Czech Technical University in Prague (2011)
8. Kapandji, A.I.: Fisiología Articular, Tomo 1, 6th edn. Maloine (2006)

Context-Awareness and Uncertainty: Current Scenario and Challenges for the Future

Leandro O. Freitas[(✉)], Pedro R. Henriques, and Paulo Novais

ALGORITMI Centre, Department of Informatics, University of Minho, Braga, Portugal
leanfrts@gmail.com, {prh,pjon}@di.uminho.pt

Abstract. One of the main aspects of Ambient Intelligence (AmI) refers to its capacity of act autonomously in benefit of human beings. This implies in a hard challenge to overcome and of enormous responsibility. AmI is directly related to other fields of knowledge such as Smart environments, which aim to improve user experience through the development of context-aware applications. In this paper we present the current scenario of context-aware systems with some conceptual metrics to be followed. We highlight the problem of dealing with uncertain context information, e.g. incomplete, out-dated or nebulous data, seen as one of the main obstacle in this area.

Keywords: Intelligent environments · Context-Awareness · Uncertainty

1 Introduction

Intelligent environments should be sensible, allowing the system to identify the current state of entities and react under different situations, considering specially the user [5]. In such environments the user is the main focus for the development of services [2]. Considering this, Ambient Intelligence (AmI) is directly related to the People-Centric Computing paradigm, which aims to increase the participation of the user inside environments full of smart devices. According [6], this paradigm supports the idea of users acting not only as final client, but also as contributors. Their actions are monitored and used as source of information to improve the system functioning.

Intelligent environments should have the ability of learning and adapting not only from users' needs or requirements, but also, for the natural evolution of their preferences or desires. This kind of domain learns by observing the user and their apprenticeship allows the development of dynamic environments [9].

Context-aware systems use events to evidence changes in the environment. When the system identifies new states of entities, new events are created. Actions are defined as response to events and are used to characterize the behaviour of applications or services. To develop context-aware systems we have to adapt the concept of event, allowing it to be programmed or manipulated by applications [10].

Context-aware systems use information provided by different types of sensors. The role of these sensors is to monitor the environment and detect relevant data. Due to the dynamicity of smart environments, there are times where the data collected is not enough

© Springer Nature Switzerland AG 2019
P. Novais et al. (Eds.): ISAmI 2018, AISC 806, pp. 174–181, 2019.
https://doi.org/10.1007/978-3-030-01746-0_20

for the system to deal with it properly. In other words, the set of collected data does not reflect, precisely, what is happening in the environment. In cases like this, the system may build a computational model with incorrect services that instead of assist users, it might disturb them.

Considering this scenario, there are, in literature, several approaches trying to minimize the uncertainty of information. In this paper we discuss some of them in order to contribute with the consolidation of smart environments and Ambient Intelligence.

The paper is structured as follows Sect. 2 we discuss concepts of context-awareness pointing some metrics of design, detection and modelling information. In Sect. 3 we highlight some issues about dealing with uncertain data of smart environments. At last in Sect. 4 we present our final considerations and future directions of research.

2   Context Awareness

Context awareness represents a connection between the environment and the computational structure. It refers to identification of current state of users and other entities and how they can influence the behaviour of applications. Such identification can be achieved through the use of different types of sensors like location monitoring, vital signs, level of stress or fatigue, among others.

According to [11], context-awareness has three main features that must be taken into account. The first one refers to the presentation of specific information to the user based on the context. The personal device of the user should be able to present information based on his location. For example, if he is at work, it could show his tasks for that day. Another critical feature of context-aware systems is the automatic execution of services. For this to happen, the system should allow a machine-to-machine communication and exchange of information. It is imperative that information used as basis to execute a service should be completely correct to avoid miss interpretations and execution of wrong services. It is known that a context-aware system should have different types of sensors monitoring and collecting data from the environment. However, if the system does not know what to do with the data, they are useless. Thus, the last main feature is the context needs, which are the goals of the services. The context needs are essential for the collection, fusion and analysis of right information.

The popularization of the use of context data in information systems brings challenges that must be overcome. There will be scenarios where the system will have to process much more data than used to do with only direct inputs. Besides that, smart environments should be full of sensors monitoring objects, users and other entities. Probably, more than one sensor would be capable of collect the same information. Considering that sensors have specific objectives, the system must be able to decide from which sensor it should use the data collect, according to the current context.

2.1 Detection of Context Information

According to [11], there are several techniques for the detection of relevant information of smart environments. The first one refers to the responsibility [12], where the context-aware system makes a request to the sensor. In other case the sensor sends information detected to the system every time it detects something. In both cases the periodicity of the requests or detection may vary between instant detection or periodically. The former characterizes events that occur only once and the acquisition of information must be at the time it occurs. The latter characterizes events that tend to repeat within a period of time. In order to acquire this information, the sensor needs to detect more than once and analyse it, in order to find a pattern.

Nowadays there are several types of sensors available that can be applied to smart environments to monitor and detect context information. According to [7], they can be categorized in physical, virtual and logical sensors. The first is capable of detecting data without any external intervention. There are plenty of devices equipped with this kind of sensor. Information generated is considered as low-level context, i.e., they are more trivial [11]. On the other hand, the virtual sensors do not detect context information without intervention. They search information from other sources, through web services, and process it as sensor data. Logical sensors use physical and virtual sensors to produce, through web services, richer context information.

2.2 Context Modelling

Context-aware systems must use context models that reflect the knowledge of a smart environment. According to [15], these models can be classified as static, with limited and previously defined set of information, or dynamic, discovering and manipulating new information over time. The modelling process needs to consider aspects like heterogeneity of entities, mobility of users, relations among entities, periodic events, uncertainty, reasoning, usability and efficiency of the context [3, 11].

The context modelling starts with the definition of entities with their characteristics (attributes, restrictions and relations). This should be performed using a previous known context as basis. It is important to highlight that one of the main issues on context modelling refers to the subjectivity. The same set of information detected in an environment may be valid in one situation and not in another. So, the more complete is the information about a context, the more accurate will be the model built based on that. After modelling a new context, it should be integrated with rest of context information in a repository.

The definition of services of context-aware system is directly related to requirements about the current state of entities. Although, in many times the signs collected by sensors do not provide enough data to fill such requirements. That happens because, under the computational perspective, the context information is naturally dynamic, uncertain or incomplete. The result is an inexact connection between the service and what is provided by the environment [14]. The description of the context and services requirements uses similarity comparison, which analyses the necessary and provided information individually, and also global comparison, which analyses the context as a whole [14].

Considering this, we can verify that one of the main challenges of modelling context is to handle with uncertainty. A context-aware system should be able of obtaining enough data to build complete contexts, i.e., to identify clearly all information present in the environment. This is crucial for a correct orchestration of events with context-aware services. If the sensors are not able to identify a situation or if the context models do not represent real scenario of the environment, the services that would be executed may not be the more appropriate. In the next section the problem of uncertainty will be discussed.

3   Handling Uncertainty

Due to the dynamic nature of AmI, we need to find ways to correctly analyse and handle incomplete information, once it could result in the construction of uncertain contexts. Uncertainty, in smart environments, can be defined as any incomplete, contradictory, vague or outdated information [16].

Information collected could be partially correct or even absent, but still should be considered for reasoning. This way, Multiagent Systems provide a generic model with the necessary flexibility, with different levels of autonomy and dependency for all components of smart environments.

The dynamism of a domain centred in the user is a complex subject, under different perspectives. We should consider aspects related to their state of mind, including level of stress, fatigue and personal preferences, and also external factors like temperature and time of the day. Such factors can influence the user causing changes in their acting patterns within a short space of time. The result is that the knowledge base that is used by the computational system starts to work with a high level of uncertainty, without the assurance that the mapped context is correct and up-to-date.

The uncertainty originated in a smart environment could be result of variability phenomenon, i.e., due to changes in the user behaviour, making them act unexpectedly [8]. Another cause refers to the lack of complete information about specific situations, increasing the complexity of the system's learning process. Besides the subjectivity generated from these two sources, the quality of the computational service must remain the same. The uncertainty identified in dynamics environments should not be ignored. We should perform an analysis of user behaviour seeking alternatives to fill information gaps evidenced during the process of building contexts [13].

Context information is naturally incomplete and uncertain, and could be result of different causes, e.g., interpretation problems of the signs sent by sensors [1]. Different manufacturers may use specific concepts and identifiers to name the same entity or relation. Also according to [1], we still have challenges to overcome regarding the uncertainty in smart environments. We should guarantee a correct modelling of the available services with formal description of their goals and requirements for execution. Besides that, we should guarantee that the service invocation is capable of verifying the reliability of the output information. At last, we should extend the conventional composition models verifying levels of probability for each service related to the detected context. The correct definition of each of these steps contributes to a more appropriate service orchestration, even if working with uncertain context information.

According to [4], the information is partial when only some of the queries in the reasoning layer can be answered. However, to build a context model aiming to assist users, these answers should be as complete as possible. Besides that, the information could also be not as trustable as it should or generate a conflicting result. In this case we have different types of sensors capable of generate the same kind of information from an entity, but, to be used in different scenarios. The data collected from them could produce conflicted information to the context model, where the same data represents more than one kind of information, depending on the situation [4]. Uncertain information can be represented in numeric models through approximation values, within an established error margin [4]. Independent of the chosen methodology, usually, a relation between two instances has external aspects that must be taken into account, once they may be relevant for the right understanding of the situation.

A good alternative to face situations of incomplete set of data is to speculate information about the future. According to [17], Speculative Computation refers to the use of defaults values as input for processing and generation of an output with previsions of future situations. These default values can be known data from a set or, in case of this data is not available, it can be defined by the user, as being a valid value for that attribute. In case of context-aware systems, by using these default beliefs as input, speculative computation helps to improve the reasoning over a context model, reducing the time of decision making and execution of applications.

Information used in any context-aware system must be represented to be computed. According to [4], this representation must have characteristics of similarity with what it represents in the real world and also, should be as simple as possible, avoiding any ambiguous meaning. To make a decision based on a specific context model, the system must analyse all the possible scenarios that match with it, to decide and orchestrate the execution of services. To ensure that the services will be the most appropriate, the information brought to the system must be complete.

4   Identification of Problems and Approach

The main focus of the proposed research is to tackle the problem of uncertainty handling in context-aware computing. Thus, the following problems were identified:

- **Lack of accurate data:** due to different sources of problems, many times a context-aware system cannot build a computational model that represents the knowledge of a real-world domain. Thus, the probability of mistaken analysis tends to increase considerably and, consequently, the decision of what action to take will not be the most appropriated front that specific situation;
- **Misinterpretation of contexts:** the analysis of situations performed by software agents is based on identification of patterns of behaviour of users. However, unexpected behaviour can happen and even if the system has enough context data to build a model, it will not understand why the user and his/her surroundings is behaving

like that. This happens because the system learned that, in that moment of time, the user used to do specific things, but his/her behaviour changed.

A good alternative to deal with these problems is to speculate information about the context. According to ~\cite{Tiago}, Speculative Computation refers to the use of defaults values as input for reasoning filling gaps of data. These default values refer to valid data attributes in previous similar situations. Thus, if what happened in the past is true, the probability of the same situation to be true in the future tend to be high. These default values help software agents to speed the reasoning over a context once they do not need to wait for all context data to start the processing. Speculative Computation is composed by two phases, Process Reduction and Fact Arrival. The former, represents the normal processes of reasoning over context data, using default values, if necessary. The latter, represents the arrival of new context data to be added to the processes. Figure 1 presents a generic architecture for Speculative Computation.

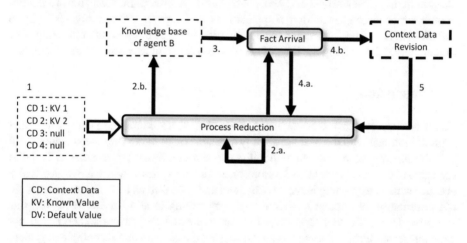

Fig. 1. Speculative computation

The figure above is composed by steps to help the understanding the flow of it. In *step 1*, a set of data is received by the software agent A that will process the context. There are gaps of information within this set, which represents missing context values. In other words, for some reason the data was not able to be sent to the system. In *step 2*, the process of reasoning over the context data starts, referring to the first phase of Speculative Computation. For this, the software agent A fills the gaps of data with default values (*step 2.a.*). The agent uses the known values that were valid in previous situations as basis. At the same time, agent A queries other agents about the missing data (*step 2.b.*). If the reasoning finishes before the agent A receives the answers, the default values are taken as being true. However, if the other agents answer the queries before the reasoning is finished, the flow of the Speculative Computation passes to the second phase, *Fact Arrival*, (*step 3*). In this case, the active processes are suspended, and agent A has to analyse the context data comparing them to the default values that were used until this point. If the context data from the answers are consistent with the default values, the suspended processes are resumed, and the reasoning continues until it is finished (*step 4.a.*). However, if they are not consistent, the default values must be ignored and replaced by the new context data from the answers of the queries (step 4.b.). In this case, the suspended processes are ignored, and a *Context Data Revision* is performed, replacing the missing values with the arrived context data. After that, a new phase of process reduction is started (*step 5*), to reason over the set of context data. This happens every time that software agents answer queries from agent A with new context data.

The main advantage of Speculative Computation to deal with uncertainty in intelligent environment is that the process of reasoning over a context can start even with gaps of data, with the use of default values to fill these gaps. This allows the system to act faster in different scenarios.

Nowadays, several approaches of context-aware system discard the missing values making the system to ignore them during the reasoning. The problem of these approaches is that this makes the system to use less data to build models to be processed. Therefore, the system may not clearly identify what happens in the environment and the probability of orchestrate a wrong execution of applications tend to grow considerably. Using these default beliefs as input, Speculative Computation helps to improve the reasoning over a context model, reducing the time of decision making and execution of applications.

5   Conclusion

Context-aware systems are, gradually, becoming more common. Smartphones that naturally interact with the user are already available on market. These new devices will help the consolidation of Ambient Intelligence, through Smart Environments. To build such kind of domains we need to develop context-aware systems, which have the premise of sense what is happening in the environment and act based on that. The system should use information of context to execute services aiming to assist users. However, this reasoning is not a trivial task, once the data collected by sensors sometimes is not complete as should be. Due to the dynamicity of the domain and its complexity, there will be times where the system will not be able to build a complete scenario due to the incompleteness of information or its reliability. This uncertainty generated must be handled appropriately, once the main idea of context-aware systems is for them to work autonomously. Considering this, we can conclude that one of the main of issues of context-aware systems is the study of approaches of handling uncertainty and what to do in situations where the system is not able to build complete context model due to information vague, partial or out-dated.

In this paper we presented some approaches of context-aware systems, with its design, detection and modelling principles as well as concepts of how to deal with uncertainty. We intend to focus on aspects like the analysis of existing models of representation and define metrics to classify levels of uncertainty of smart environments.

Acknowledgements. This work has been supported by COMPETE: POCI-01-0145-FE\-DER-007043 and FCT – Fundação para a Ciência e Tecnologia within the Project Scope UID/CEC/00319/2013.

References

1. Amdouni, S., et al.: Handling uncertainty in data services composition. In: International Conference on Services Computing, 11th edn., pp. 653–660. IEEExplore, Anchorage (2014)
2. Aztiria, A., et al.: Learning Frequent behaviors of the users in intelligent environments. IEEE Trans. Syst. Man Cybern. Syst. **43**(6), 1265–1278 (2013)
3. Bettini, C., Brdiczka, O., Henricksen, K., Indulska, J., Nicklas, D., Ranganathan, A., Riboni, D.: A survey of context modelling and reasoning techniques. Pervasive Mob. Comput. **6**, 161–180 (2010)
4. Bhatnagar, R.K., Kanal, L.N.: Handling uncertain information: a review of numeric and non-numeric methods 1986. In: Kanal, L.N., Lemmer, J.F. (eds.) Uncertainty in Artificial Intelligence. Elsevier Science Publishers B.V. (2014)
5. Cook, D.J., Augusto, J.C., Jakkula, V.R.: Ambient intelligence: Technologies, applications, and opportunities. Pervasive Mob. Comput. **5**(4), 277–298 (2009)
6. Delmastro, F., Arnaboldi, V., Conti, M.: People-centric computing and communications in smart cities. IEEE Commun. Mag. **54**(7), 122–128 (2016)
7. Indulska, J., Sutton, P.: Location management in pervasive systems. In: Proceedings of the Australasian Information Security 2003, vol. 21, pp. 143–151 (2003)
8. Mokhtar, S.B., et al.: Efficient semantic service discovery in pervasive computing environments. In: International Middleware Conference (MIDDLEWARE 2006), Proceedings of the ACM/IFIP/USENIX, 7th edn. Bordeaux, France, pp. 240–259 (2006)
9. Novais, P., Carneiro, D.: The role of non-intrusive approaches in the development of people-aware systems. Prog Artif Intell, cidade, Berlin, vol. 5, n. 3, pp. 215–220 (2016)
10. Nugroho, L.E.: Context-awareness: connecting computing with its environment. In: International Conference on Information Technology, Computer, and Electrical Engineering (ICITACEE), 2nd edn. Semarang, Indonesia (2015)
11. Perera, C., Zaslavsky, A., Christen, P., Georgakopoulous, D.: Context aware computing for the internet of things: a survey. IEEE Commun. Surv. Tutor. **16**(1), 414–454 (2014). (First Quarter)
12. Pietschmann, S., Mitschick, A., Winkler, R., Meissner, K.: Croco: Ontology-based, cross-application context management. In: Third International Workshop on Semantic Media Adaptation and Personalization, SMAP 2008, pp. 88–93, December 2008
13. Samia, B., Allel, H., Aicha, A.M.: Handling preferences under uncertainty in recommender systems. In: IEEE International Conference on Fuzzy Systems (FUZZ-IEEE), 1st edn. Pequim, China, pp. 2262–2269 (2014)
14. Vanrompay, Y., Pinheiro, M. K., Berbers, Y.: Context-aware service selection with uncertain context information. In: International Discotec Workshop on Context-Aware Adaptation Mechanisms for Pervasive and Ubiquitous Services (CAMPUS 2009), 2nd edn. Lisboa, Portugal: Proceedings of the Second International DisCoTec Workshop on Context-aware Adaptation Mechanisms for Pervasive and Ubiquitous Services, vol. 19 (2009)
15. Yanwei, S., Guangzhou, Z., Haitao, P.: Research on the context model of intelligent interaction system in the internet of things. In: 2011 International Symposium on IT in Medicine and Education (ITME), vol. 2, pp. 379–382, December 2011
16. Yaghlane, A.B., Denoeux, T., Mellouli, K.: Elicitation of expert opinions for constructing belief functions. In: Uncertainty and Intelligent Information System, pp. 75–89 (2008)
17. Oliveira, T.J.M.: Clinical Decision support: knowledge representation and uncertainty management. Ph.D. thesis, University of Minho - Doctoral Programme of Biomedical Engineering, R. da Universidade, 4710-057, Braga, Campus de Gualtar, 4 (2017)

Urban Mobility: Mobile Crowdsensing Applications

João Simões[1], Rui Gomes[1,2], Ana Alves[1,2(✉)], and Jorge Bernardino[1,2]

[1] Polytechnic Institute of Coimbra – ISEC, Coimbra, Portugal
a21220126@alunos.isec.pt, {rui.gomes,aalves,jorge}@isec.pt
[2] Centre for Informatics and Systems of the University of Coimbra – CISUC, Coimbra, Portugal

Abstract. Mobility has become one of the most difficult challenges that cities must face. More than half of world's population resides in urban areas and with the continuously growing population it is imperative that cities use their resources more efficiently. Obtaining and gathering data from different sources can be extremely important to support new solutions that will help building a better mobility for the citizens. Crowdsensing has become a popular way to share data collected by sensing devices with the goal to achieve a common interest. Data collected by crowdsensing applications can be a promising way to obtain valuable mobility information from each citizen. In this paper, we study the current work on the integrated mobility services exploring the crowdsensing applications that were used to extract and provide valuable mobility data. Also, we analyze the main current techniques used to characterize urban mobility.

Keywords: Urban mobility · Ubiquitous systems · Mobile crowdsensing

1 Introduction

Today, mobility is one of the toughest challenges that cities face, with 54% of the world's population living in urban areas and by 2050 this number is expected to reach 67% [1]. Automobile sales are expected to increase from 70 million a year in 2010 to 125 million by 2025 and some analysts says that with the existing growth rate car fleet could double by 2030 [2]. These facts combined with the circumstance that cities today are responsible for more than 70% of global carbon dioxide emissions [3] demands mobility integrated services capable to provide solutions for these problems. The existing urban infrastructure cannot support such an increase in vehicles on the road. Congestion is already close to be intolerable in many cities due a lot of time lost in traffic, wasted fuel, and increased cost of doing business. The World Health Organization estimated in 2014 that seven million premature deaths were attributed to air pollution, and a significant share is the result of urban transit [3]. These are not the only problems that exist in urban mobility. Mobility has also a big impact on quality of life of each citizen which is directly influenced by the frustration, pollution, time lost, and noise of traffic congestion [4]. Understanding human mobility is also crucial to provide solutions capable of improving emergency services. Defining the better route to arrive quickly as possible can sometimes do the difference. A lot of use cases can be applied on human mobility and concerning these enormous challenges that urban mobility and transport policies are facing, fresh and

© Springer Nature Switzerland AG 2019
P. Novais et al. (Eds.): ISAmI 2018, AISC 806, pp. 182–189, 2019.
https://doi.org/10.1007/978-3-030-01746-0_21

valid data of different kinds are extremely important. To deal with the problems regarding urban mobility, cities need to offer a set of mobility integrated services capable of extracting data from various sources with different granularity like smartphone sensors or social networks. Due to the significant growing number of mobile phones, the study of human mobility has significantly changed. Currently, 68% of the world's population owns a mobile phone, and by 2019 this figure is expected to be about 72% [5]. Smartphone sensors became in the latest years a good valid source of data to analyze human behaviors. With them, we can know where we are, what we are doing and in some cases, they can tell us why we are there or doing something. Accompanying the growing number of sensors, mobile crowdsensing has also been turned a popular way to share data collected by sensing. In this paper, we explore the use of crowdsensing applications to capture valuable information from citizen mobility and also describe the actual techniques and frameworks used. This paper is organized as follows. In Sect. 2 we present mobile crowdsensing applications and frameworks used to obtain valuable mobility data. Section 3 shows a comparative analysis between the applications and frameworks. Finally, Sect. 4 presents our conclusions.

2   Mobile Crowdsensing Applications

Mobile Crowdsensing is a promising way to obtain valuable information from mobility of each citizen. We can classify mobile crowdsensing in two types: opportunistic and participatory [6]. Opportunistic, if the data is collected and shared automatically without user intervention and in some cases, even without the user's explicit knowledge. Participatory, if the users voluntarily participate in surveys with the goal of providing useful information's. The research methodology applied on this paper was the following: as we want to use SenseMyFeup to collect crowdsense data in the metropolitan area of Porto, we start our search gathering the most recent applications developed on the current repositories such ResearchGate[1] and DBLP[2], falling two sub-categories: Opportunistic and Participatory. Our goal was to provide examples of applications that are included in this two types and present the actual techniques used for both.

2.1   Opportunistic Crowdsensing

SenseMyFeup
SenseMyFeup consists of a mobile application and a scalable backoffice developed with the aim of studying human processes in a large urban area. The app is responsible to collect data from embedded sensors, like GPS, WiFi, Accelerometer, Gyroscope and also external sensors like Bluetooth sensors. After collecting data, the mobile application is responsible for sending it to the gathering units that in turn, send it to main servers. SenseMyFeup uses two mechanisms that make the mobile application non-intrusive to user [6]. SenseMyFeup uses an algorithm that can detect movement without any

---

[1] https://www.researchgate.net/ [Accessed: 11-Dec-2017].
[2] https://dblp.uni-trier.de/ [Accessed: 11-Dec-2017].

intervention. This technique decreases the impact of the application on battery life, saving smartphone resources more efficient. When a new activity or location is received the incoming sensor, data is analysed to classify a user movement as "moving", "stopped" or "undefined", when is not possible to conclude nothing. A decision algorithm is used to analyze the results, starting a session if any of the classifiers returned "moving", and stopping a session when every classifier agrees with "stopped" state. To classify an activity and see if the user is effectively moving, SenseMyFeup uses the Google Activity Recognition API to sense the user physical activity. The tool also uses a location classifier to determine if the current movement of the user is relevant to capture. When we have the goal of extracting useful data to understand the user mobility we don't need to collect data from an indoor movement unlike daily commute trips or a walk in the street, that are interesting facts that can be useful to understand the daily routine. The other mechanism implemented in this tool is the opportunistic WiFi data upload. The Synchronization that is made between client and server can be done using whatever connectivity is available or using WiFi connections only. This option is more useful for the user, because is more desirable to establish user's confidence by not using data plan and consuming less battery.

SWIPE

SWIPE is an open-source platform, for sensing, recording and processing human dynamics using smartwatches and smartphones. SWIPE is available under MIT license (http://github.com/sfaye/SWIPE). Faye et al. [7] have decided to use SWIPE to study a methodology to obtain user data in motion. According with the researcher characterizing human mobility typically depends solely on data retrieved by accelerometers, or GPS and user inputs. So, in this study, they define a methodology where different sensor groups and other combinations are used to understand user mobility. The smartwatch offers the possibility to measure stress level that can be a promising way to identify the regions with more traffic congestion on the city. Moreover, they offer in this study the possibility to use other devices like smartwatches to study user mobility. Combining those devices, offers new perspectives that help understanding user mobility and a large amount of data can be obtained. The sensing system proposed in the article is an android application that collects data simultaneously on a smartwatch and a smartphone. The smartwatch regularly sends the data that it has collected to the smartphone. Smartphone receives the data from smartwatch and also collects data from their sensors. It is also responsible to serve as a gateway to access an online platform over the Internet. The metrics collected are: Maximum average acceleration (m/s^2), pedometer (number of steps taken by the user, provided by phone and watch using accelerometer function), heart rate, ambient sound (from 0 to 100, provided by microphone sensor from smartphone), WiFi Aps (number of anonymized BSSIDs of Wi-Fi access points, collected by the smartwatch), Bluetooth devices (number of Anonymized BSSIDs of Bluetooth devices collected by the smartwatch), mobile network data state and speed (km/h) [7]. All the data is collected following two variables, sampling rates and recording. Sampling rates is used to define when the devices should acquire data from sensors and recording is used to indicate the frequency at which a metric is saved.

CitySensing Framework

CitySensing Framework is a framework that consists of mobile application components and visualization/analytics components organized in a distributed architecture. It supports both opportunistic and participatory methods of crowd sensed data collection. On the client side, the mobile app is for capturing physical, virtual and social sensors data. Physical data can be gathered using the mobile sensors such as GPS, microphone, camera, ambient light sensor, accelerometer, gyroscope, compass, proximity sensor and in some advanced smartphones the temperature and humidity. Social data is detected by the interactions that the user makes in their social networks such tags, like, publications, tweets or photos. This kind of data can be extracted through the appropriate API's. In this model, the data is sent to the server in a distributed approach. After collection, the data is processed and analyzed at the mobile devices and high level contextual and mobility information are sent. On the server side is where all data are collected and processed for analysis. To analyze the data, CitySensing server components are based on distributed processing frameworks, such as MapReduce/Hadoop and Spark, for processing offline crowd-sensed data stored in a distributed file system over a cloud/cluster. For real-time processing and analysis of large crowd sensed data streams appropriate data stream processing frameworks are employed, such as Apache Storm and Spark Streaming. This analysis is used to detect aggregated mobility patterns, trajectories, group activities and behaviour's that can be reflect a daily mobility routine for a citizen [8].

Future Mobility Survey

Future Mobility Survey (FMS), is a smartphone based prompted-recall travel survey system developed and tested in Singapore [9]. The main role of the smartphone devices in this project is to act as data loggers. This application was implemented to allow other types of devices to upload data to the server, such as dedicated GPS loggers. The data is collected from the smartphone application and uploaded into a central server. On the server side, the data is mapped, automatically cleaned, analysed and made accessible to the user from the project website, where the user is asked to provide more travel information via prompt recall survey. So, as we previously explained we easily subdivide this application in three different modules: client side (smartphone app), server database side and Web interface side. On the client side, the main challenge for this kind of application is the battery consumption due to the GPS sensor. The application uses a mechanism to minimize this problem, GPS data is collected in an intensive way during some time (collection time) and after that, no data is collected during another time (sleeping time) and the new GPS localization is predicted with accelerometer data. This was the main approach followed to improve the problem regarding with battery consumption. On the server data base side, several phases are included to transform data in to valid information: process raw data, stop inference, mode inference, activity inference and learning from user validations. The user also can validate the activities that was detected by the system analysis (Learning from user validations phase). The validation by the users will improve the system analysis that will use them for future decisions.

Travel Mode Detection

Shafique et al. [10] propose a model that uses data collected by smartphone sensors to infer human activities. The application extracts data from smartphone sensors such as accelerometer, orientation sensor (pitch and roll) and GPS, but in this case the data collected is used only for validation. In addition to resultant acceleration, six features were further extracted from resultant acceleration namely standard deviation, skewness, kurtosis, maximum resultant acceleration, average resultant acceleration and maximum average resultant acceleration [10]. The model was tested with 50 participants from Kobe, Japan. They contributed to the data collection phase using an android smartphone over a month. When they start a trip, the participants select an option on the smartphone application to start the recoding and only input the travel mode. When they have reached their destination, the user selects on the smartphone an option to stop the recording. At the end of the day, a recall survey would be conducted to check the reliability of the collected data. In this phase of validation, the data extracted from GPS sensor was used to generate routes maps were users can easily reconfirm the starting and ending times of various trips as well the modes of transportation model. In this study, six activities were detected: walk, bicycle, car, bus train and subway.

2.2   Participatory Crowdsensing

PublicSense

PublicSense is a mobile crowdsensing framework for urban sensing and public facility management [11]. Few studies pay attention to the citizen call data coming from various complaint platforms. The data generated from complaint calls can give a crowdsourcing way to know the problems, events and urban dynamics happening in the city and also understanding citizen mobility. Crowdsourcing call data is distinctive and allow us to achieve refined urban sensing. Potential applications that can use this data can be public facility management, transportation services and emergency situation monitoring.

The input of the framework is heterogeneous data, including data that come from complaint platforms, road networks, POIs (points of interest) and social media. Using only complaint call data without connecting them with geographic information may not be useful to achieve interesting conclusions. So, the PublicSense framework has a geo location recognition module responsible to obtain the GPS missing information of the call data. As case of study they use the complaint call data from Xi'an city in Japan. They create a dictionary containing the street and road names of Xi'an and leverage NLPIR[3] to obtain the street involved in the complaint. After this, they use the Geocoding Javascript API[4] to extract the correct positions (longitude and latitude) of the street involved in the complaint call. After getting all of the data, the next step was the generation of a heat map with a purpose to find interesting conclusions such what is the areas that has more complaint calls. This approach can also be used to improve traffic congestion identifying the regions with more complaint call data.

---

[3] http://www.nlpir.org/ [Accessed: 11-Dec-2017].

[4] https://developers.google.com/maps/documentation/javascript/geocoding   [Accessed: 11-Dec-2017].

Questionnaire-Based Human Mobility

Satomi et al. [12] developed a study where questionnaire-based human mobility data was indexed and merged with GPS logger-based mobility data in order to analyze a long-term human mobility pattern associated with socio-economic development between 2000 to 2015. In this study, human mobility was used as a parameter for extracting local interaction under rapid socio-economic and environmental transformations. First, they start by conducting an interview to obtain mobility-related information. For this, the data extracted includes travel origin, destination, distance, time and transportation mode. Besides questionnaires to obtain user mobility information, this study uses GPS Log Records to improve the system. Wearable GPS devices such as "i-gotU USB Travel & Sports Logger—GT-600" were used to log human mobility and validate the mobility data obtained from questionnaire survey. This device is lightweight (37 g) and small with an automatic motion detector which can be worn on the waist or clipped to clothes [12]. Next step was the conversion of mobility data from non-spatial (data collected by questionnaires) to spatiotemporal data. This data was merged with GPS logger's data and coordinates as latitude and longitude was used. All of the information merged was visualized in a GIS platform, in order to obtain important conclusions. Researchers concluded that when urban area is increased, the travel distance of local population also increases. This can be explained with the fact that when urban area is growing other economics activities are created. With new work opportunities available the number of people that travel do that urban area increases. This approach can be used for understanding changes of livelihood and its monitoring in other places where topological representation is greatly required.

3   Comparative Analysis

In this section, we provide a comparative analysis between projects and frameworks presented on the previous section. In Table 1 we provide an overview of the crowdsensing projects previously presented showing the main advantages and disadvantages of each one.

In the table above, we can see some important characteristics of crowdsensing applications. Any participatory crowdsensing requires directly an action from the user. However, in certain situations the participatory crowdsensing applications can be more easily to develop, because they do not apply any machine learning technique to learn with the user the current movements. On the other side, opportunistic crowdsensing does not request interaction from users and needs to implement techniques capable to learn the user context. In both ways, any crowdsensing application needs to focus on the use of privacy data.

Table 1. Comparative Analysis between crowdsensing projects

Crowdsensing Type	Framework Name	Advantages	Disadvantages
Participatory Crowdsensing	Public Sense	A crowdsourcing way to know the problems, events and urban dynamics and citizen Mobility	Needs user input
	Questionnaire-based human mobility	Less complex to implement comparative with other tools and methodologies	Needs user input
Opportunistic	CitySensing Framework	Social data and techniques that detect aggregated mobility patterns, trajectories, group activities and behaviors reflecting a citizen's daily mobility routine	High battery consumption on client-side. No techniques to face battery consumption problems
	Future Mobility Survey (FMS)	Concerns with battery consumption	Data collected is sent in to server database. Application needs to learn with the user to predict activities
	Travel Mode Detection	Reduce frequency from sensors to improve battery consumption., using Random Forest	Battery consumption
	SenseMyFeup	Focus battery consumption: only sent data when Internet connections are available. Detect movement without any user intervention. Using a second server as a database improves the machine learning prediction. Gathering process and prediction process do not collide.	Can be improved using techniques to infer conclusions with mobility data
	SWIPE	Combine multiple sources of gathering data (smartphone, smartwatch). Use of two parameters to improve battery consumption: recording samples and recording.	Too dependent of smartwatch to gathering the physical state of the user

4   Conclusions

Human mobility is a real challenge that cities face today. Understanding how people move is essential to create solutions capable of improving quality of life. Identifying personal travel patterns is the main step to plan sustainable urban transportation systems. To do this effectively and timely, urban and transportation planners need a dynamic way to profile the movement of people and vehicles. In this study, we provide an overview of the current literature on mobility data collection, presenting and describing some projects and methodologies that can be applied in the mobility context. The design of a future crowdsensing application should have in mind principally concerns about battery consumption, privacy politics and non-intrusive actions for the user. As future work, we intend to use SenseMyFeup in the metropolitan area of Porto as part of URBY.Sense Project. We aimed acquiring data describing transportation related resources and also public data about places and events from social networks.

Acknowledgements. URBY.Sense is co-financed by COMPETE 2020, Portugal 2020 - Programa Operacional Competitividade e Internacionalização (POCI), FEDER and FCT.

References

1. UN-Habitat, Urbanization and Development: Emerging Futures (2016)
2. Dargay, J., et al.: Vehicle ownership and income growth, worldwide: 1960-2030. Energy J. **28**(4), 143–170 (2007)
3. World Health Organization, World Health statistics 2014 (2014)
4. Becker, R.A., et al.: COMMUNICATIONS human mobility characterization from cellular network data. Commun. ACM **56**(1), 74–82 (2013)
5. The Statistics Portal. http://www.statista.com/statistics/274774/forecast-of-mobile-phone-users-worldwide. Accessed 11 Dec 2017
6. Rodrigues, J.G.P., et al.: Opportunistic mobile crowdsensing for gathering mobility information: Lessons learned. In: Proceedings of the IEEE Conference ITSC, no. 978, pp. 1654–1660 (2016)
7. Faye, S., et al.: Characterizing user mobility using mobile sensing systems. Int. J. Distrib. Sens. Networks **13**(8), 1–13 (2017)
8. Stojanovic, D., et al.: Mobile crowd sensing for smart urban mobility. In: Capineri, C., Haklay, M., Huang, H., Antoniou, V., Kettunen, J., Ostermann, F., Purves, R. (eds.) European Handbook of Crowdsourced Geographic Information, pp. 371–382. Ubiquity Press, London (2016)
9. Pereira, F., et al.: The Future Mobility Survey: overview and preliminary evaluation. In: Proceedings of the Eastern Asia Society for Transportation Studies, vol. 9 (2013)
10. Shafique, M.A., Hato, E.: Travel mode detection with varying smartphone data collection frequencies. Sensors (Switzerland) **16**(5), 716 (2016)
11. Zhang, J., et al.: Public sense: refined urban sensing and public facility management with crowdsourced data. In: Proceedings of UIC-ATC-ScalCom, Beijing, pp. 1407–1441(2015)
12. Kimijima, S., Nagai, M.: Human mobility analysis for extracting local interactions under rapid socio-economic transformation in Dawei. Sustain. **9**(9), 1598 (2017)

Sensorized Toys to Identify the Early 'Red Flags' of Autistic Spectrum Disorders in Preschoolers

Marco Lanini[1], Mariasole Bondioli[1(✉)], Antonio Narzisi[2], Susanna Pelagatti[1], and Stefano Chessa[1]

[1] Department of Computer Science, University of Pisa, Largo Pontecorvo 3, 56127 Pisa, Italy
mariasole.bondioli@di.unipi.it
[2] IRCCS Fondazione Stella Maris, Calambrone, Italy

Abstract. The observation of the games in children affected by autistic spectrum disorders is an important and widely used diagnostic method, which, at the state of the art, is conducted in a clinic with the direct observation of the specialist. This paper presents a prototype of a sensorized toy that can detect movements and orientation of the toy in real-time so to allow for an indirect observation of the young patients. This will open to new ways of performing such tests in a non-invasive way and in the own environments of the patient. The prototype is based on a low-power platform embedding accelerometers, magnetometer and gyroscope, and on a server that provides different views to the specialist in real-time.

Keywords: E-health · Internet of things · Sensors · Autism · Smart toys

1 Introduction

The observation of the play in children affected by autistic spectrum disorders (ASD) is an important and widely used diagnostic method, especially when used in an early stage on young children [1, 2]. This because a child with such disorders lives in different ways the first learning phases of his/her childhood. In fact, the game modes of these subjects are often different than those of their peers: they often perform actions such as carrying toys around without ever playing and they often lack shared, social play and imagination, as in the case of games of fiction not rigid and adequate to age, while afterwards the insistence is in games with very fixed rules. Of course, such an observation with diagnostic purposes is not easy to implement: it requires specialist's knowledge and it is conducted in clinics. Consequently, it suffers of some limitations: it is conducted in an unusual environment for the child, with objects that are not his/her own.

A new perspective on all this may be opened by the recent evolution of information technology, especially by the internet of things (IoT). In particularly, with IoT is possible to embed sensors and "intelligence" within any object [3]. In our case, the introduction of devices capable of sensing, data processing and wireless communication within the toys would allow the specialist to conduct their observations remotely, even when the child is in his/her own environment and with his/her own toys. From a technological point of view, the adoption of IoT devices for this specific application poses significant challenges in the selection of a suitable set of sensors in the algorithms for the analysis

© Springer Nature Switzerland AG 2019
P. Novais et al. (Eds.): ISAmI 2018, AISC 806, pp. 190–198, 2019.
https://doi.org/10.1007/978-3-030-01746-0_22

of sensor data (the data fusion process) [4] and in the protocols and platforms form the management of the data [5].

In this work, we address the problem of indirect observation of the child, by means of IoT devices embedded in toys and with the capability of providing an accurate instantaneous measurement of the orientation, direction and force applied to the Toy. Our design implements all the layers of a typical IoT application, from the device to the remote server providing storage and data presentation by means of web interfaces. We present the hardware and software architecture and the realization of two prototypes (embedded in two toys, namely a truck and an airplane), along with a preliminary experimental study aimed at validating the approach, which was conducted at the hospital of the IRCCS Stella Maris in Calambrone (Italy). It should be noticed that, although the small scale of the experimentation does not allow a medical assessment of the approach, was indeed very valuable from the technological point of view to validate the entire system and to show the validity of the approach to the specialists.

2 Related Works

Previous research has established the connection between the way in which children interact with objects and the potential early identification of children with autism. Those findings motivate our own work to develop "smart toys," objects embedded with wireless sensors that are safe and enjoyable for very small children, that allow detailed interaction data to be easily recorded. These sensor-enabled toys provide opportunities for autism research by reducing the effort required to collect and analyze a child's interactions with objects.

The ways in which infants and toddlers play with objects can be indicative of their developmental progress. Depending on their age, a child's object play activities can display simple physical milestones such as placing objects in their mouth to sophisticated cognitive tasks such as symbolically using a banana as a telephone receiver. In fact, observing the subtleties surrounding the way in which infants play with toys may highlight early indicators for developmental delays, such as autism. To better quantify how play may serve as an early indicator for autism, researchers have conducted studies examining the differences in object play behaviors among infants. However, such studies are laborious, requiring that researchers repeatedly inspect videos of play often at speeds much slower than real-time to indicate points of interest.

Some studies have focused specifically on atypical use of objects employing direct assessment rather than retrospective methods (home movies or parent report). The work in [6] demonstrated that lateral glances toward objects were significantly more common in a group of children with autism (mean age 44 months) than children with typical development. Another work [7] found that children with autism (mean age 33 months) who spent a large proportion of time in restricted object use showed poorer joint attention, imitation, and social engagement abilities. Finally, [8] reported group differences in repetitive movements with objects (defined as 'at least 3 consecutive movements with objects, such as taps, spins, bangs, lines up, rubs, twirls, rolls, collects') between toddlers with ASD and comparison toddlers with either typical or delayed development.

From a technological point of view, there has been a growing interest for using technology to ease several aspects of diagnosis and treatment of ASD people [9]. Automatic detection of stereotypical movements in children with autism using sensors has been investigated by Goodwin et al. [10]. Their experiment involved six children and each child wore three wireless accelerometers, one on the left wrist, one on the right wrist, and one around the chest, with no restriction in movement. Session were taken at school or during therapy in situations that usually triggered stereotypical behavior. Each session was video recorded, and specialists annotated them both online and offline using an annotation software. A classifier algorithm was trained to automatically detect repetitive behaviors from the accelerometer readings with good results. Plötz et al. [11] used a commercially available device with a 3D accelerometer and a microcontroller combined with machine learning techniques to detect and classify anomalous behaviors. Other approaches [12, 13] used Kinect to recognize stereotypical movements. In [13], the Kinect camera was used to film 12 actors performing three separate stereotypical motor movements each. Then a software recognizing these gestures was developed and instructed using machine learning approach. Manual grading was used to confirm the validity and reliability of the software. Our approach differs from the above ones, since we aim at designing a device that is not invasive (does not requires wearable devices or cameras), and can be used even in the private houses, without the supervision of specialists. To this purpose our design exploits a state of the art IoT architecture, which is not present in the above works.

3   Reference Scenario

The motion capture system described in this article aims to be an innovative IoT based support for the specialists interested in a different observational approach to the ASD early diagnosis, during their regular clinical activities. The sensorized toys are, in fact, designed to be easily introduced during the traditional therapy session, in the clinic environment, to permit to the doctors the improvement of a new non-invasive ASD diagnosis method based on a constant data collection of the child's playing motions. The system has been developed to be used in diagnosis clinical process, concomitantly with the direct observation of the therapist but also in remote control, living the child to interact only with the toy without the interfering presence of other people, if necessary. In this perspective, the system reference scenario is based on the interaction steps between the toys moved by children, the server collecting and elaborating the data of the children's actions and the user interface, where the server gives the data back to the user.

We can resume the scenario of the motion capture process as follow: (1) The user has access to his interface through the browser, connecting to the server IP; (2) To start a new data collection session, the user selects on his interface a patient and the related toys and press the "start" button; (3) The corresponding devices automatically acquire from the sensors (and send to the server) the data of the children's movement during the play session; (4) The motion sensors detect the shift of the toy, elaborate the data of the specific movement and, through wireless, send it to the server; (5) The server communicates the informations to the application web pages, where the data are visualized and elaborated for the diagnosis goals.

During this process, the server has to provide for a punctual data storage in a database, constantly receptive to the entry information from the software of the motion capture devices, and to return the elaborated information to the user through the User Interface. At the same time, the specialist has to have the control of all the data in the databased and of the main activities of the motion capture devices. In this sense, the UI should provide for all the functionalities that connect the user to the system. Through this interface, the user could in fact create a new patient and add him to the database, remove a patient from the database (by the patients' menu) or start a new capture session connect a patient with a related toy (by means of session's menu).

Moreover, the application should permit the availability of all the data of the active and past sessions, to give to the user a complete overview of the movement registrations of each patients. For the same reason, in the database page the user could print or download all the results obtained from the sensorized devices, filtered by patient, showed by the system in an intelligible graphic form.

4   The Sensorized Toys and the Data Fusion Algorithm

As discussed in the previous section, the main purpose of a sensorized toy is to measure the direction of the movement and the force applied to the toy by a child. Implicitly, this also gives also information about if, when and how long the child plays with the toy. Due to the need of avoiding invasive solutions, the natural choice is to embed in the toy a sensor with transducers of acceleration and direction. Specifically, we used a three-axis accelerometer, a magnetometer and a gyroscope. Note that, even with these transducers, measuring the direction of movement and the force applied to the toy is not immediate, since the output of the transducers is rather noisy, the accelerometer is disturbed by the gravity force and the gyroscope diverges over time. For this reason, the system implements data filtering and fusion algorithms. A prototype of the system, which was used for the preliminary experiments, has been implemented by using two sensors based on the Particle Photon[1] (an Arduino-based platform) and one server on a Raspberry PI device[2]. The sensors have been embedded in two toys, a truck and a plane "super wings", suitable for 3 years old babies.

The system is designed to manage an (arbitrary) number of sensorized toys and it implements a typical IoT design: the sensor(s) interacts with a remote server by using the MQTT protocol, and the processed data are stored in a database and presented to the medical specialists by a web-based dashboard. Each sensor is powered by a power-bank of 2600 mAh, and it can be recharged easily through an USB port. The sensor is hidden and placed carefully in the toy so that the X-axis is parallel to the line of sliding of the wheels of the toy, the Y-axis is lengthwise, and the Z-axis is oriented vertically (Fig. 1 shows a detail of the installation of the sensor in the truck and the orientation of the accelerometer in the airplane). The sensor reads at 25 Hz data from the accelerometer,

---

[1] https://www.particle.io/.

[2] https://www.raspberrypi.org/.

magnetometer and gyroscope, and implements a data fusion algorithm. The result of the data fusion is encoded in a JSON record and it is sent to the server for the storage. To the purpose of communication with the server, the sensor implements an MQTT client based on Mosca[3] and communicates with the server via Wi-Fi.

a)   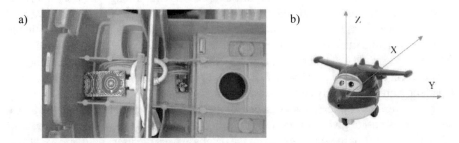   b)

Fig. 1. (a) detail of the installation of the sensor in the truck; b) orientation of the three-axis accelerometer in the airplane.

The server executes implements the MQTT broker (still based on Mosca) and a web server. The MQTT broker and the web server communicate by means of a REST API. The graphical user interface of the web server provides functionalities for the management of patients and of the toys (in case a new toy is added or a toy is removed to/from the pool of available toys), for the configuration and initiation of a diagnostic session (which associates a toy with a patient), and for the visualization of the results of the current and past diagnostic sessions. In particular, the interfaces for the visualization can show the movement diagrams of a diagnostic session, and for the 3D visualization in real time (by using the Three.js library) of the toy used in the current session.

The objective of the data fusion algorithm (Fig. 2) is to extract from the 3-axis accelerometer, gyroscope and magnetometer time series two information for the user, which are the estimation of the orientation of the toy (this is used for the 3D visualization of the toy in real time) and the estimate of the intensity and direction of the movements of the toy (this is used by the medical staff to assess the use of the toy by the child and thus make their diagnosis). The two information in output are produced for each new tuple of data samples in input and hence form two time series.

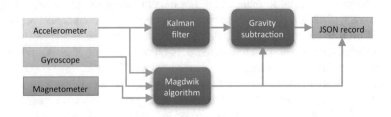

Fig. 2. The data fusion algorithm

───────────────
[3] http://www.mosca.io/.

It should be observed that each sensor taken individually (either accelerometer, gyroscope or magnetometer) is insufficient to compute either the orientation or the movement of the toy, for several reasons. The 3-axis accelerometer measurements have limitations in the detection of position in dynamic conditions and it is also impossible to estimate the orientation of the horizontal plane of the azimuth with reference to Earth's gravity (to which this sensor is subject). The estimate of the orientation can be improved by using the gyroscope, which, however, suffers of drifts in its measurements. To overcome the limits of the gyroscope we thus also the use a magnetometer (whose measurements refer to a global reference system given by the Hearth's magnetic field). The estimate of the toy orientation (in quaternion representation) is performed by fusing all these data by means of the Madgwik algorithm [14, 15]. The estimation of the movement is obtained from the accelerometer data, from which is detracted of the contribution of the hearth's gravity along each axis. In turn, this contribution is estimated on the bases of the orientation of the toy computed by the Madgwick algorithm. To make this estimation more stable, the accelerometer data is preprocessed with a standard Kalman filter. Both data of orientation and movement estimation are stored in a JSON record structured in two parts. The first contains the quaternion representation of the orientation, the raw acceleration along the three axes and the estimation of the power of the acceleration signal (this is used in the 3D real-time visualization), and the second part contains the orientation of the toy expressed in the Eucledian angles of roll, pitch and yaw, and the movement estimation in terms of magnitude and angles of direction (this is used for the storage and for the visualization of the movement diagrams).

5   Preliminary Experiments

The system has been tested during the therapy sessions of two ASD children, a nine and an eight-years-old boys (respectively named child1 and child2), in a period from May to July 2017. The experiment has been conducted by a therapist, in the clinic of the department of "Neuroscienze dell'età evolutiva", at the Stella Maris institute. After some preliminary meetings with the doctor, child 1 and child 2 received both two different sensorized toys (Truck and Super Wings), one toy at a time, during their therapy session. The specialist has collected, through the application, the first motion capture data set, for each child and the corresponding session's toy. At the end of the first experimental cycle we detected from the data collected a significative difference between the movements of the two children play sessions.

Figures 3 and 4 show the output of two 12 min sessions conducted on the 5th July 2017 with two children. The graphs report all the parameters obtained from the device, concerning its the Eucledian angles of roll, pitch and yaw, and the angle and magnitude (parameter sFactor) of its motion acceleration. From the figures it is seen that the children demonstrate a different interest in the play activities proposed. In particular:

- Child1: the toy object has been moved from the child1 for the 13% of the session, with sudden and long movements and sFactor peaks up to 0.8, but with an average of 13,6% during the motion capture time;

- Child2: the toy object has been moved from the child2 only for the 4% of the session, with more moderate and shorter movements and sFactor values with very fewer peaks up to 0.9% an average of 12,5% during motion capture time.

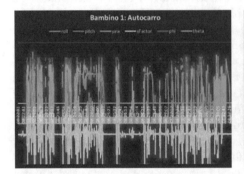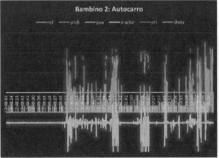

Fig. 3. Child1 plays 12 min with the truck   **Fig. 4.** Child2 plays 12 min with the truck

With the Super Wings toy sessions we could observe very similar results. In conclusion, as mentioned before, the attitude to the toys really diverges from Child1 to Child2. Child1 session has been characterized by a more compulsive behavior with a high interest to the toy, while Child2 has shown less attention to the play session with little activation face to the toys. Moreover, even with a limited testing sample and a short observational period, the preliminary experiments have revealed the efficiency of the system in the constant collection and elaboration of very detailed information about the children movements. As shown in Fig. 5, the punctual graphic visualization of the data has allowed the analyst to prepare his future clinical observations basing on a quick and efficient detection of any change of perceived movement of each child session and, at the same time, any activation/deactivation of a different toy.

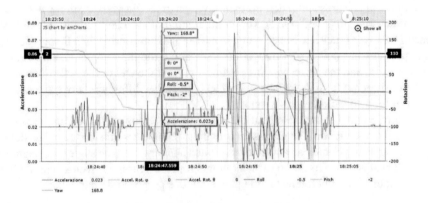

Fig. 5. Graphic representation aircraft toy flight

For instance motion capture elaboration has clearly detected and efficacy represented from the database the difference between the flight activity of the toy aircraft (Fig. 5).

6   Conclusions

The sensorization of toys will open a new perspective in the identification of the 'red flags' of the autistic spectrum disorders already in an early stage, due to their usefulness as part of the clinical and in-home assessment tools. In particular, sensorized toys will enable an earlier observation of infants in their own environment and for long periods. Although the experiments with two prototypes conducted in our work so far did not provide enough data to produce a satisfying clinical diagnosis, they showed indeed the feasibility of the approach and the quality of the data that can be produced in a realistic clinical protocol. On the base of these results we are now planning an extensive experimental campaign in the hospital of Fondazione Stella Maris in Calambrone for medical assessment and validation of the approach and medical protocol.

References

1. Baranek, G.T., et al.: Object play in infants with autism: methodological issues in retrospective video analysis. Am. J. Occup. Therapy **59**(1), 20–30 (2005)
2. Ozonoff, S., et al.: Atypical object exploration at 12 months of age is associated with autism in a prospective sample. Autism **12**(5), 457–472 (2008)
3. Gubbi, J., et al.: Internet of Things (IoT): A vision, architectural elements, and future directions. Futur. Gener. Comput. Syst. **29**(7), 1645–1660 (2013)
4. Amoretti, M., et al.: Sensor data fusion for activity monitoring in ambient assisted living environments. LNICST, vol. 24, pp. 206–221 (2010)
5. Fagerberg, G., et al.: Platforms for AAL applications. Lecture Notes in Computer Science (including subseries Lecture Notes in Artificial Intelligence and Lecture Notes in Bioinformatics), LNCS, vol. 6446, pp. 177–201 (2010)
6. Mottron, L., et al.: Lateral glances toward moving stimuli among young children with autism: early regulation of locally oriented perception? Dev. Psychopathol. **19**(1), 23–36 (2007)
7. Bruckner, C.T., Yoder, P.: Restricted object use in young children with autism: definition and construct validity. Autism **11**(2), 161–171 (2007)
8. Wetherby, A.M., et al.: Early indicators of autism spectrum disorders in the second year of life. J. Autism Dev. Disord. **34**(5), 473–493 (2004)
9. Cabibihan, J.J., et al.: Sensing technologies for autism spectrum disorder screening and intervention. Sensors **17**(46) (2017)
10. Goodwin, M.S., et al.: Automated detection of stereotypical motor movements. J. Autism Dev. Disord. **41**(6), 770–782 (2011)
11. Plötz, T., et al.: Automatic assessment of problem behavior in individuals with developmental disabilities. In: Proceedings of the ACM UbiComp, Pittsburg, USA, pp. 391–400 (2012)
12. Goncalves, N., et al.: Automatic detection of stereotyped hand flapping movements: two different approaches. In: Proceedings of the IEEE International Symposium on Robot and Human Interactive Communication (RO-MAN), Paris, France, pp. 392–397 (2012)
13. Kang, J.Y., et al.: Automated tracking and quantification of autistic behavioral symptoms using microsoft kinect. Stud Health Technol. Inform. **220**, 167–170 (2016)

14. Sebastian, O.H., et al.: Estimation of IMU and MARG orientation using a gradient descent algorithm. In: IEEE International Conference on Rehabilitation Robotics, ETH Zurich (CH) (2011)
15. Madgwick, S.O.H.: An efficient orientation filter for inertial and inertial/magnetic sensor arrays. Dep. of Mechanical Engineering, University of Bristol, 30 April (2010)

Making Ambient Music Interactive Based on Ubiquitous Computing Technologies

Yukiko Kinoshita and Tatsuo Nakajima[(✉)]

Department of Computer Science and Engineering, Waseda University, Tokyo, Japan
{yukiko-kinoshita,tatsuo}@dcl.cs.waseda.ac.jp

Abstract. Although music offers various functionalities in our contemporary urban life, the potential of making music with widely available computing technologies may not be well investigated. Ambient music is a category of music that focuses on the places where people are and puts an emphasis on tone and calmness over traditional musical structure or rhythm. This study reports on an initial effort to investigate ambient music's potential by enhancing ambient music with ubiquitous computing technologies. We have developed a prototype system named DESI (distributed embedded sound information) to realize the potential of making ambient music interactive. DESI records stereophonic sound using a piano and visual markers and then pastes the sound on everyday objects like tables and chairs. The pasted sounds construct music according to their locations. Then, we investigate the possibilities and potential pitfalls through user studies.

Keywords: Ambient music · Ubiquitous computing · Stereophonic sound

1 Introduction

Why do people want music? Why do people spend money to listen to music for hours? Does music have a social meaning? In the research in the social-personality psychology field, music has been shown to perform various functions such as emotion control, self-expression, and social combination in our daily lives [10]. In addition, music has been changed in accordance with social and technological changes, and the social functions of music have also changed. As a result, music functions socially by managing ourselves, interpersonal relationships, and the atmosphere surrounding us, etc. [4]. Therefore, using music will become more important in our everyday lives.

Ambient music is one of the most important categories of music that we come into contact with on a daily basis. Ambient music is a music genre that emphasizes calm sounds and atmosphere rather than the musical structure and rhythm. Brian Eno is a pioneer of ambient music. With the progress of synthesizers and recording technologies in the 1940s, it became easier to create sounds that cannot be heard naturally. In the 1970s, it became easy to listen to music anytime and anywhere. In such an environment, Brian Eno thought about making music that represents place and emotion. Ambient music is different from general background music. While background music has defined music structure and makes the foreground distinct, ambient music is designed to enhance the surrounding atmosphere and to keep the place calm by exploiting uncertainty about

© Springer Nature Switzerland AG 2019
P. Novais et al. (Eds.): ISAmI 2018, AISC 806, pp. 199–207, 2019.
https://doi.org/10.1007/978-3-030-01746-0_23

the music structure. Therefore, ambient music needs to relate to the place. The user should not be aware of the ambient music, and it should not capture his/her attention [3].

In ambient music, the relationship with music tends to be passive such that people accept the music unconsciously or listen to it carefully. To investigate interactivity in ambient music, basic music knowledge is necessary. In addition, other musical composition supports and tools to apply music effects require explicit conscious work, such as watching a waveform or using a specific machine. This requirement makes people conscious of being involved with the music. However, as mentioned above, ambient music needs to be seamlessly embedded into our everyday lives, and it is not appropriate for making listeners explicitly conscious that they are engaged with the music.

This paper reports on an initial effort toward the realization of interactive ambient music that involves the enhancement of ambient music using widely available computing technologies. We have developed a prototype system called DESI (distributed embedded sound information) that records stereophonic sound using a piano and markers and pastes the sound on everyday objects like tables and chairs; and the sounds pasted to the things construct music according to their locations. As a preliminary assessment of the utility of our approach, we demonstrated our system at an event for college for high school students. Additionally, to investigate the possible patterns of the interaction with the surrounding sounds, we obtained insights by conducting a design exploration workshop on interactive changes of sounds by using nearby everyday objects. This paper explores the new possibilities of music that is enhanced through ubiquitous computing technologies and opens a new research direction toward realizing a more mindful society.

2   Related Work

Normally, two ways of interacting with music are possible: a person can both listen to and compose music. In typical music sessions, music is changed by adding new melodies to existing sounds. A disc jockey plays music live, and she/he interactively changes music by applying effects to existing sounds. In addition, in the field of human-computer interaction, computer technologies have attracted attention with the interactivity of the music. For example, reacTable [6], a new music interaction method, combines visual information and sound information by changing the levels and effects of sound by moving objects on the table that are projected by a projector and connecting lines among these objects. Additionally, in terms of music composition, various composition support tools have been developed [9, 11].

Live coding is a new approach to creating music whereby a player changes minimal musical materials in music by interacting with program code [2]. This is an artistic performance that reproduces music in real time by writing information on music, such as tone color, code, and tempo, as programing codes. Each set of codes represents a small piece of musical sound that is repeatedly played so that the entire work of music reproduced by multiple codes is similar to the musical atmosphere of ambient music. Even if players do not actually have instruments, they can express their own music if they have musical knowledge, and they can make further changes to the sounds they hear. However, this is unsuitable for ambient music because it requires knowledge of music, and skills to write program codes that are not familiar to ordinary audiences.

An interactive system called music room utilizes the relationship between space and the input and output of music [7]. In this system, a couple who enters one room interacts with the music in collaboration. As the distance between the two people shrinks, the musical sound changes will be more pleasant, the moving speed changes to the strength of the musical sound. This makes it possible for them to interact with the music without having knowledge of the music, and it is possible to find the relationships with music by comparing their movements with the changes in the music. Thus, this approach is intuitive and easy to understand. In this system, two people need to explicitly collaborate each other, so they need to be conscious of the music. If the people change the tune or tempo by moving things using the positional relationship of physical objects in our daily life, we may be able to establish a new interaction method as a way of interacting with music and everyday life. In this system, since two people collaborate, it is necessary to face the music and be conscious. However, when the positional relationship of physical objects controls the music, the tune or tempo will be changed by moving the objects. It may be possible to establish a new interaction method as a way of interacting with everyday life.

3   Interactive Ambient Music Concept

Ambient music is inspired by *"Musique d'ameublement,"* which was composed by Erik Satie during the first half of the 19th century. Satie's music aimed to not disturb everyday life as if it was furniture in our surroundings. Composers could only create music by actually playing sounds with musical instruments in this era. Later, composers became able to make unnatural sounds and could produce sounds from speakers rather than using only classical musical instruments, and the concept of ambient music was born. Ambient music is similar to minimal music in that it is a form of art music that employs limited or minimal musical materials.

In interactive ambient music, we inherit the concept of ambient music, namely, the concept that people should not be aware that they are explicitly engaged in music. Nevertheless, people can interact with music in a manner similar to live coding in which they do not change a program explicitly but instead interact with music through everyday objects and ubiquitous computing technologies. Interactive ambient music is also considered as an approach to realize alternative reality experiences that refine the meaning of the current real space with virtuality [5].

The goal of interactive ambient music is to reduce people's awareness that they are listening to music. For this purpose, we believe that it is important to reduce the presence of physical objects that make people conscious that they are listening to music, such as speakers, instruments, DAW (Digital Audio Workstation: tool for Desktop Music) screens, and specific devices, as much as possible. In the current research, we focus on how to interact with music in an unconscious way in terms of the output aspect of sounds. The input aspect of sounds will be a future issue, so we will not deeply investigate it in this paper. With the concept of music that does not interfere with everyday life in mind, we imagined producing interactive ambient music in a way that music will come from what people touch in their daily lives and not from speakers installed in their rooms.

To realize the interactive ambient music concept, we use VR sound, which is the latest technology in the music and sound fields. Stereophonic sound is the source of the VR sound and provides sound localization, which controls the direction and position of the sound by filtering the sound with a function of HRTF (head-related transfer function). Combining the head tracking technology currently used in HMDs (head-mounted displays) with head tracking techniques to acquire the head movement with a sensor, it is possible to compose sound sources in absolute positions, that are not relative to head orientation, which are called VR acoustics [1].

Figure 1 provides an overview of how people interact with interactive ambient music using the VR acoustic technology and illustrates the idea that a sound can be present in everyday objects, and people can influence musical sounds by changing the location of those things. First, interactive ambient music does not use traditional musical instruments, but furniture is used in everyday life and can be used to play sounds in interactive ambient music. In the first scene in Fig. 1, a bookshelf, a vase, a PC, and a coffee cup produce musical sounds. To add a new music sound to a new everyday object, a person plays a piano to produce a short musical sound, records it and pastes it to that object. In the second scene, the person is recording a short musical sound, and in the third scene, she is pasting that sound to a phone. In the fourth scene, the phone is playing the music sound. By moving the location of an everyday object, the person can bring her favorite sounds nearby or influence the atmosphere of the music. In the fifth scene, she moves a vase and a PC, and in the sixth scene, she is satisfied with the music influenced by the locations of the objects. Additionally, other people sharing the same space can interact with interactive ambient music by changing the location of everyday objects in that space. In the seventh scene, another person records a short musical sound, and in the eighth scene, the second person is pasting the sound to a clock. Then, in the ninth scene, he moves the coffee cup and the phone. Finally, in the tenth scene, both people enjoy the ambient music that is played from the surrounding objects. People who are in the same space can add sounds and change the atmosphere of the music by interacting with everyday objects in the space.

Since the prototype system described in the next section uses VR sounds, the surrounding sound is blocked by taking on headphones. In recent years, bone conduction headphones have started avoiding the risk of blocking surrounding sounds. Thus, it is relatively easy to present music on the surrounding sounds. As a result, it is possible to preserve the concept of interactive ambient music by transparently embedding music into surrounding daily sounds.

Fig. 1. Interactive ambient music

4   System Implementation

This section contains a brief overview of the DESI prototype system that is used to realize interactive ambient music. The software structure of DESI is shown in Figs. 2 and 3 shows photos showing persons using the prototype. First, in order to recognize the relationship between everyday objects and their positions, the system uses two types of markers. A user puts a triangular marker (the head tracking marker depicted in Fig. 3) on his/her head for head tracking, and an LED light marker (the object tracking marker in Fig. 3) to track the everyday object. The triangular marker is attached to the user's head with the sharp tip in the direction of the user's face. The LED light marker recognizes different sound sources through the combination of colors. The everyday object to which this marker is attached produces a musical sound as if the thing itself is playing the sound. In the current prototype system, LED light markers contain three different colored lights. The number of everyday objects that users interact with can be easily increased by increasing the combination of colors. These markers are detected using a web camera hung on the ceiling of the space.

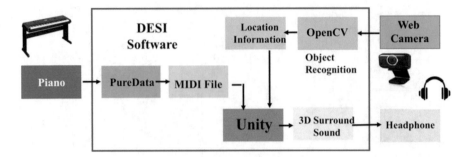

Fig. 2. Software structure of DESI

Fig. 3. Demonstrating interactive ambient music

Next, we will explain how the system works by referring to Fig. 2. For object recognition, as shown in green, a web camera and OpenCV are used, and the relative distance between the object and the user's head is calculated. Sounds are input by recording MIDI files with a piano, shown in blue, and Puredata, which is an open source visual programming language for multimedia and is shown in red. The MIDI file allows the system to record the height, length, and strength of a sound so that the system can easily change the tone of the sound. The position information of an everyday object is recognized by a marker. The MIDI file is converted to stereophonic sound on Unity and played as shown in yellow. When converting and reproducing a MIDI file in stereophonic sound, the system uses C# Synth for playback and Unity's Oculus Audio SDK for the conversion. The stereophonic sound can be heard by a user through headphones. Additionally, the buttons for starting and stopping the recording, sound reproduction, sound changes, and changing markers are implemented as a simple GUI on Unity, as shown on the screen of the PC in Fig. 3. The position information is sent from OpenCV to Unity every 0.5 s, and Unity sends instructions to Puredata to play the MIDI files.

5   User Studies

To generate insights and clarify potential pitfalls of interactive ambient music, in the first user study, we actually asked users to use the prototype system and we elicited some of their opinions. Next, we conducted a design exploration workshop as our second user study to generate ideas for possible enhancements to the interactive ambient music experience and how to use surrounding objects for inputs and outputs of respective music sounds.

5.1   User Study 1: Feasibility Demonstration

We demonstrated our prototype for two days at a university-hosted event for high school students, as shown in Fig. 3. Nearly eighty people, mainly high school students and their parents, attended the demonstration. The number of LED light markers used in the demonstration was limited by the space. As shown in Fig. 3, when placing a display showing a GUI screen and a piano in an L shape and sitting toward the piano, the everyday objects that markers are attached to are placed behind and to the left of the participant. The participants could freely move the everyday objects as they desired. Before the demonstration, we explained the background setting of the prototype and how to use it, then we actually performed the demonstration with each participant. Finally, they freely used the prototype to interact with music.

Afterwards, we asked the participants whether they felt that the sound actually originated from then everyday objects, then they discussed their impressions of using the prototype. First, when questioning the thirty-two participants about whether the sounds could be heard from the everyday objects, twenty-nine of them responded that "*The sound is heard from the direction of the things.*" In addition, in the free discussion, participants described the interactive ambient music experience as "*Mysterious feeling,*" "*Fun,*" "*Impressive,*" "*New,*" "*Interesting in changing the sound direction in real time.*" Additionally, there were some negative opinions including "*I have a difficulty in playing without piano experience,*" "*It is hard to get expected sounds when I play by myself,*" "*I want to use only my favorite music without playing a piano by myself,*" and there was one participant who was not aware of the relationship with the surrounding sounds.

5.2   User Study 2: Design Exploration Workshop

The design exploration workshop included eleven male (average age: 22.7) participants and one of the authors who served as a facilitator. First, the facilitator explained the interactive ambient music concept and why it is important to be unconscious of music when interacting with it. Next, she showed the participants a film that demonstrated how to use this prototype and how to interact with the music by moving surrounding objects. In the first phase of the discussion, she asked the participants to think about "*What kind of occasion, what kind of scene you want to listen to, and what kind of media you want to use?*" to extract opinions about the concept of interactive ambient music that participants possessed in the workshop. Similarly, she also asked them "*Where you like to interact with music and what kind of music is desirable as interactive ambient music for sharing each other's thoughts?"* In the second phase of the discussion, she divided the participants into three groups and asked each group to generate ideas about how to use surrounding things for the input and output of musical sounds, and what kinds of interactions can be used to change the sounds.

For the first question, participants answered *"While working, sleeping and not wanting to listen to surrounding sounds, or "While staying in offices and public spaces like department stores, I want to listen to ambient music."* For the second question, some of participants claimed *"The purpose of ambient music is lost, when the explicit interaction with music becomes conscious for me."* In the second phase discussion, the facilitator asked to each group to focus on how a user's activity influences musical sounds in ways that are not through explicit interactions. The following ideas were proposed from the groups: *"For increasing human concentration ability, detecting concentration loss from body information change the tempo and sound density of the music," "When pouring liquid into a cup, warm or cold music is played according to the liquid's temperature,"* and *"Music sounds are influenced according to the number of people and the degree of congestion."*

One finding from the workshop is that music is not ambient when the user is conscious of their explicit interaction with the music. Similarly, music does not become ambient if a user predicts how their actions will influence the sounds in the music. People easily predict changing music sounds in music. On the other hand, people have trouble noticing sounds that change due to the changes of the number of people or the degree of congestion. Therefore, we must consider that the nature of ambient music can be preserved if the interaction with music remains implicit. For example, by automatically assigning different musical sounds to moving objects and people in a current space, the musical sounds can be ambiently influenced according to movement.

One interesting future direction is to use ambient music as an ambient notification tool similar to ambient display [8]. For example, if interactive ambient music is used in an airport, when things and people play different music sounds according to their movements, the music is ambient for most people, but for a person who knows what the sounds mean the occurrence of an event, the music becomes meaningful notifications for him/her.

6   Conclusion

Ambient music is one-sided music that is only passively listened to, and emphasis is placed on blending it into a place so that it can be heard unconsciously. By adding interaction to ambient music without raising consciousness, we have proposed a new interactive ambient music concept. We developed DESI to realize this concept, and investigated the feasibility through user studies.

References

1. Chiba, K.: A world of dazzling VR acoustics. Sound & Recording Magazine, no. 11, pp. 20–23 (2016)
2. Collins, N., McLean, A., Rohrhuber, J., Ward, A.: Live coding in laptop performance. Organ. Sound **8**(3), 321–330 (2003)
3. Eno, B.: Ambient music. In: Cox, C., Warner, D., (eds.) Audio Culture: Readings in Modern Music, pp. 94–96 (2004)
4. Hargreaves, D.J., North, A.C.: The functions of music in everyday life: redefining the social in music psychology. Psychol. Music **27**(1), 71–83 (1999)
5. Ishizawa, F., Sakamoto, M., Nakajima, T.: Extracting intermediate-level design knowledge for speculating digital–physical hybrid alternate reality experiences. In: Multimedia Tools and Applications, 42 p. (2018)
6. Jordà, S., Geiger, G., Alonso, M., Kaltenbrunner, M.: The reacTable: exploring the synergy between live music performance and tabletop tangible interfaces. In: Proceedings of the 1st International Conference on Tangible and Embedded Interaction, pp. 139–146 (2007)
7. Morreale, F., De Angeli, A., Masu, R., Rota, P., Conci, N.: Collaborative creativity: the music room. Pers. Ubiquit. Comput. **18**(5), 1187–1199 (2014)
8. Nakajima, T., Lehdonvirta, V.: Designing motivation using persuasive ambient mirrors. Pers. Ubiquit. Comput. **17**(1), 107–126 (2013)
9. OtoComplete. https://www.youtube.com/watch?v=zWfGsEITpH8. Accessed 26 Dec 12 2017
10. Rentfrow, P.J.: The role of music in everyday life: current directions in the social psychology of music. Soc. Pers. Psychol. Compass **6**(5), 406–412 (2012)
11. WaveDNA, Liquid Music. https://dirigent.jp/product/wavedna/liquid-music/. Accessed 26 Dec 12 2017

Perceptions of Shared Leadership Through ICT in Secondary Schools in the City of Melilla

Antonio Campos Soto, Inmaculada Aznar Díaz,
Antonio-Manuel Rodríguez-García[✉],
and Carmen Rodríguez Jiménez

University of Granada, Granada, Spain
acaso4@hotmail.com

Abstract. This article analyzes the perceptions about the distribution of leadership and implemented by the use that the ICT present the governing bodies and didactic coordination of the secondary schools of the Autonomous City of Melilla. For this, we studied the Distributed Leadership (LD) and ICT dimensions of the doctoral thesis: Analysis of Leadership distributed in multicultural contexts and its incidence in Pedagogical and organizational Autonomy in secondary schools in the city of Melilla through ICT.

A mixed methodology has been used. In the quantitative research, we have worked with an invited sample (population) of 181 teachers, of which 164 (90.60%) have agreed to collaborate in the study, but finally have answered 145 (80.11%). The research was completed with 8 semistructured interviews with teachers from management teams and educational project coordinators.

The results indicate that the management teams distribute the functions of educational leadership, even though ICTs are hardly used to implement such functions.

Keywords: Leadership · ICT · Secondary education · Teaching practice

1 Introduction

Nowadays, Information and Communication Technology (ICT) and Internet, together with the globalization of education, either at a formal or informal level, are promoting methodological change which are the result of continuous methodological and curricular updating (Morales Capilla et al. 2015) which have triggered the implementation of new teaching and learning styles, resulting in the long-awaited methodological and networking change (Kocolowski 2010) in all sectors of the educational community and schools/education centres that the society demands, which redound in a change at an organizational, pedagogical and management level (Torrance 2013b; Ottestad 2013; Cox 2013; Kalinovich and Marrone 2017). With the change of the teaching model distributed leadership can foster cooperative learning communities in which teachers can work together in order to integrate ICTs in their learning and teaching (De Valenzuela del Aguila 2013). Thus, the leadership model in schools affects the curriculum and ICTs training (Voogt et al. 2013).

© Springer Nature Switzerland AG 2019
P. Novais et al. (Eds.): ISAmI 2018, AISC 806, pp. 208–215, 2019.
https://doi.org/10.1007/978-3-030-01746-0_24

In Spain, the Organic Law for the Improvement of the Educational Quality (LOMCE in Spanish) has set the stage for strengthening the management capacity of educational centre leadership, conferring the head teachers, as representatives of the educational administration in the school and as managers of the educational project, the opportunity to exercise greater pedagogical and managerial leadership. In this respect, according to authors like Corbella (2016), (Vermeulen et al. 2017) or Voogt et al. (2013), we attend a school culture in which the school leader promotes the ICT in order to programme, develop and share the teaching practice, where pedagogies focused on students and leadership actions, which foster digital competence between the staff, predominate.

For this reason, in recent years an interest focused on a leadership management has changed towards a wide spectrum leadership in which actions are focused on the whole of the school community. We speak of distributed leadership, of shared character, facilitating, inclusive, motivating and founding (Torrance, 2013a) thus fostering teacher engagement in stimulating an atmosphere which facilitates students' learning autonomy, something that will redound in an improvement of the results and the organizational aspects of the educational centre (Säljö 2010; García et al. 2011).

Starting from this base, this article focuses on analyzing the relationship of the distribution of educational leadership between management teams and teacher coordinating bodies in the secondary schools of Melilla (Spain) and the ICTs as tools which must implement such leadership in the pedagogical and organizational sphere (Valverde-Berrocoso and Sosa-Díaz 2014; Hauge et al. 2014). This allows us to study the dimensions of the distributions of the leadership such as communication, collaboration, trustworthiness, commitment, management, cultural change, innovation and the ICTs integration in teaching and learning processes.

2 Methodology

2.1 Objectives and Design

The present work has as its purpose the following objectives:

1. To ascertain a clear perception of the distribution of the leadership that management teams, project coordinator and department managers of the secondary schools of the autonomous city of Melilla have.
2. Analyzing to what extent the use of ICTs promote the distribution of the leadership of secondary schools of the autonomous city of Melilla.
3. Suggesting guidelines, if applicable, that facilitates the integration of the ICTs in the processes of the distribution of leadership in the secondary schools of the autonomous city of Melilla.

2.2 Instruments

A questionnaire of 35 items of Likert-type scale of four values was developed. The values were: 1 (None), 2 (Low), 3 (Moderate), 4 (High) which are distributed in three dimensions: *Academic and personal information* (API), *Distributed Leadership* (DL) and *Information and Communication Technology* (ICT) (Table 1).

Table 1. Dimensions, sub dimensions and items of the survey questionnaire

Dimensions	Sub dimensions	Items
Academic and personal information (API)	Personal information	1, 2, 3, 4
	Academic information	5, 6, 7, 8
Distributed Leadership (DP)	Collaboration	9, 10
	Trustworthiness	11, 12
	Commitment	13, 14
	Management	15, 16, 17, 18
	Educational quality	19, 20, 21
	Communication	22, 23
Information and Communication Technology (ICT)	Innovation	24, 25, 26, 27, 28
	Web 2.0 tools	29, 30, 31, 32
	PLE	33, 34, 35

With the purpose of verifying the validity of the tool, it was submitted to the judgement of nine experts. With the purpose of measuring the degree of reliability and internal consistency of the instrument, the Cronbach's Alpha was calculated, leading to a value of (,934) for a total of 27 items which is considered as excellent according to George and Mallery (2003) with which the questionnaire has high internal consistency and hence the results obtained are highly reliable.

2.3 Population and Sample

The whole of the target study population, which is composed of 181 members (N = 181), being the total acceptor sample to participate in the research of 164 (90,60%) answering it 145 (80,11%). Of those subjects who exercise their teaching and educational management work in nine centres of the city of Melilla. 51% are men and 49% are women.

The ages between 41 and 50 years are the most numerous in the research, representing 37,95% followed by the 27,6% of the age category between 31 and 40 and 24,8% of teachers belonging to the age category between 51 and 60 years.

3   Results

After the pertinent recount of the frequencies obtained from the data analysis which are derived from this research, the following table has been developed, where you can observe the arithmetic averages and their respective standard deviations (Table 2).

It can be observed that the average scores of the majority of the items fluctuate between 2,14 and 3,30, which explains the unanimity between management teams,

Table 2. Arithmetic average and standard deviation of the results obtained after the completion of the questionnaire.

Items	None		Low		Moderate		High		Analysis of measures		Total	
	N	%	N	%	N	%	N	%	M	σ_X	N	%
SMT	7	4,8	24	16,6	73	50,3	41	28,3	3,02	,803	145	100
MTP	7	4,8	28	19,3	70	48,3	40	27,6	2,99	,816	145	100
MTR	10	6,9	18	12,4	55	37,9	62	42,8	3,17	,898	145	100
MTP	10	6,9	31	21,4	59	40,7	45	31,0	2,96	,897	145	100
EPC	7	4,8	23	15,9	64	44,1	51	35,2	3,10	,836	145	100
MTC	7	4,8	29	20,0	66	45,5	43	29,7	3,00	,833	145	100
MTN	15	10,3	26	17,9	72	49,7	32	22,1	2,83	,890	145	100
TMT	7	4,8	26	17,9	70	48,3	42	29,0	3,01	,816	145	100
TTT	2	1,4	28	19,3	89	61,4	26	17,9	2,96	,655	145	100
TTB	8	5,5	40	27,6	73	50,3	24	16,6	2,78	,786	145	100
TRS	13	9,0	11	7,6	69	47,6	52	35,9	3,10	,888	145	100
CIP	10	6,9	27	18,6	52	35,9	56	38,6	3,06	,922	145	100
MTA	5	3,4	20	13,8	47	32,4	73	50,3	3,30	,834	145	100
TCC	9	6,2	31	21,4	57	39,3	48	33,1	2,99	,894	145	100
TMT	12	8,3	32	22,1	58	40,0	43	29,7	2,91	,920	145	100
TDP	11	7,6	34	23,4	68	46,9	32	22,1	2,83	,858	145	100
ATB	13	9,0	52	35,9	60	41,4	20	13,8	2,60	,837	145	100
IID	8	5,5	44	30,3	55	37,9	38	26,2	2,85	,877	145	100
BTI	6	4,1	43	29,7	61	42,1	35	24,1	2,86	,830	145	100
ICA	7	4,8	28	19,3	66	45,5	44	30,3	3,03	,845	145	100
TTC	8	5,5	30	20,7	70	48,3	37	25,5	2,94	,827	145	100
TFP	10	6,9	39	26,9	65	44,8	31	21,4	2,81	,852	145	100
SNM	50	34,5	33	22,8	41	28,3	21	14,5	2,23	1,078	145	100
PLE	34	23,4	60	41,4	44	30,3	7	4,8	2,17	,842	145	100
PLR	36	24,8	62	42,8	38	26,2	9	6,2	2,14	,863	145	100
PLM	29	20,0	59	40,7	44	30,3	13	9,0	2,28	,887	145	100
TFE	7	4,8	41	28,3	65	44,8	32	22,1	2,84	,822	145	100

project coordinators and the department managers of Melilla in relation to the statements detailed in the measurement tool (M = 2.84).

Since the variability of the results of the standard deviation [0.655, 1.078] are very close to the unit, the dispersion in relation to measures is reduced to the minimum, giving a high degree of reliability to the tool, and, thus, of internal consistency to the results.

Firstly, according to the gender of the participants, there is only a significant difference in the variable YWE "years of work experience", where the male representation in the director position is 66,7% against the female 33,3%. In the case of the head teacher or similar we find ourselves with 58,1% with more years of experience in these positions, against 41,9% female. Regarding the secretarial position the entire representation is male. In terms of heads of department there is a higher female representation (56%). Finally, as the ICT coordinator or similar there is also a higher female representation with 52,9%.

Depending on the years of work experience, 42,8% are teachers with between 11 and 20 years of experience. Nearly 20% have experience between 21 and 30 years, the 18,6% between 1 and 10 years, 15,9% between 21 and 30 years and around 3% for more than 40 years.

Once cross-matching data is done between the variable "school centre" (SS) and the items of the DL dimension, we observe that regarding "decision making in a collective way from the part of the management team" (SMT), it calls attention to the fact that 21,7% of the teachers participating in the secondary schools Rusadir consider that the decision making is not done in a collectively way. In reference to whether "the management team of the secondary schools promote the participation of the different sectors of the educational community...", more than 50% of school teachers in the Leopoldo Queipo, Enrique Nieto, Juan A. Fernández Pérez y Miguel Fernández secondary schools, affirm it positively, which is 48,3% of the participants. Regarding whether "the management team fosters a climate of confidence and respect between all the staff of the centre" (MTR), it is observed that in more than three secondary schools more than 80% of the teachers value it positively. It also highlights negatively this same fact in the Juan A. Fernández Pérez and Rusadir secondary school, with 30%. In terms of whether "the management team inspires professionalism to the rest of the teachers because of its way of working and addressing the issue" (MTP), it was observed that, on an overall level, 71,7% of the total teachers staff receive it positively. Results with regard to "the management team chooses the collaborators depending on their initiative and personal worth" (CIP), 80% of teachers in Enrique Nieto secondary school and 82% of teachers in Virgen de la Victoria school centre received this positively. In general, all centres resemble each other in term of the perception of teachers' staff, except Rusadir secondary school that evaluated it negatively with 60%.

Finally, concerning if "the management team fosters smooth and clear communication with all members of the community" (TCC), the majority of teachers had a positive view thereon, with each secondary school passing 60% with a rating between *moderate* and *high*. It should be noted that the school centre Rusadir has 55,6% in the value of *none*, and 19,4% in the *low* one, which are meaningful percentages in comparison with the rest of the schools.

Then, we show the crosses which present a statistically significant difference between the variable "school centre" (SS) and the corresponding ones in the *Information and Communication Technology* (ICT) dimension.

With regard to "the use of ICTs for the distribution of teaching assignments between members in the Tutors Board" (ATB), Leopoldo Queipo and Juan A. Fernández Pérez secondary schools obtained a higher teacher representation in the use of ICTs. In connection with "ICT innovation development in management teams for the educational integration and the educational cross-cultural action" (IID) this is an aspect valued in a positive way on the part of teachers in Enrique Nieto y Leopoldo Queipo secondary schools. In reference to if "beginner teachers tender to innovate with ICT in the learning and teaching process" (BTI), teachers' appreciations fluctuate equally for and against in Leopoldo Queipo school centre, which contrasts with 80% of teachers in Enrique Nieto secondary school, who evaluate positively (*moderate and high)*. Slightly higher data can be found in Virgen de la Victoria secondary school with *moderate* (63,6%) and *high* (27,3%). In terms of "the use of PLE (Personal Learning Environments) for the implementation of the teaching coordination with the rest of the management team and/or department members" highlights the fact that 35% in Enrique Nieto secondary school use them, and 45,5% of teachers in Juan A. Fernández Pérez secondary school do it in a *moderate* degree. Regarding if "the integration of ICT in the curriculum foster students' school success", teachers from Enrique Nieto (45%), Juan A. Fernández Pérez (54,5%), Miguel Fernández (45,7%), La Salle (50%) and Rusadir (47,8%) evaluate it positively. On the contrary, 75% of teachers in Buen Consejo school centre evaluate it in a negative way.

4 Conclusions

The administration of the management teams of the secondary schools of Melilla is, in general, highly valued from the point of view of their teachers except for one of the centres (Rusadir secondary school). With respect to the administrations of the department managers of those schools, the general perception is positive, although in Miguel Fernández and Rusadir secondary schools this is perceived negatively.

From the analysis of the sub dimensions like trustworthiness (MTR, MTP), communication (MTA, TCC), education quality (TRS, CIP) or the collaboration (SMT) it emerges that the distribution of the leadership of the bodies with organizational and governing responsibilities in the secondary schools of Melilla is positive in all the public centres and very positive in the private ones.

With regard to the use of ICTs in the centres of Melilla, the inclusive compensating nature of inequalities that present the use of ICTs from the part of the students is valued. However, urgent action needs to be taken aimed at using ICTs from the part of teachers both at curricular and classroom level. In this sense, specific training courses need to be offered in schools and oriented to the use and management of ICTs through effective practices.

All those who have answered the questionnaire appreciate positively that the management teams foster the collaboration and participation in the use of ICTs. Nevertheless, some teachers have misgivings regarding the implementation of ICTs because of a lack of time.

Finally, it can be assumed that the leadership is distributed between the management teams and the didactic organizational bodies of the secondary schools of Melilla, co-operating with all the staff, developing a positive climate of participation, communication and trustworthiness which redound in an improvement of the organization and the management at an academic, social and inclusive level. However, the use of ICT does not implement the management and didactic coordinating bodies leadership distribution, because its use remain in early stages.

5   Prospective

With this research we have revealed the educational reality in secondary schools of the autonomous city of Melilla, in terms of the use and development of ICT, both in an organizational and teachers' coordination levels, as well as curricular level. Thus, we give answers to our initial concerns about detecting strengths and deficiencies in the use and development of ICT in secondary education in the city of Melilla. Research whose importance we consider as appropriate if we start from the base that in this educational stage and in this city, together with Ceuta the higher index of early school leaving in Spain is found.

The purpose of these conclusions is double: in one hand, reporting the educational administration the results obtained in order to apply improvements in future education policies. On the other hand, act as starting point for future researches that will be focused on the analysis of the relation between the early school leaving in secondary school stage in Melilla, employment and the development of ICT in the teaching and learning processes of students and teachers, and the degree of acquisition of Spanish linguistic competence on the part of students.

References

Corbella, M.R.: Liderazgo y responsabilidad educativa: el necesario liderazgo de directores y profesores en la educación. Rev. Fuentes **14**, 85–104 (2016)

Cox, M.J.: Formal to informal learning with IT: research challenges and issues for e-learning. J. Comput. Assist. Learn. **29**(1), 85–105 (2013)

De Valenzuela del Aguila, A.: Herramientas TIC que potencian el trabajo en equipo. Recuperado a partir de (2013). http://repositorio.ual.es:8080/jspui/handle/10835/2098

García, C.H., Serrano, J.L.E., Ruiz, T.B., Sánchez, M.G., González, M.D.T., Postigo, A.C., Katarain, J.C.G.: El liderazgo educativo: proyectos de éxito escolar. Ministerio de Educación (2011)

George, D., Mallery, M.: Using SPSS for Windows step by step: a simple guide and reference (2003)

Hauge, T.E., Norenes, S.O.: Collaborative leadership development with ICT: experiences from three exemplary schools. Int. J. Leadersh. Educ. **18**(3), 340–364 (2015)

Hauge, T.E., Norenes, S.O., Vedøy, G.: School leadership and educational change: tools and practices in shared school leadership development. J. Educ. Change **15**(4), 357–376 (2014)

Kalinovich, A.V., Marrone, J.A.: Shared leadership: a primer and teaching recommendations for educators. J. Leadersh. Educ. **16**(1), 205–215 (2017)

Morales Capilla, M., Trujillo Torres, J. M., Raso Sánchez, F.: Percepciones acerca de la integración de las TIC en el proceso de enseñanza-aprendizaje de la universidad. Pixel-Bit. Revista de Medios y Educación, (46) (2015)

Säljö, R.: Digital tools and challenges to institutional traditions of learning: technologies, social memory and the performative nature of learning. J. Comput. Assist. Learn. **26**(1), 53–64 (2010)

Torrance, D.: Distributed leadership: challenging five generally held assumptions. Sch. Leadersh. Manag. **33**(4), 354–372 (2013a)

Torrance, D.: The challenges of developing distributed leadership in Scottish primary schools: a catch 22. Educ. 3–13 **41**(3), 330–345 (2013b)

Valverde-Berrocoso, J., Sosa-Díaz, M.J.: Centros educativos e-competentes en el modelo 1: 1. el papel del equipo directivo, la coordinación TIC y el clima organizativo. Profesorado. Revista de Currículum y Formación de Profesorado **18**(3), 41–62 (2014)

Vermeulen, M., Kreijns, K., Van Buuren, H., Van Acker, F.: The role of transformative leadership, ICT-infrastructure and learning climate in teachers' use of digital learning materials during their classes. Br. J. Educ. Technol. **48**(6), 1427–1440 (2017)

Voogt, J., Knezek, G., Cox, M., Knezek, D., ten Brummelhuis, A.: Under which conditions does ICT have a positive effect on teaching and learning? A call to action. J. Comput. Assist. Learn. **29**(1), 4–14 (2013)

A Multi-agent System Using Fuzzy Logic Applied to eHealth

Afonso B. L. Neto[✉], João P. B. Andrade[✉], Tibério C. J. Loureiro[✉], Gustavo A. L. de Campos, and Marcial P. Fernandez

Universidade Estadual do Ceará (UECE), Fortaleza, Ceara, Brazil
afonsoblneto@gmail.com, joaopbern7@gmail.com,
tiberiocj@gmail.com, {gustavo,marcial}@larces.uece.br

Abstract. As time is passing by, life quality is becoming one of the most concerns for people who are getting old. According to studies involving countries in Europe, older people tend to live alone or, at most, with one other person. The technology currently available considering health products helps those people to achieve their goals. Taking that into account, this article proposes a multi-agent system architecture that uses IoT devices to catch patients' heart signals and, using fuzzy logic process, estimates the level of hypertension, considering systolic pressure, diastolic pressure, age and body mass index. We used information of 768 patients obtained from a public database and evaluate the performance of the presented fuzzy logic model. The results of such fuzzy logic were compared with an evaluation made by accredited nurses, reaching a 94.40% of accuracy in the diagnosis.

Keywords: Fuzzy logic · Multi-agent system · Health · IoT

1 Introduction

In 2050, world population may rise nearly 20% comparing with 2017, as published by the European Commission Research and Innovation report [1], having the majority living in small households - alone or with just one person. Everyone desires to have an optimal quality of life. Thus, researches have been conducted in order to provide an environment on which technology can improve people's health and welfare.

The new concept of Medical Internet of Things (MIoT) [2] involves wearable devices that monitor vital signals on real time. Application areas for the IoT resides on medical and health care [3]. Lots of sensors and medical devices have been developed constituting a core part of the IoT.

Considering the lack of knowledge of most people in perceiving any imminent cardiac problems, in this article is proposed a solution with Multi-Agent System that sends typed information of heart signals using IoT, processes remotely the information, and infers diagnosis of high blood pressure, showing the result on a remote device. The solution proposed uses fuzzy logic, embedded into a multi-agent system. In order to evaluate the performance of the presented fuzzy logic model, was used information of

© Springer Nature Switzerland AG 2019
P. Novais et al. (Eds.): ISAmI 2018, AISC 806, pp. 216–223, 2019.
https://doi.org/10.1007/978-3-030-01746-0_25

768 patients obtained from a public database. The results of such fuzzy logic were compared with an evaluation made by accredited nurses.

This article is organized with background on Sect. 2, the architecture of the MAS on Sect. 3, Sect. 4 presents the application and results of the solution proposed, and Sect. 5 concludes the article and brings future works.

2 Background

This section will present the fundamental concepts for understanding the proposal.

2.1 Sensors, Controllers and Mobile App

In order to get information of heart activity, it is necessary to use devices where such information could be gathered either automatically or manually. The market has several types of shields available, such as the AD8232 Heart Rate Monitor [4], to get ECG. There are mobile apps that get information as well, and other devices that send that information to another place. Some of these devices are shown on Fig. 1.

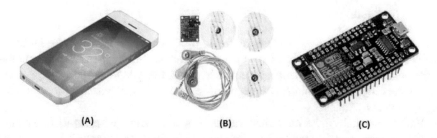

Fig. 1. (A) MQTT App, (B) AD8232 and (C) NodeMCU

The increase of calculation power and computational intelligence in the last few years made possible for IoT devices to process real-time data and adjust to changing conditions, as proposed by Toader [11].

2.2 Publish/Subscribe Pattern

The publish/subscribe pattern is part of message pattern, which describes how two different parts of a message passing system connect and communicate with each other. As shown in Fig. 2, the publishers do not program the messages to be sent directly to specific subscribers, but instead categorize published messages into classes on a broker, without knowledge of which subscribers. Any message published to a topic on the broker is immediately received by all the subscribers to the topic.

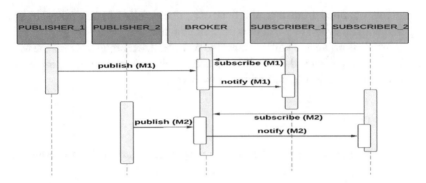

Fig. 2. Publish/Subscribe pattern

As data exchange performs a vital role in healthcare, a Publish/Subscribe based Architecture moderated through a web service that can enable an early exchange of healthcare data among different interested parties e.g. doctors, researchers, and policy makers was proposed by Wadhwa et. al [9].

2.3 Fuzzy Logic

Zadeh [5] presented the context of the theory of fuzzy sets, that is used to introduce Fuzzy logic. A fuzzy set designates a measure of membership, typically a real number in the interval [0, 1], to elements of a universe. Thus, fuzzy logic purposes to be a way that models logical reasoning with vague or imprecise statements like "John is short (ugly, cool, etc.)".

Variables usually take numerical values in mathematics, whereas in fuzzy logic applications non-numeric values are often used to facilitate the expression of rules and facts. A linguistic variable such as size may accept values such as short and its antonym long. As a fuzzy value scale cannot always be represented by natural languages, it can modify linguistic values with adverbs or adjectives [6]. Fuzzy logic process consists in four main steps: Fuzzification Module (Fuzzifier), Knowledge Base (Rules), Inference Engine(Intelligence) and Defuzzification Module (Defuzzifier), as shown in Fig. 3.

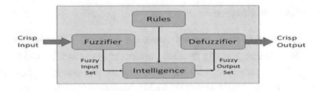

Fig. 3. Fuzzy logic overview

Membership functions are used to quantify linguistic term and represent a fuzzy set graphically. A membership function for a fuzzy set A on the universe of discourse X is defined as

$$\mu A{:}X \rightarrow [0, 1] \tag{1}$$

Each element of X is mapped as one membership function that gives it a value between 0 and 1. It is called membership value or degree of membership and quantifies the degree of membership of the element in X to the fuzzy set A:

- x axis represents the universe of discourse.
- y axis represents the degrees of membership in the [0, 1] interval.

2.4   Multi-Agent Systems (MAS)

In artificial intelligence research, agent-based systems technology has been hailed as a new paradigm for conceptualizing, designing, and implementing software systems. According to Russel [7], an agent is "anything that can be viewed as perceiving its environment through sensors and acting upon that environment through actuators.". Figure 4 shows the elements of an agent and a Multi-Agent System.

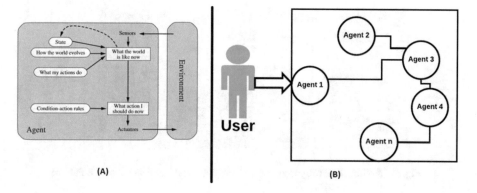

Fig. 4. (A) Elements of an agent and (B) Multi-Agent System(MAS).

Increasingly, multiple agents that can work together are required by applications. A multi-agent system (MAS) is a network of software agents that is loosely coupled and interact to solve problems that are beyond the individual capacities or knowledge of each problem solver [8].

3   MAS Solution Architecture Using IoT

As stated by Russel [9], the task environment involves the PEAS (Performance, Environment, Actuators, Sensors) that should be described when designing an agent. The architecture proposed consists in three agents, as described on Table 1.

Table 1. PEAS description of the task environment

Agent type	Performance measure	Environment	Actuators	Sensors
Mobile	Correct signals	Home	Pub/Sub Topic	ECG sensors/ Mobile APP
Processing	Diagnosis accuracy	Cloud	Pub/Sub Topic	Pub/Sub topic
Monitoring	Availability time	Monitor centre	Mobile Phone Screen/Monitor Screen	Pub/Sub Topic

The embedded multi-agent architecture that we propose for our MAS/IoT solution using Publish/Subscribe Pattern is resumed in Fig. 5.

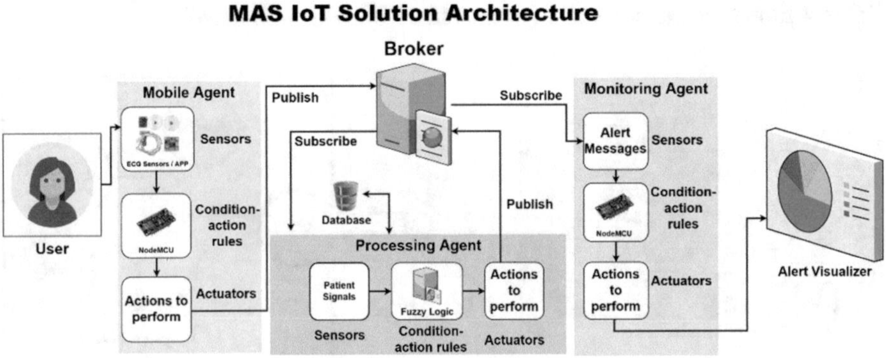

Fig. 5. Architecture proposed - MAS (3 agents) and Publish/Subscribe Pattern

The Mobile Agent that is encapsulated on a physical device containing ECG sensors and NodeMCU board or mobile APP which the user can interact. The Processing Agent can process information through fuzzy logic and infer how the blood pressure of the user is going, based on systolic blood pressure, diastolic blood pressure, age and body mass index, and indicates some abnormality. A database stores received and processed data for analytics purpose. The Monitoring Agent gets the process result which is shown on an output device. Publish/Subscribe Pattern "connects" this environment.

In this article we proposed a MAS/IoT solution, that consists in:

1. Publish/Subscribe environment using MQTT broker;
2. MySQL Database;
3. Mobile Agent: consists in a mobile MQTT client app used to get the patient's systolic and diastolic pressure, body mass index and age, validate them, and publish on a topic named "measurements" on MQTT broker.
4. Processing Agent: There is a subscriber service developed in python that gets those data and save on a MySQL database. As proposed in [13], we developed a fuzzy expert system using R [16], for the management of hypertension (High Blood

Pressure), that classifies the hypertension risk as Mild, Moderate or Severe. The results are stored on the database and published on a topic named "alerts" on MQTT broker;

5. Monitoring Agent: The results can be shown in two ways: NodeMCU with a Led, that is a subscriber of a specific patient pre-configured topic, and a website that access the database and show the results according to available filters.

The code of all services and applications are available on GitHub at https://github.com/afonsoblneto/eHealth.

4   Application and Results

The solution has been used in two ways: first, inserted data from a specific patient using a mobile app with MQTT client feature, allowing that all three agents could be used. Second, we imported the UCI Machine Learning Repository Database [14, 15] into MySQL, allowing the processing agent to be used in its fullness, using real data from 768 distinct patients.

Using the first way, the user introduces data using a mobile MQTT client app. The data is validated and published on MQTT broker. After that, the published values are taken by Processing Agent and the information is processed using fuzzy logic. When is detected some abnormality, the inferred diagnostic is published on MQTT broker into an alert topic. Finally, the Monitoring Agent has a feature as subscriber of that alert topic and when any alert is read, the NodeMCU turn its LED on, meaning that some abnormality was detected.

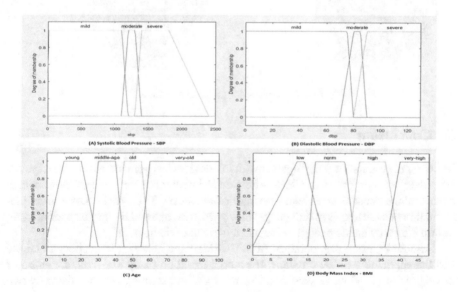

Fig. 6. Membership functions

Some adjustments were made on classifiers to keep them aligned to 7th Brazilian Director of Blood Hypertension [10]. The membership function for input parameters is shown in Fig. 6.

The membership functions considered the following:

- Input
  - Systolic: Mild, Moderate, Severe
  - Diastolic: Mild, Moderate, Severe
  - Age: Young, Middle-Aged, Old, Very Old
  - Body Mass Index: Low, Normal, High, Very High
- Output
  - Blood Pressure: Mild, Moderate, Severe

Considering the second way of use the application, in order to validate the applied fuzzy logic, the data stored into MySQL database was processed by the Processing Agent, and the results were stored into another table of the database. The results were accessed and checked separately by two accredited nurses. 560 patients have Mild/Normal blood pressure, and 120 patients have Moderate and 47 patients have Severe blood pressure. The results showed a positive predictivity of 94.40% (725 results calculated by fuzzy logic match with results calculated by nurses) in High Blood Pressure Diagnosis, as Fig. 7 shows.

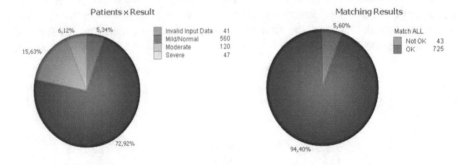

Fig. 7.   Achieved results and accuracy.

5   Conclusion

Multi-agent systems working together with IoT devices has been increasing in last years. Moreover, applying a certain level of artificial intelligence on some of those agents makes things become better, considering fast diagnosis. This article shows an implementation of such agents in order to get blood pressure, age and BMI information, using affordable hardware devices, and the due treatment of them.

In addition, using data obtained from UCI Machine Learning Repository Database and the application of the implemented model based on fuzzy logic, the diagnosis of blood pressure level was interesting. The results of our solution were validated by two accredited nurses and achieved a positive predictivity of 94.40% in High Blood Pressure Diagnosis, where 725 results of 768 available match.

As a future work, we intend to use PPG and ECG signals, as proposed by Kumar et al. [4], process them and infer diagnosis, as well as automate the integration of this model, turning all elements into a single solution.

References

1. European Union, "Research and Innovation". Publications Office of the European Union, Luxembourg (2014)
2. Haghi, M., Thurow, K., Stoll, R.: Wearable devices in Medical Internet of Things: scientific research and commercially available devices. US National Library of Medicine National Institutes of Health, January 2017
3. Pang, Z.: Technologies and architectures of the Internet-of-Things (IoT) for health and well-being. KTH Royal Institute of Technology, Kista, Sweden, xiv, 75 p. (2013)
4. Kumar, S., Ayub, S.: Estimation of blood pressure by using electrocardiogram (ECG) and photoplethysmogram (PPG). In: Fifth International Conference on Communication Systems and Network Technologies 2015, pp. 521–524 (2015)
5. Zadeh, L.A.: Fuzzy sets. Inf. Control **8**, 338–358 (1965)
6. Tamir, D.E., et. al.: Fifty years off fuzzy logic and its applications. In: Studies in Fuzziness and Soft Computing, vol. 326. Springer (2015)
7. Russell, S.J., Norvig, P.: Artificial Intelligence: A Modern Approach, 3rd edn. Prentice Hall, Upper Saddle River (2010)
8. Shoham, Y., Leyton-Brown, K.: Multiagent Systems: Algorithmic, Game-Theoretic, and Logical Foundations, rev. 1.1 (2010)
9. Wadhwa, R., Mehra, A., Singh, P., Singh, M.: A Pub/Sub based architecture to support public healthcare data exchange. In: Net Health Workshop, COMSNETS 2015 (2015). http://ieeexplore.ieee.org/document/7098706/
10. 7th Brazilian Director Of Blood Hypertension, vol. 107, N° 3, Supl. 3. Brazilian Society of Cardiology, September 2016
11. Toader, C.G.: Multi-agent based e-health system. In: 2017 21st International Conference on Control Systems and Computer Science (2017). http://ieeexplore.ieee.org/document/7968635/
12. The MQTT (2017). http://mqtt.org/
13. Chandra, V., Singh, P.: Fuzzy based high blood pressure diagnosis. In: International Journal of Advanced Research in Computer Science and Technology - IJARCST 2014, vol. 2, issue 2, Ver. 1, 2014
14. Sigillito, V.: Pima Indians Diabetes Data Set. The Johns Hopkins University
15. Lichman, M.: UCI Machine Learning Repository. University of California, School of Information and Computer Science, Irvine, CA (2013). http://archive.ics.uci.edu/ml
16. The R (2018). https://www.r-project.org/

Medical Digital Library Tool

T. Pereira[1], C. Martins[1(✉)], A. Almeida[1], N. Fonseca[1],
L. Faria[1], and J. A. Lopes-Santos[2]

[1] GECAD - Knowledge Engineering and Decision Support Research Center,
Institute of Engineering – Polytechnic of Porto (ISEP/IPP), Porto, Portugal
`tiago.pereira@eu.ipp.pt`, `{acm,amn,nuf,lef}@isep.ipp.pt`
[2] Hospital Center of Porto, Porto, Portugal
`jlopesdossantos@gmail.com`

Abstract. The approach for the definition and implementation of a platform-independent medical digital library, using only open-source tools, will be described. As a first test, the library will be used in the development of a tool aimed to aid doctors in otoplasty candidates' evaluation. This tool was been developed and tested first on a mobile platform, so the potential of the developed library can be tested on the most adverse situation. Also, the resulting library will be license free, making easier the collaboration of others in development and integration with other software, translating in a better spread of the library in medical practice.

Keywords: Medical imaging · Mobile devices · Digital medical tool

1 Introduction

There is still great potential for investigation and development around image processing and open-source, especially when directed to the medical area. Scientific research and investigation can be easily found around these areas, but these are usually independent of one another [1]. The majority of tools on the market for medical imaging are commercial.

Presently mobile devices like smartphones and tablets are present in the hands of millions of people, and this includes health care professionals. These devices have increasingly more processing, sensor and connection capabilities, making them good candidates to become a portable medical auxiliary tool. Having in mind that the cameras can be considered sensors, creating a medical auxiliary tool based on modern mobile devices with these sensors can prove to be useful.

Mobile Medical applications seem to contribute to help patient to involve in health promoting behaviours outside the clinical environment [2]. Also, mobile devices can be a complementary part of assessment and intervention for health problems [2]. Care providers who treat patients with chronic or mental diseases recognise the importance of maintaining contact with their patients outside of the exam room [2].

Healthcare mobile apps can be used to deliver medication alerts, patient education material and human interactions to gauge a patient's current mental state [2].

P. Novais et al. (Eds.): ISAmI 2018, AISC 806, pp. 224–230, 2019.
https://doi.org/10.1007/978-3-030-01746-0_26

Furthermore, mobile tools can contributes to improve independent living and support the prevention of health and welfare of each person [2]. But, there are a number of challenges it faces.

Despite the progress that many other industries have made, healthcare is likely to be the one market where imaging processing using artificial intelligence techniques can truly have an impact and positively affects medical diagnosis.

The need to create a medical auxiliary tool using medical imaging and open source has already been studied [1], with the conclusion that there is still work to be done in order to make the tools really useful.

It is expected that the development of an open-source medical imaging library help the conversion of future academic research in real-life applications.

2   Medical Imaging and Auxiliary Diagnosis Tools

Establishing a diagnosis is a very significant part in the medical profession. Medical auxiliary diagnostic tools to collect the information needed to analyze and infer about the patient's condition, for this effect the doctor uses various methods and tools. Usually the primary method is to make questions to the patient regarding his present status, his health history, family and analyzing his social groups. Besides this, the doctor also relies on auxiliary tools, as stethoscope, otoscope, ophthalmoscope, thermometers, which are known as medical auxiliary diagnostic tools, to collect information about the patient's organs [3].

A disadvantage of typical big diagnostic devices is that these have to be operated inside the hospital. Having in mind underdeveloped countries with doctors working on remote villages or emergency scenarios where on-site treatment tents have to be installed, portable diagnostic devices are an increasingly studied field. Examples of this effort are ultrasound machines, which started by being a fixed machine, evolved to a transportable machine, and presently it is typically the size of a laptop, existing already a smartphone version of this tool [4, 5].

The use of computers as an auxiliary diagnostic tool presently a key device in the work of a doctor. Medical imaging investigation using computers is therefore presently one of the main investigation fields [6, 7].

Nevertheless, there is still a great deal of fragmentation and a low number of conversions of the studies into practical applications. Fritzsche et al. [1] explain very well the reasons for this on a previous work:

"Two important reasons for this are (a) a lack of open source tools and standardization among them that would allow the community to incrementally build upon existing solutions and better handle the increasing complexity in the current state of the art, and (b) a lack of powerful visualization and interaction mechanisms in many of the current tools that would allow the physician to better understand and interpret the obtained results. Once a clinical application is ready for use, the resulting experience is a powerful driver for further refinement of the tools, algorithms, and state of the art."

This statement can be easily understood with the help of Fig. 1, where the drivers and the refinement propellers are shown.

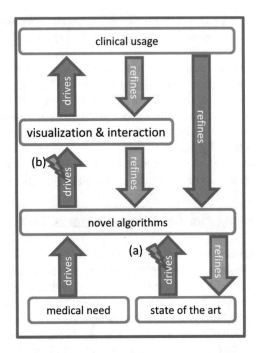

Fig. 1. Present state of medical imaging development [1]

Here it can be understood that medical needs drive novel algorithms, which in turn should drive the creation of new visualization & interaction methods that would then drive the clinical use of the application. And in the other direction, it is shown that clinical usage demands the refinement of visualization & interaction and novel algorithms. In case the drive between novel algorithms and visualization & interaction is ignored, as in the case of (b) in Fig. 1, it is easily understood that novel algorithms shall never reach clinical use.

3   Medical Imaging Library Tool

To fill the gap where there is a lack of open-source tools, standardization, and the visualization & interaction is rarely developed due to the extra effort required, a library using OpenCV will be proposed.

OpenCV is open-source and has the confidence of having been created by Intel, and runs on the major current operating systems, like Windows, Linux, OS X, OpenBSD, FreeBSD, NetBSD, Android and iOS. In addition to this, it has bindings for most popular programing languages like Java, C#, Ruby or Python, when programming in its native language (C/C++) is not possible.

The project will rely on the collaboration of doctors from pediatric surgery of a Hospital.

Being that, OpenCV has built-in interfaces and methods to access filesystems, cameras and other relevant hardware, this becomes beneficial, as researches do not need to waste time studying ways to do this. There is an abstraction that takes care of this, allowing the researches to focus on the most important task of their work: the investigation and algorithm development.

Also, the proposal is to try and create an easy and standard interface between what we can consider the core (the OpenCV block) and the algorithms the researches are developing. This way, there will be no need to adapt the methods for different platforms, the same code block can be used independently of the operating system or device. The structure of the library can be seen on Fig. 2.

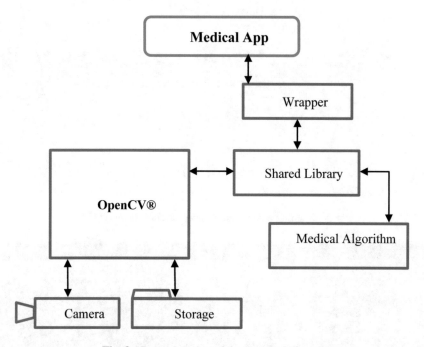

Fig. 2. Representation of the proposed library

All this will benefit cross-platform development, allowing the research projects that achieve success, to rapidly be adopted on a wide range of devices, and with this lower costs and reach a much broader range of people.

4   Medical Mobile Tool

The Frontal Aesthetics Translation Index for Measurement of Amplitude of the Ears (FATIMAE) was described and published by Portuguese investigators as a novel tool for the digital classification of prominent ears, thus a more objective diagnosis of a disease that affects about 5% of the Pediatric Population. Its wide implementation in clinical practice, however, will require the development of mobile applications that will allow to perform prompt measurements in a pediatric consultation [9]. Therefore, the objective of this work was to test the applicability of a mobile device in a clinical setting for instant calculation of this index.

A mobile app [8] (Fig. 3) was developed in an Android platform using our medical digital library, for the purpose of calculating the FATIMAE in a clinical setting. This developed mobile tool is currently available in Google Play for download and use under the name "Constantindex FATIMAE Calculator" [10].

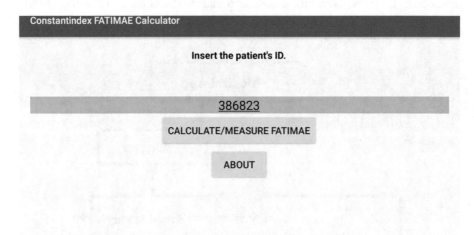

Fig. 3. Mobile App first screen (Patient ID input) [10]

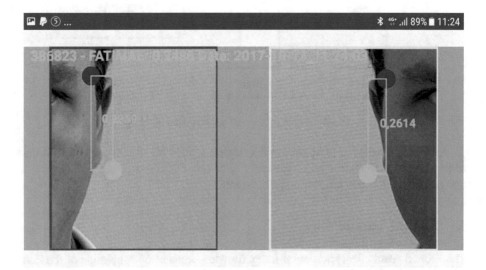

Fig. 4. Screenshot example of exact marking of the region of interest [8, 10]

For such task, an automatic and as precise as possible ear position system was be developed, which shows the index value to evaluate if the patient's ears can be classified as "prominent ears" or not, and by that decide if such patient is a otoplasty surgery candidate or not [9].

The challenge was to detect and mark as accurately as possible the region of both the patient's ears (Fig. 4), allowing Doctors to easily define manually the exact area of each ear, giving after that the calculated index [9], based on the Doctoral work of Master José Lopes dos Santos, using only the medical digital library developed [9].

5   Evaluation and Conclusion

A work regarding preliminary results for this mobile app was presented at the CMIN Summit in the year 2017 [11], were fifteen patients with complaints of protruding ears, aged from 4 to 17 years of age, were prospectively evaluated before being submitted to corrective otoplasty. After proper informed consent, frontal facial clinical photographs were obtained, with the head in the Frankfurt plane, and the FATIMAE was calculated using the desktop program ImageJ. Index measurements were also performed using the newly developed app and both index assessments were then statistically compared.

The work reported a positive correlation was demonstrated between the measurements performed by the traditional desktop ImageJ program and the mobile application.

Also, each FATIMAE calculation with the novel android App took less than one minute to perform. Furthermore, occasional glitches were reported and the program was continuously improved during the study.

The lack of open-source tools and easy methods to visualize and interact with the data, preventing the conversion of academic research into real-life applications, have been a big obstacle for the creation of digital medical diagnostic and/or auxiliary tools, limiting the access to the latest scientific developments in real life.

A library that helps the transposition of academic and scientific research to useful applications and that also facilitates researchers and scientist to develop and achieve results, will enhance and promote global access to digital medical tools. Such library would also lower costs and focus on a wider range of devices, being able to run even from smartphones or tablets, independently of the operating system and brand, providing better healthcare access to wider regions the world.

References

1. Fritzsche, K.H., Neher, P.F., Reicht, I., van Bruggen, T., Goch, C., Reisert, M., Nolden, M., Zelzer, S., Meinzer, H.-P., Stieltjes, B.: MITK diffusion imaging. Methods Inf. Med. **51**(5), 441–448 (2012)
2. Fonseca, N., Almeida, A., Faria, L., Martins, C., et al.: Development of a hybrid application for psychotic disorders self-management. In: De Paz, J., Julián, V., Villarrubia, G., Marreiros, G., Novais, P. (eds.) Ambient Intelligence–Software and Applications – 8th International Symposium on Ambient Intelligence (ISAmI 2017). ISAmI 2017 (2017)
3. Jones, R.: Oxford Textbook of Primary Medical Care. Oxford University Press, Oxford; New York (2004)

4. Schleder, S., Dendl, L.-M., Ernstberger, A., Nerlich, M., Hoffstetter, P., Jung, E.-M., Heiss, P., Stroszczynski, C., Schreyer, A.G.: Diagnostic value of a hand-carried ultrasound device for free intra-abdominal fluid and organ lacerations in major trauma patients. Emerg. Med. J. **30**(3), e20–e20 (2013)
5. Smartphone Ultrasound, MobiUS SP1 | Mobisante. http://www.mobisante.com/products/product-overview/. Accessed 29 Mar 2017
6. Sriniva, M.B.: Portable diagnostic device can help save lives. Microsoft Research (2008)
7. Doi, K.: Computer-aided diagnosis in medical imaging: historical review, current status and future potential. Comput. Med. Imaging Graph. Off. J. Comput. Med. Imaging Soc. **31**(4–5), 198–211 (2007)
8. Pereira, T.: Definição de uma biblioteca para apoio à decisão de avaliação de orelhas proeminentes. MSc. Thesis, in Computer engineering, ISEP, Portugal (2016)
9. Lopes-Santos, J.A., Martins, C., La Fuente, J.M., Costa-Carvalho, M.F.: A novel approach for classifying protruding ears is easy to calculate and implement in daily clinical practice. Acta Paediatrica (2017)
10. FATIMAE mobile App. https://play.google.com/store/apps/details?id=com.joselopesdossantos.fatimae_calculador
11. Lopes-Santos, J.A., Rocha-Pereira, T., Martins, A.C., La Fuente, J.M., Costa-Carvalho, M.F.: Mobile devices as an emergent tool in clinical medicine: the present and future of the FATIMAE digital index. Birth Growth Med. J. **26**(Supp II) (2017)

Find_Me: IoT Indoor Guidance System

Stuart Martinho, João Ferreira[✉], and Ricardo Resende

Instituto Universitário de Lisboa (ISCTE-IUL), Information Sciences,
Technologies and Architecture Research Center (ISTAR-IUL), Lisbon, Portugal
{scmoa,jcafa,jrpre}@iscte-iul.pt

Abstract. This work presents an approach to combine location information from beacons and local building information to give real-time location and guidance to a user inside a building. This information can help users orientation inside unknown buildings and the data stored from different users can provide useful information about users movements inside a public building. Beacons are installed on the building and emit signals that give a geographic position with an associated imprecision, related with Bluetooth's range. This uncertainty is handle by building layout and users' movement in a developed system that maps users position, gives guidance and store user movements. This system is based on an App (Find Me!) for Android OS (Operating System) which captures the Bluetooth Low Energy (BLE) signal coming from the beacon(s) and shows, through a map, the location of the user 's smartphone and guide him to the desired destination.

Keywords: Location indoor · Mobile APP · BLE · iBeacon · Path Finding · A*

1 Introduction

Most of modern life in developed countries is spent inside buildings, which are frequently large and complex. The reasons behind this complexity are poor architecture layout, an accumulation of interventions or poorly planned expansions. As a result, users get lost easily: they don't know where they are or they know where they are but don't know how to reach their destination. On the other hand, it is very difficult for building planners to implement guiding systems that cover all possible destinations and are easy to read. The only option to users is to wander, ask for directions, often multiple times. This situation is usual in hospitals, airports, shopping centers or museums. This problem can be easily simplified with Indoor Location System (ILS).

ILS or Indoor Positioning Systems have the purpose of finding a user's location inside a building, typically with a small, smart device like a phone, tablet or watch. The method for finding the device can be implemented via Bluetooth, infrared, magnetic fields, RFID, WIFI. When the smart device connects with some Indoor Location electronic device, the system can store that data and take many conclusions. An example of an Indoor Location technology is Beacon Technology. A beacon is a device that broadcasts a Bluetooth Low Energy (BLE) signal in a predefined range. This signal can be interpreted as the Location of the user inside of a building, without the need of an internet connection.

© Springer Nature Switzerland AG 2019
P. Novais et al. (Eds.): ISAmI 2018, AISC 806, pp. 231–238, 2019.
https://doi.org/10.1007/978-3-030-01746-0_27

Most people have their own smartphone these days [1], so the best and easiest way to help with the orientation is through a mobile App such as the Find Me! App, here proposed. This software's function is to show, through a map, where the person/user is and how can he get to the intended destination, guiding him until he reaches it. The most important and crucial parts of Find Me! App are the user's current location (from the nearest beacon) and the desired destination. Having these two locations as input, a Path Finding Algorithm (A* Search Algorithm Type [2]) calculates the shortest way between these points and draws it on a map with the objective of guiding the user to his destination. When the user intersects another beacon region, the drawn path is updated with a new user's current location.

2   State of Art

The navigation space in large buildings such as schools, public buildings or shopping centers is frequently complex, comprising rooms and corridors, stairs, escalators, lifts and ramps. Traditional architectural drawings such as floor and site plans, elevations and cross sections and more natural perspectives can be used to aid orientation.

A floor plan, as shown in Fig. 1, is a rectangular projection from above showing space arrangement of one level. It displays anything that could be seen below a certain threshold (typically 1.2 m): floor, walls, windows and door openings, stairs up to the section level, sometimes furniture. The public is used to this kind of drawing, but a considerable part of users have difficulty in reading.

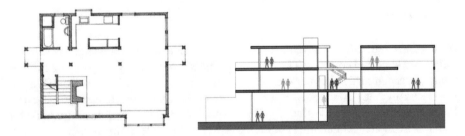

Fig. 1.   Floor plan of an apartment (left) and a cross-section of a multilevel building (right).

Elevations and cross sections are rectangular projections in the horizontal direction. Elevations display the exterior, while cross sections, as shown in Fig. 1 show the interior, as seen as if a vertical plan cuts open the building. They illustrate the relationship between levels of the building, but are not very often used in navigation.

A more sophisticated category comprises three-dimensional projections. These are more easily read by lay users since they present a more natural way of showing the interior of buildings, walls, doors, and windows, passages, and furniture. Finally, first and third-person views, as popularized in action games, allow users to relate what they see on the screen with reality. They are increasingly being employed, but are much harder to implement since they demand precise location, detailed 3D description of the building and more powerful, 3D, graphics processing capabilities on the devices.

The choice of which elements are represented and their level of detail is important in navigation. Most structural and construction details are not usefull, but prominent decorative features, floor or wall colors or lining materials are, as well as human figures, shadows, transparency or animations provide a better grasp of the environment.

Pathfinding or pathing can be described as a computer Algorithm that finds the shortest route/path between two points. To achieve the main objective, the Pathfinding needs to receive three mandatory parameters: current point, destination point, and a map. The A star algorithm, or A*, holds the objective of calculating lowest route/path cost from a current/initial node to the destination node out of one or more possible nodes. It is based on an evaluation function: $f(n) = g(n) + h(n)$. The $h(n)$ is optimal path cost estimate from node n to the destination node, and $g(n)$ is described as the current cost from the current/initial node to any node n, with other words, the optimal path cost finding. As A* traverses the map it follows the path with the lowest cost while keeping alternative nodes in a sorted priority queue. If a node being traversed has a higher cost than another encountered node at any point, it discards the node with the higher-cost and traverses the lower-cost node instead. This process continues until the destination is reached. The base map can be configured with two types of components: nodes and obstacles. A cost value is set to each node but for obstacles no cost is provided so the algorithm doesn't keep it in its queue. Figure 2 describes, in an objective way, the basic concept of A* algorithm in a flow chart, adapted for our mobile usage.

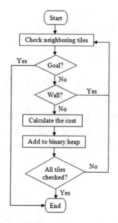

Fig. 2. Flow chart of the A* pathfinding algorithm

Fig. 3. 3D section views of the BIM model of ISCTE-IUL's Building 1 – Sedas Nunes.

BIM – Building Information Models – are 3D descriptions of buildings which associate information with the geometry of the building and of its contents: fixings, furniture or spaces. A well parametrized BIM model is aware that a specific door is a double door, glass made, has level 1 fire-rating, connects corridor 1S to room 1S04, opens to the rooms on the left-hand-side, etc. ISCTE-IUL's facility management office has been developing a BIM model which is being used to feed maps, room listings, and locations, as well as beacon's locations. That proves a major advantage: since the BIM models are data-based they can be queried and updated with information to and from the app.

2.1 Indoor Location System

Work [3] by Lin *et al.* describes an Indoor Location System, using Beacon technology, applied to a Hospital Building, which is similar to the system described in this paper, but is applied to a Hospital. This ILS is divided into three main parts:

1   Patients' Mobile App side: patients install a mobile app on their smartphone that, depending on the patient location, intersects beacons emission and send it to the server that converts it into a Location.
2   Medical Staff's Devices side: doctors and staf use an app that besides showing patients' data (condition, name, age, address, etc.) also shows the patient's current location on a hospital map.
3   Server Side: this middle tier connects the two users': Patients and Medical Staff. It holds the Database that stores patient's data and location of the beacons. When a beacon's id is received from the Patients' App, the Server looks for that beacon in the database and returns a location, that is sent to Medical Staff's Devices.

This ILS helps doctors and probably security staff as well, in day-to-day utilization but also in emergencies.

Our ILS holds a major difference to the described system, as our system is private, as it will only show the position of the user. Also, instead of finding the current location of the user, we also gather destination and the shortest path between these two points.

3   Find Me! Architecture

The system architecture, described in Fig. 4, is based on four modules: Beacons Configuration and Georeference, Find Me! Back End – Database and Path Finding Algorithm and Find Me! Front End – Insertion of Destination and Map View. Having a local Database directly implemented/configured on smartphone storage permits this application to be independent of WIFI or 3G/4G.

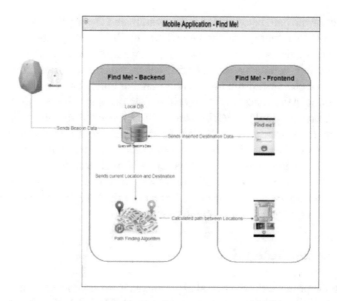

Fig. 4. System Architecture based on Estimote Beacon device.

The beacon type/make used in this ILS is Estimote [4]. It has several specifications [5] and two protocol types: EddyStone [6] and IBeacon [7]. IBeacon was developed by Apple in mid 2014. This protocol allows the broadcast of a specific format data:

- *"UUID, 16 bytes, usually represented as a string, e.g., "B9407F30-F5F8-466E-AFF9-25556B57FE6D";*
- *Major number, 2 bytes, or an "unsigned short", i.e., a number from 1 to 65,535;*
- *Minor number, 2 bytes, same format as Major.*

In this project, all the beacons are configured with the same UUID to create a proper mesh. This prevents the adition of external beacons without knowledge of the specific UUID. Major and minor fields will differentiate each beacon. With Beacons Manager, a method supplied by the *Estimote* API to scan the closest beacons, there's no need to handle the BLE connection from each beacon region that the smartphone intersects. It's through this method that beacons receive their UUID.

When the App runs within the range of a becon, it automatically receives the data it transmits and sends it to the Find Me! Backend which holds a local Database that stores the building geometry and room name and the installed Beacons information and location. If the user inserts a destination room in the app Front End – Insertion of Room, that information is also sent to the Backend. The Backend runs a query with both data and returns two locations: User's Current Location and Destination. Having this data, the system finds the floor plan maps that correspond to each location.

The pathfinding algorithm receives, as input, the current location, destination and the matching maps. These contain information on walkable regions (the cost of each area) thus preventing the algorithm from returning paths that cross walls or off-limits areas.

To calculate the path between Current Location and Destination, two hipotheses are possible. If both locations are on the same floor, the Path Finding Algorithm runs an A* method which calculates the shortest path. If the locations are on different floors the Path Finding Algorithm calculates the shortest path between the current location and the closest stairs (or elevator, if the user as choosen that option) and guides the user to the destination floor. After that, the A* algorithm calculates the shortest path between floor destination stairs and the destination location. Once the shortest path is calculated it is sent, with the maps and locations, to the Find Me! Front End – Map View.

The Map View interface is where the user observes orientations until he reaches its destination. The interface renders both current and final locations, maps, and the optimized path on the screen. As the user is on it's way to the destination and intersects other Beacons, the current location is updated. Also, Path Confirmation Photos are shown on the screen to assure the user he is on the right path.

4   Testing and Validation

To evaluate the proposed ILS, several experiments were conducted in the experimental test-bed. The experimental test-bed was located in ISCTE-IUL's Building 1, represented in Fig. 3. This building has three floors, 335 rooms and 48 installed Beacons (4 per hall per floor). Beacon transmission frequency was one packet/second, and configured signal broadcasting power was -20 dBm – a range of approximately 4 m of radius, but also dependedent of physical barriers such as slabs or walls.

A group of twenty persons/users (students, local infrastructure responsible and teachers) acted as beta testers. The users installed the Find Me! app in a Huawei P8 Lite Android smartphone [9]. One random user test was choosen for the following test

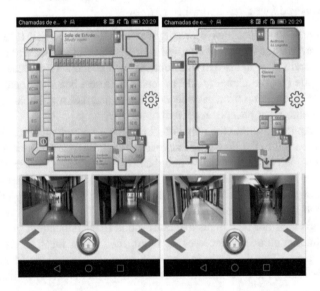

Fig. 5.   Guiding the User to nearest stairs (left), destination floor (right).

description: the user was located in floor 1. When he ran the App, he intersected a beacon, and the received beacon data was sent to the App (UUID, Major and Minor values). The user inserted the destination: room 0S2 – located in floor 0. As the current location floor was different from the destination floor the app automatically gave orientations to the closest stairs (temporary destination), as can be seen in Fig. 5 (left).

The user checked Path Confirmation Photos to make sure that he was on the right way. As he was walking, he intersected other beacons which provided the new current location, so the path was correctly updated. When floor 0 of building was reached, he entered the range of a beacon and the map and Path Confirmation Photos(calculated throught the returned shortest path from A* algorithm), were updated floor. After that, it was possible to identify the destination room on the map – red pin - on Fig. 5 (right). The user just needed to walk there.

On another test, the application was started outside beacon's range. Without beacon data, the Find Me! App couldn't calculate the current location of the users so didn't pass to the next activity, the destination insertion. The user then started to walk randomly to try to trigger the next beacon. When he intersected the first beacon, the app followed the normal flow and helped the user reach the destination. In some cases, the smartphone intersected a beacon from the floor above or bellow and compromised all the user orientations. On the next release, this is a point that needs a focus to solve it.

The main user's criticism is that the number of beacons needs to be higher in a way to get more live updates of the current location because only four beacons per floor aren't enough. After all the users experiments, it was given to each one, an online survey with the propose of evaluate the implemented ILS. In Table 1 it's possible to confirm the average values of all the filled surveys.

Table 1. Results of performed surveys

Questions	Rate(0–10)
The simplicity of the App	7
Quality of Orientations	8
The optimized path between Locations	7
Performance of the App	7
Front End Design	6

5   Conclusion

This study describes the implementation of an ILS complemented with a mobile App, to help users of a building that is frequently described as confusing – ISCTE-IUL's building 1– to receive the right orientation, not only on the beginning but also until the user gets the destination.

It was confirmed that an increase of the numbers of beacons, will lead to a more precise location and that can further combine beacon location information with available buildings maps to provide guidance. With a larger number of beacons their range can be lowered thus avoiding the app detecting beacons in different floors.

Another contribution of this research work is that we can register passively (without user identification) the movements of persons inside buildings. This important information can be analyzed to extract knowledge about users' movements inside public buildings. We can connect this beacon information to IoT system to turn on/off lights and give information about the number of persons.

This work can be easily replicated to other buildings with the introduction of new maps and calibration of beacons signal range due to different propagation conditions. Also, there is the possibility shortly the possibility of association of advertising services and information through the beacons. For example, the beacon nearer to the library will advertise library services, for canteen menus, available secretary services, general university information and others. We also intend in the near future introduce evacuation guidance for emergency situations.

References

1. Ching, F.D.K.: Design Drawing, 2nd edn. John Wiley & Sons Inc., Hoboken (2010)
2. Loong, W.Y., Long, L.Z., Hun, L.C.: A star path following mobile robot Wong. In: 4th International Conference on Mechatronics, pp. 17–19, May 2011
3. Lin, X.Y., Ho, T.W., Fang, C.C., Yen, Z.S., Yang, B.J., Lai, F.: A mobile indoor positioning system based on iBeacon technology. In: Proceedings of Annual International Conference of the IEEE Engineering in Medicine and Biology Society, EMBS, vol. 2015, pp. 4970–4973, November 2015
4. E. Inc., "Estimote Beacons" (2012). https://estimote.com/. Accessed 01 Dec 2017
5. E. Inc., "Beacon Specification". https://community.estimote.com/hc/en-us/articles/204092986-Technical-specification-of-Estimote-Beacons-and-Stickers
6. E. Inc., "EddyStone Protocol". (2016). http://developer.estimote.com/eddystone/. Accessed 01 Dec 2017
7. E. Inc., "IBeacon Protocol" (2016). http://developer.estimote.com/ibeacon/. Accessed 01 Dec 2017
8. J. Inc., "Canvas." https://developer.android.com/reference/android/graphics/Canvas.html. Accessed 07 Dec 2017
9. Huawei, "Huawei Phone P 8 Lite Specs." http://consumer.huawei.com/pt/phones/p8-lite-2017/. Accessed 10 Dec 2017

m-Health Application for Infection Control and Prevention Focused on Healthcare Professionals

Ariane Baptista Monteiro[✉], Luzia Fernandes Millão, Ícaro Maia Santos de Castro, Silvio César Cazella, Rita Catalina Aquino Caregnato, and Karin Viegas

Federal University of Health Sciences of Porto Alegre,
Rio Grande do Sul, Rua Sarmento Leite 245, Porto Alegre, RS 90050-170, Brazil
aribmonteiro@gmail.com, luziam.ufcspa@gmail.com,
icaromscastro@gmail.com, silvioc.ufcspa@gmail.com,
ritac.ufcspa@gmail.com, Kviegas@gmail.com

Abstract. Decisions about precautions and isolations are performed by healthcare professionals based on manuals, through tables and texts, not always available at the point of care. For this reason, the use of mobile application software to aid in decision making during health care is increasing. This article describes m-health application on standard precautions and based on transmission routes development to support the decision of healthcare workers while providing patient care. It is an applied research of technological production, where it was carried out prototyping, content database elaboration, creation, development and system evaluation. m-health application, called Isolation App, provides information on precautions and isolation, personal protective equipment suitable for each type of precaution, hand hygiene guidelines and contains national and international links related to the subject. Application was evaluated by healthcare workers experts in infection control and prevention. Results suggest the prototype may support health professionals, facilitating decision making at the point of care.

Keywords: Applications · e-Learning · e-Health

1 Introduction

The use of mobile devices in health care settings bring great benefits to professionals, contributing to more productive practices and avoiding misunderstandings, and providing a patient safety care [1, 2].

Despite the widespread availability of health-related applications, the widespread mobile devices use, and the urgent need to reduce and eliminate healthcare-associated infections (HAI), few support applications have been identified in HAI Prevention, but which can reduce infections through easy access to manuals, hand hygiene monitoring, and a list of step-by-step procedures for reducing infection at point of care [3].

Standard Precautions (SP) implementation is the main strategy HAI transmission prevention. Technological evolution and accessibility to healthcare have enabled the interventions expansion that favor health services users, but assistance to these patients

© Springer Nature Switzerland AG 2019
P. Novais et al. (Eds.): ISAmI 2018, AISC 806, pp. 239–246, 2019.
https://doi.org/10.1007/978-3-030-01746-0_28

is increasingly complex. Many have conditions that require precautions that prevent infections transmission, in addition to the SP that should be performed for all patients.

In such cases, associated with SP, precautions based on transmission routes are required, additional measures to stop the transmission of epidemiologically important microorganisms in the hospital environment or in another healthcare scenario [4]. Healthcare professionals should be aware of which infectious condition the patient may have and thus take appropriate precautions to reduce the transmission risk by implementing the most appropriate transmission-based precautions [5] and know when to discontinue the precaution [6].

However these measures are really effective only when used properly. In this sense, mobile applications are presented as an alternative to the professionals to make quick consultations, facilitating decision making at the point of assistance. Global Observatory of the World Health Organization for e-Health defines mobile health (mHealth) as "a practice of medicine and public health supported by mobile devices such as mobile phones, patient monitoring devices, digital personal assistants and other devices "wireless" [1]. Mobile applications can support the collection and dissemination of evidence to reduce HAI [3].

One of the study authors, due to his infection control experience, observed the healthcare professionals difficulty in identifying the appropriate precaution for each situation. In most hospitals, infection control professionals are present to guide staff, but not 24 h a day, when doubts arise, professionals need information. In addition, m-health application can be a great support tool to be used by health students, as it provides guidance in an accessible way, facilitating the choice of the most appropriate preventive measure. This article describes m-health application on standard precautions and based on transmission routes development to support the decision of healthcare professionals while providing patient care.

2   Infection Control and the Use of Mobile Applications

The concern with HAI is increasing due to the impact of these events for both patients and health institutions. International multi-center study verified HAI prevalence in Intensive Care Units and concluded these are higher in Brazil and perceived a clear relationship between infection and mortality [7].

HAI identification, prevention, and control represent prior intervention on risk in healthcare setting, before harm reaches the patient. SP include a set of practices that apply to all patients regardless of suspected or confirmed infection in any healthcare setting. In this sense, hand hygiene and the use of gloves, gown, mask, glasses or facial protector, whenever there is risk of professional exposure are imperative [8].

When the transmission route is not completely interrupted using only the SP, precautions based on the transmission route must be used. There are three categories of precautions based on the route of transmission: contact, droplet, and airborne [4, 8].

In Jordan a study analyzed the knowledge of nursing students about SP and based on transmission routes, evidenced that the students presented more knowledge regarding the SP, however a small percentage correctly answered questions regarding additional precautions [9]. The use of a card with a route-based precautions summary as a tool for healthcare workers decision-making improved patient care [10].

This scenario increases microorganisms transmission risk, especially in high complexity healthcare setting. In addition, a tool to be used at the point of care could guide which preventive measure would be most appropriate. Information and communication technologies may provide information on which precautions are most appropriate in each situation, thus preventing possible HAI.

Applications use during patient care may be an alternative. Health interventions, through mobile devices, have been designed to increase healthy consumer behaviors or to improve disease monitoring, increasing adherence to treatments [11].

Applications are often used in the healthcare environment, but to a lesser extent when it comes to HAI control and prevention. Study published in 2016 evaluated tuberculosis prevention and treatment applications and found they had features aimed at front-line healthcare professionals, and focused on tuberculosis information or in data collection [12]. In 2015 there were found 17 applications on health care infection prevention and control [3], and integrative review [13] on mobile applications developed for health in Brazil, met 27 published articles, between 2006 and 2013, none of them focused on precautions and isolations. Tortorella and Kinshuk [14] in 2017 described a context-aware learning system to reduce the transmission of pathogens directed towards the cleaning of the environment and surfaces, but did not address standard precautions and based on transmission routes.

3   Materials and Methods

This is an applied research on the production of a prototype of an application with guidelines on disease prevention, based on the Center for Diseases Control and Prevention (CDC) manual [7]. Development of an application mobile to assist choose the type of precaution most appropriate for each patient condition. Personal protective equipment (PPE) that should be used by health professionals to carry out precautions, hand hygiene guidelines and referential links related to the subject.

4   Isolation App Prototype

Development of the m-health application including: prototyping, database elaboration with the content, creation, development of the system and evaluation. In the application elaboration, it was first performed the requirements survey for the system, according to the use case diagram presented in Fig. 1.

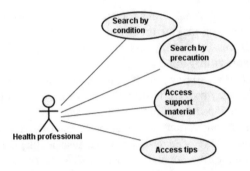

Fig. 1. Diagram of use cases of the prototype

In the diagram shown in Fig. 1 each ellipse represents a feature offered to healthcare professionals in the application. Search by condition: The actor can search the system for information about the condition/infection by its name. Search by precaution: The actor can filter your search for conditions according to the type of precaution. Access support material: the stutter to access to related links that are used as reference in Brazil and in the world. Access tips: the actor can access tips on care to be taken, such as hand hygiene and the correct use of Personal Protective Equipment – PPE.

The application database was prepared using Mysql and contemplated the following data: (1) condition/infection, (2) type of precaution and the duration of the same, (3) the necessary personal protective equipment, (4) peculiar additional information to each condition/infection, according to the CDC list. The modeling of the database is presented in Fig. 2.

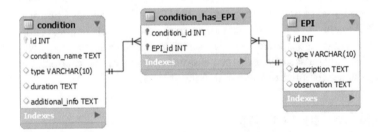

Fig. 2. Relationship entity diagram of the database.

The condition table is used to store the data for each condition/infection. This table contains the following fields: Id: unique identifier of the condition/infection; Condition_name: condition/infection name; Type: Precaution Type (Standard, Contact, Droplet and Aierborne) according to Center for Diseases Control and Prevention; Duration: duration of precaution according to infection/condition; Additional_info: Additional information about the condition (peculiarities of each condition relating to precautionary types).

The PPE table PPE is used to store data-on each type of PPE. This table contains the following fields: Id: unique identifier of PPE; Type: Precautionary type; Description: stores the description of the PPE; Observation: Notes on PPE.

After the database was completed, it was exported to the SQL (Structured Query Language) format to be used in SQLITE locally in the mobile application.

The mobile application intittled as Isolation App was designed for android and has five interfaces, the first screen, in the central part is the logo of the prototype, an paragraph of presentation of the App, explaining what can be searched in the tool, at the top there is a bar containing a menu that allows the user perform searches for infectious conditions, access screen precautions, access the bibliographic references screen, or click on hand hygiene during health care. Figure 3 shows the Initial Interface of the application menu with guidance on the application.

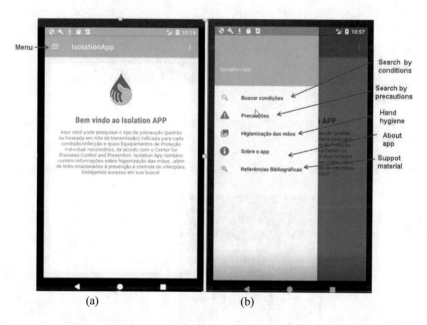

Fig. 3. (a) Isolation App's initial interface and (b) Interface with the application menu.

Figure 4 shows search interface, where the user can type the condition that he wants to consult, to access information of a specific condition, or double-click the search space, accessing the list of conditions for the Centers for Disease Control and prevention (CDC), type and duration of precaution, plus additional information on each condition (if any).

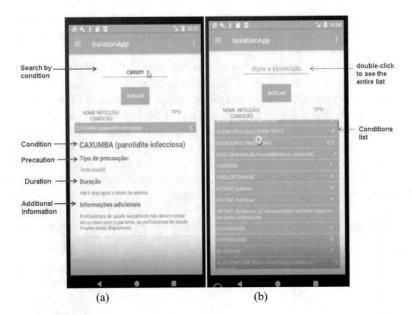

Fig. 4. Search interface (a) search for a specific condition/infection and (b) access to the list of conditions/infections.

Figure 5 presents the interface that provides guidance on each precaution type.

Fig. 5. Interface with guidelines on the types of precautions.

Figure 6 presents interfaces on hand hygiene, which provides information on when, how and why to hand hygiene during health care, bibliographical references, which contains links with scientific evidence used to support the guidelines provided by the prototype.

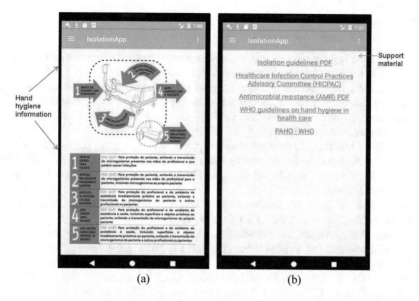

Fig. 6. (a) Interface with information on hand hygiene and (b) Interface with national and international guidelines links on standard precautions and based on transmission routes.

5 Evaluation

Prototype evaluation was conducted through a questionnaire, applied to health professionals with expertise in infection control. The questionnaire addressed the following characteristics: Functional Adequacy, Accuracy, Reliability, Usability, Apprehensibility, Operationality, Layout, Clarity, Aesthetics, Sensitivity to time, Efficiency, Accuracy, Consistency and Information Quality [12]. The participants evaluated the application positively, this suggest the prototype may support healthcare workers, facilitating decision making at the point of care.

6 Conclusions and Future Work

This article describes the development of a mobile application that can be updated and customized according to the literature and epidemiological institution profile where it will be used. The tool is informative, offers guidelines on standard precautions and based on transmission routes and presents the content in a clear and easy to understand. The application, therefore, can be used during the patient care offering information based on scientific evidence to health professionals and students, in any place and at any time. One point to emphasize in the work developed is the novelty, since none was found article on the development of a mobile application that provides information on standard precautions and based on transmission routes for health professionals of various categories, or national application for this purpose, after a search conducted on the web until July 2017.

References

1. World Health Organization, mHealth: New horizons for health through mobile Technologies, Global Observatory for eHealth series, vol. 3 (2011)
2. Cazella, S.C., Feyh, R., Ben, Â.J.: A decision support system for medical mobile devices based on clinical guidelines for tuberculosis. In: Ramos, C., Novais, P., Nihan, C., Corchado Rodríguez, J. (eds.) Ambient Intelligence - Software and Applications. Advances in Intelligent Systems and Computing, vol. 291, pp. 217–224 (2014)
3. Schnall, R., Iribarren, S.J.: Review and analysis of existing mobile phone applications for healthcare-associated infection prevention. Am. J. Infect. Control **43**, 572–576 (2015). Author, F.: Contribution title. In: 9th International Proceedings on Proceedings, pp. 1–2. Publisher, Location (2010)
4. Patrick, M.R., Hicks, R.W.: Implementing AORN recommended practices for prevention of transmissible infections. AORN J. **98**(6), 610–624 (2013)
5. Harding, A.D., Almquist, L.J., Hashemi, S.: The use and need for standard precautions and transmission-based precautions in the emergency departament. J. Emerg. Nurs. **7**(4), 367–373 (2011)
6. Banach, D.B., Bearman, G., Barnden, M., Hanrahan, J.A., Leekha, S., Morgan, D.J., et al.: Duration of contact precautions for acute-care settings. Infect. Control. Hosp. Epidemiol. **39**(2), 127–144 (2018)
7. Da Collina, G.A., et al.: Oral hygiene in intensive care unit patients with photodynamic therapy: study protocol for randomised controlled trial. Trials **18**, 385 (2017). PMC. Web, 20 March 2018
8. Siegel, J.D., Rhinehart, E., Jackson, M., Chiarello, L.: Guideline for isolation precautions: preventing transmission of infectious agents in healthcare settings. CDC, Atlanta (2007)
9. Darawad, M.W., Al-Hussami, M.: Jordanian nursing students' knowledge attitudes towards, and compliance with infection control precautions. Nurs. Educ. Today **33**, 580–583 (2013)
10. Russel, C.D., Young, I., Leung, V., Morris, K.: Health care worker's decision-making about transmission-based infection control precautions is improved by a guidance summary card. J. Hosp. Infect. **90**, 235–239 (2015)
11. Free, C., Phillips, G., Galli, L., Watson, L., Felix, L., et al.: The effectiveness of mobile-health technology-based health behaviour change or disease management interventions for health care consumers: a systematic review. PLoS Med. **10**(1), e1001362 (2013) https://doi.org/10.1371/journal.pmed.1001362. Accessed 20 Apr 2017
12. Iribarren, S.J., Schnall, R., Stone, P.W., Carballo-Diéguez, A.: Smartphone applications to support tuberculosis prevention and treatment: review and evaluation. JMIR Mhealth Uhealth **4**(2), e25 (2016)
13. Tibes, C.M.S., Dias, J.D., Zem-Mascarenhas, S.H.: Mobile applications developed for the health sector in Brazil: na integrative literature review. Rev. Min. Enferm. **18**(2), 471–478 (2014)
14. Tortorella, R.A.: A mobile context-aware medical training system for the reduction of pathogen transmission. Smart Learn. Environ. **4**, 4 (2017). https://doi.org/10.1186/s40561-017-0043-9

Vessel Trajectories Outliers

Tomás Machado[1], Rui Maia[1], Pedro Santos[1], and Joao Ferreira[2(✉)]

[1] INOV, Lisbon, Portugal
[2] Information Sciences, Technologies and Architecture Research Center
(ISTAR-IUL), Instituto Universitário de Lisboa (ISCTE-IUL), Lisbon, Portugal
jcafa@iscte-iul.pt

Abstract. In this work we describe our first steps towards our H2020 project MARISA participation, where we intent to develop a tool-kit towards the identification of outliers in Vessel trajectories based on electronic data regarding position and time. These outliers can correspond to illegal activities that could be related with illegal immigration, drugs transshipment among others. We developed process tools that based on any electronic Vessel position systems, like Automatic Identification System (AIS) data, it is possible to extract routes in an unsupervised approach. At the same time identify non-conformities based on AIS data signal lost and to identify situation when two or more Vessels are approaching close to each other, called the rendezvous.

Keywords: Vessel · Trajectory · AIS · Outliers · Rendezvous

1 Introduction

Approximately 90% of the global trade being carried by the international shipping industry, turning the Ocean vital for the World's economy. Nowadays there are approximately 50,000 merchant ships trading internationally, and with the current demand this number tends to increase[1]. Although this efficient way of transportation presents threats that prevail in the maritime domain (i.e. piracy, trafficking of drugs, illegal immigration, arms proliferation, illegal fishing etc).

Automatic Identification System (AIS) is an automated tracking system, that broadcasts information via mobile maritime radio band aiding Vessels in collision avoidance. Imposed by the International Maritime Organization (IMO) every passengers ship must be equipped with an AIS device. Autonomously broadcast AIS messages contain Vessels kinematic information (including ship location, speed, heading, etc.) and static information (including ship name, ship Maritime Mobile Service Identity (MMSI), etc.), which can be transformed into useful information for maritime traffic manipulations (e.g. Vessel path prediction and collision avoidance). The introduction of AIS in the maritime domain, increased the volume of Vessel trajectory data exponentially, making human

---

[1] ICS Shipping and World Trade, www.ics-shipping.org/shipping-facts.

© Springer Nature Switzerland AG 2019
P. Novais et al. (Eds.): ISAmI 2018, AISC 806, pp. 247–255, 2019.
https://doi.org/10.1007/978-3-030-01746-0_29

analysis and evaluation of this data extremely inefficient. Therefore, new effective ways of automatically mining this data show a great contribution for the future of nautical surveillance. However, mining maritime trajectory data present several challenges, such as: (1) Maritime trajectory data possess the data uncertainty typical of moving object trajectories. Geo-referenced locations of trajectory positioned by location sensing techniques may be collected with spatial uncertainty due to computational error and signal loss or degradation associated with the positioning device, [1]; (2) maritime traffic is not constraint to roads; Vessels are free to navigate in open waters, increasing the complexity of anomaly detection drastically.

Vessel Anomalous Behavior can be subdivided into: (1) Kinematic behaviors relate to the motion of ships including the routes taken and speed of travel; (2) AIS transmission behaviors include the switching on or from AIS systems and changing the Vessels name or other details, [2].

Current work is under an H2020 project MARISA - Maritime Integrated Surveillance Awareness[2], where is created a tool-kit that provides a suite of services to correlate and fuse various heterogeneous and homogeneous data and information from different sources, including Internet and social networks. In the context of this project current work provides an analysis and synthesis of the traffic spatio-temporal data streams provided by the AIS cooperative self-reporting system requires a suitable degree of automation and efficiency to detect and characterize inconsistencies, anomalies, ambiguities and ultimately transform this information into usable and actionable knowledge. It provides the activity at sea as contextual information and patterns of life, referred as maritime routes and summarizes the maritime traffic over a given period of time and a given area.

2   State of Art

Vessel Behavior is as considered as a baseline in which abnormal behavior can be found. This baseline can occur as normal trajectories are various and constant, producing a normalcy model of Vessels dynamics, that Machine Learning Techniques can learn. A vast number of frameworks in which Vessel behaviour analyses with the purpose of anomaly detection are fully defined as integrated systems. The authors in [3], suggested the framework MT-MAD (Maritime Trajectory Modeling and Anomaly Detection), in which a given set of moving objects, the most frequent movement behaviour are explored, evaluating a level of suspicion hence detecting anomalous behaviour.

TREAD (Traffic Route Extraction and Anomaly Detection) is a framework in which an Unsupervised Route Extraction is used to create a statistical model of maritime traffic from AIS messages, in order to detect low-likelihood behaviours and predict Vessels future positions, [4].

A framework focused on Vessel interaction and Rendezvous, is proposed by [5], which uses a logically connected 3-phase process, reducing the volume

---

[2] Marisa Project - www.marisaproject.eu.

of data that is processed by the sub-sequential phases, therefore prioritizing critical scenarios, that request human intervention.

A Partition-and-Detect framework, in which trajectories are partitioned into a two-level of granularity, achieving high efficiency or high quality trajectory partitions is proposed in, [1]. Accessing the performance of these frameworks, is an arduous task, as there is no defined benchmark labelled sets where test assessment can be performed, [6]. Although a solution for constructing an AIS database, with the potential value for being used as benchmark database for maritime trajectory learning, is proposed in [7].

As the volume of positional AIS data exponentially increases, it's important to find methods in which raw trajectories data can generate knowledge.

Trajectory learning is the process of learning motion patterns from trajectory data using unsupervised techniques, mainly clustering algorithms [8]. Morris and Trivedi [9], further categorize trajectory learning as a three-step procedure: (1) Trajectory Preprocessing; (2) Trajectory Clustering; and (3) Path Modelling. Pallotta proposed a method that enriches the raw Vessels trajectories with a description of the ship movements, labeling the raw trajectories as 'Stationary' or 'Sailing' [4].

A way to discretize a trajectory is to represent a trajectory into a spatial grid in which a cell represents a geographical area with a defined size. Analysisng this grid is allows a effective discover of frequent regions, [3]. The authors in, [9], propose a method to dynamically analysing a trajectory, with the emphasis on the learning of AP (Activity Path) and the discovery of POIs (Point of Interest), which can indicate common Vessel destinations (e.g. frequent fishing zones, ports, etc.).

3   Methodology

Figure 1, represents our development defined as a process, towards the contributions specified above. Our process is partitioned into 3 major parts, Data preprocessing, Route extraction and Anomaly Detection.

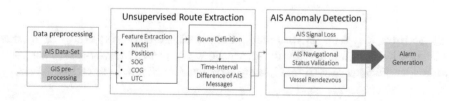

Fig. 1. Implemented process towards our goals of Vessel Rendezvous and AIS signal loss.

3.1   Data Preprocessing

The open-source AIS data type file is a raw database file(.dbf), that permits low efficiency, data manipulations. Thus, it becomes important to transform this data format into a more efficient Data Wrangling format. By the means of an open-source Geographic Information System[3], we transformed the raw data into csv format. As dbf to csv transformations are time consuming, and a decent sized data-set is achieved with just one area. The chosen area 10, represents data whose longitude is from -120 to -126 and latitude is from 30 to 50 .

3.2   Unsupervised Route Extraction

We developed a efficient method for Unsupervised Route Extraction, based on [4], work. Our method can be fed with either a AIS Data Stream, an AIS Data Base, or Unprocessed AIS Data Base.

Feature Extraction. By extracting only the relevant features, related to the Vessels Kinematics, and Navigational Status, we are able to reduce significantly the amount of data that we processed. Thus, increasing the effectiveness of our Anomaly Detection methods.

Route Definition. A simple, but not as effective way to represent a trajectory, as more effective ways are proposed in the literature, is to represent a trajectory with no compression. This can be easily achievable as Vessel are identified with unique ID, and are obliged to broadcast their AIS information in semi-continuous rates. Thus, we consider a Vessel Trajectory as the aggregation of all its broadcasts, defined as a set of multidimensional-points represented as:

$$TR_{MMSI} = p1, p2, p3, p4, \cdots, pn \tag{1}$$

where each multidimensional point p is defined as:

$$p = [t, x, y, SoG, CoG] \tag{2}$$

Time-Interval AIS Broadcast. Our, definition of Trajectory, is further enriched by determining the difference between every t, t_{-1}, transmission, this allows an efficient way to manipulate data, so AIS Signal Loss can be found.

3.3   AIS Anomaly Detection

Vessel anomaly detection is a research field which poses an immense level of complexity. We constraint our work, to some Vessel anomalies, which are described as requirements in the MARISA project.

---

[3] QGIS Geographic Information System, http://qgis.osgeo.org.

AIS Navigational Status Validation. AIS Navigational Status, describes the Vessel periodic navigational activity, according a set of static Status such as: under-way by engines, at anchor, moored, aground, engaged in fishing, underway by sail, etc. A detailed list and description all this status is found in[4].

AIS Signal Loss. Blocking AIS is a process that Vessels responsible use, to hide their position, as an anti-piracy defense, or possibly for illegal activities. This is done by simply turning off AIS equipment or by block antenna signal. We developed an real-time Heuristic based process that generates alarms, for Vessels that do not broadcast AIS information, for a period longer, than a set threshold.

Vessel Rendezvous. A requirement imposed by the MARISA project was the development of services, able to detect and generate alarms when two or more Vessels are approaching close to each other. In the maritime world this can be considered as a possible anomaly, which is called rendezvous. The concept of rendezvous in the Maritime world is quite complex, as there are numerous legislation. For this work, and because the emphasis is on the alarm generation, a simplification of this Vessel interaction is assumed. Thus, Vessel Rendezvous is considered as the interception or closeness of two or more Vessels, in a configurable Time Period.

3.4   AIS Data-Set

The sources of AIS data our work, are derivative from two different types: (1) Public sources this includes data from Australia[5] and USA[6]; and (2) Confidential AIS and S-AIS, data sources stemming from Military Forces via the MARISA project. From the USA open-source data, we created a Data-Set of 1659 different Vessels, representing approximately 12,3 Million AIS broadcast, representative of the year 2017. Although, for the Experiments conducted in Section X, we have used a small subset composed of 38 Fishing Vessels, which represents approximately 280.000 AIS messages (Table 1).

4   Experience

AIS Navigational Status Validation. In our experiment, we discovered that a large number of Vessels do not keep their Navigational Status updated. A plausible cause; as the Navigational Status is manually introduced and updated, it can lead to expected Human error. Nevertheless, the wrong use of a Navigational Status can lead to fines by the Maritime Authorities, thus being considered an Anomaly (Table 2).

---

[4] Solas Chapter V Annex 17 AIS - www.mcanet.mcga.gov.uk.

[5] Vessel Tracking Data, www.amsa.gov.au/Spatial/DataServices/DigitalData.

[6] MarineCadastre Vessel Traffic Data, www.marinecadastre.gov/ais.

Table 1. Example of AIS data transmitted by Vessel, MMSI: 636081210; first 4 transmissions from a total of 6256 transmissions.

MMSI	X	Y	SOG	COG	Time
636081210	−125.993218	48.355773	14.3	73.300003	2014-02-27 13:33:02
636081210	−125.985303	48.357340	14.5	73.800003	2014-02-27 13:34:23
636081210	−125.979437	48.358500	14.6	73.099998	2014-02-27 13:35:23
636081210	−125.973353	48.359692	14.7	73.000000	2014-02-27 13:36:25

Table 2. 32 Vessel subset AIS Navigational Status description.

AIS Navigational Status (Number - Name)	Count	%
0 - Under way using engine	150465	54
15 - Default	97264	35
7 - Engaged at Fishing	24444	9
5 - Moored	5462	2
8 - Under way Sailing	2138	0,7

By determining the actual movement characteristics of a Vessel, we are able to deduce if either a Vessels is Stopped or Moving. This simple extrapolation permits the validation, of most used Navigational Status, we further our Stopped/Moving method with the detection of AIS Signal Loss. In this work we do not Validate if a Vessel is Engaged at Fishing, as it presents a certain level of uncertainty, and cannot be solved with a simple Stopped/Moving validation, a possible solution for this case is presented by the authors in [10].

Vessel Rendezvous. Algorithm 1, describes the implemented process in a algorithm way, using as input distance D_{tresh}, and a interval time-group T_g. For every single Vessel Track, the Track is partitioned into defined time-groups (e.g. a T_g of 5 min). If two or more Vessels are on the same T_g, the distance between these Vessels is calculated, using the Haversine, (3). If this distance is less than the D_{tresh}, an alarm is generated for these two Vessels.

Haversine Distance is used, as latitude and longitude features are represented in a spherical coordinate system, using on the following equation, as d represents the distance between the 2 points:

$$d = 2rsin^{-1}(\sqrt{sin^2(\frac{lat_2 - lat_1}{2}) + cos(lat_1)cos(lat_2)sin^2(\frac{long_2 - long_1}{2})}) \quad (3)$$

5   Results

Results evaluation were based on two developed process: (1) AIS Navigational Status Validation; and (2) Vessel Rendezvous.

Algorithm 1. Vessel Rendezvous Method

Input   : A set of defined AIS routes, S_{tr}; Time-Group, T_g; and a Distance
threshold, D_{tresh};
Output: A set of two or more Vessels with a distance in NM and Time,
$RES : [MMSI_{set}; Timestamp; D_{NM}]$;
1 **Initiation:** Partition S_{tr}, into sub-groups SG_{Ttr} ,defined by T_g.
2 **foreach** SG_{Ttr} in S_{tr} **do**
3   **if** SG_{Ttr} contains more than 2 routes **then**
4   Calculate HarversineDistance in NM over all $(SG_{Ttr})C2$; **if**
             HarversineDistance $> D_{tresh}$ **then**
5                $RES = [MMSI_{set}; Timestamp; D_{NM}]$
6   **end**
7   **end**
8 **end**

From AIS Navigational Status validation experience, we were able to deduct
that Vessels tend to not update their Navigational Status. Thus, by labeling the
AIS data, with Stopped and Moving labels, and we are able to determine how
many mislabeled Broadcast were sent (Table 3).

Table 3. Results for AIS Navigational Status and AIS Signal Loss occurrence count,
with 32 Vessel subset

Broadcast AIS Status	Total count	Moving count	Stopped count	Error count - %	Error %
0 - Under way using engine	150465	54734	95731	95731	64
5 - Moored	5462	1038	4424	1038	19
8 - Under way Sailing	2138	1118	1020	1020	47
15 - Default	97264	57616	39541	*	*
AIS Signal Loss	411	-	-	-	-

The accuracy for the 15 - Default Navigational, cannot be measure as the
other Statuses, this Status represents that Vessels have kept the default AIS
Status for the whole Trajectory. Thus, our results presents the number of AIS
Default Status, for Moving and Stopped kinematics characteristics.

Vessel Rendezvous test was conducted using a 2-min time group, and 1,5
Nautical Miles as distance Threshold. Table 4, represents the collection of alarms
generated, after conducting the Vessel Rendezvous test in the 32 routes sub-set.
The table shows 10 occurrences, ordered by distance, from 4 different Vessels.
Multiple occurrences can occur from the same two Vessel routes if these Vessels
were close to each other more than the defined time-group parameter.

In Fig. 2, the two Vessels, MMSI 413104010 and 432000385, are represented.
These Vessels represent the Vessels that were close to each other, for the longest
time-period, resultant from the Vessel Rendezvous test. Although, while Vessels

Table 4. Results obtained for Vessel Rendezvous, with the 32 Vessel subset.

TimeSlot	MSSI A	MMSI B	Distance in NM
2014-02-15 11:32:00	357103080	370000802	1.012450
2014-02-01 21:30:00	413104010	432000385	1.160439
2014-02-01 21:36:00	413104010	432000385	1.160971

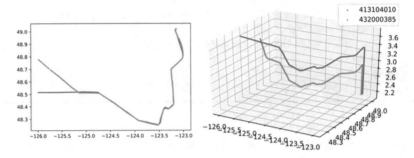

Fig. 2. Routes of MMSI: 413104010 and 432000385; axes representing (lat.,long.)[Left] and (lat.,long.,time)[Right]

are able navigate freely in open waters, it is common that certain routes are shared by Vessels, which can be caused due Maritime traffic. Which could, make normal the fact the Vessels were close to each other for the whole route. Although what could be considered abnormal, is the fact that when the Vessels are the closest to each other's, is the moment in which the Vessels change directions, which is shown in Fig. 2(Right).

6   Conclusion

In the current research work, we developed tools to handle Vessel electronic data, mainly oriented to AIS and extract Vessel routes. From these routes, missing positions were identified and rendezvous situations. These processes generate alerts towards responsible control entities with the mission of checking these possible non-conformities in real-time. Until present moment, we are only able to validate Navigational Scenarios based on Stopped or Moving kinematics, the next step is the identification of fishing and constrained by her draught real scenarios, and further validation with MARISA marine partners.

References

1. Lee, J.-G., et al.: Trajectory outlier detection: a partition-and-detect framework. In: 2008 IEEE 24th International Conference on Data Engineering (2008)
2. Lane, R.O., et al. Maritime anomaly detection and threat assessment. In: 2010 13th International Conference on Information Fusion (2010)

3. Lei, P.-R.: A framework for anomaly detection in maritime trajectory behavior. Knowl. Inf. Syst. **47**(1), 189–214 (2015)
4. Pallotta, G., et al.: Vessel pattern knowledge discovery from AIS data: a framework for anomaly detection and route prediction. Entropy **15**(12) (2013)
5. Shahir, H.Y., et al.: Maritime situation analysis: a multi-vessel interaction and anomaly detection framework. In: 2014 IEEE Joint Intelligence and Security Informatics Conference (2014)
6. Laxhammar, R.: Anomaly detection for sea surveillance. In: 2008 11th International Conference on Information Fusion (2008)
7. Mao, S., et al.: An automatic identification system (AIS) database for maritime trajectory prediction and data mining. In: Proceedings in Adaptation, Learning and Optimization, Proceedings of ELM-2016 (2017)
8. Le Guillarme, N., Lerouvreur, X.: Unsupervised extraction of knowledge from S-AIS data for maritime situational awareness. In: Proceedings of the 16th International Conference on Information Fusion (2013)
9. Morris, B.T., Trivedi, M.M.: A survey of vision-based trajectory learning and analysis for surveillance. IEEE Trans. Circuits Syst. Video Technol. **18**(8), 1114 (2008)
10. Natale, F., et al.: Mapping fishing effort through AIS data. Plos One **10**(6) 2015

Testing a New Methodology for Accelerating the Computation of Quadratic Sample Entropy in Emotion Recognition Systems

Arturo Martínez-Rodrigo[1]([✉]), Beatriz García-Martínez[2],
Antonio Fernández-Caballero[1,3], and Raúl Alcaraz[1]

[1] Instituto de Tecnologías Audiovisuales,
Universidad de Castilla-La Mancha, 16071 Cuenca, Spain
Arturo.Martinez@uclm.es
[2] Instituto de Investigación en Informática,
Universidad de Castilla-La Mancha, 02071 Albacete, Spain
[3] Centro de Investigación Biomédica en Red de
Salud Mental (CIBERSAM), Madrid, Spain

Abstract. Emotion recognition has become an important area of study for the development of human-machine interfaces able to recognize and interpret human emotions. In order to construct emotional systems, signals from physiological variables have to be registered and processed rapidly to provide a fast emotional response from the computer system to the user. In this regard, several studies have claimed that nonlinear methodologies applied to electroencephalographic signals can provide relevant information about emotions recognition. However, given the multimodal nature and nonlinear behaviour of that signals, the data processing is often very slow to give a fast response, producing an important delay between feeling an emotion and receiving the adequate response from the emotional system. In order to overcome this difficulty, this work computes a modification of quadratic sample entropy accelerating the computation by exploiting vectors with dissimilarity.

Keywords: Electroencephalography · Quadratic sample entropy Analysis · Vectors with dissimilarity

1 Introduction

Today, machines are characterized by a lack of capability to modulate their actions according to the human's mood. Consequently, affective computing has emerged as a technological paradigm to provide computers the ability to feel, infer and interpret human emotions [1]. Recognizing emotions from different human activities may help to develop better human-machine interfaces (HMIs) and perform adequate responses from computers depending on the human's affective reactions. Considering the massive use of HMIs in the society, it seems

© Springer Nature Switzerland AG 2019
P. Novais et al. (Eds.): ISAmI 2018, AISC 806, pp. 256–264, 2019.
https://doi.org/10.1007/978-3-030-01746-0_30

appropriate to make efforts in developing advanced computational models to endow machines with a more humanized behavior.

Developing emotional models demands an intensive mathematical computation because of the heterogeneity of the different processes conforming the emotional system. Indeed, multimodal physiological signals are firstly monitored, amplified and discretized before undergoing to thorough signal processing. In this regard, different variables related to the autonomous nervous system can provide relevant information about the human emotional state. Some of the most analyzed physiological variables are the electro-dermal activity, electromyogram, electrocardiogram, or skin temperature, among others. However, the electroencephalography (EEG), which measures the electrical activity of the brain directly on the scalp, is the physiological variable that has reported the most promising results [2]. The main reason is that EEG signals represent the first bodily response against any external stimulus, while the rest of physiological variables are secondary effects of brain processes [3].

In this regard, different features involving temporal, statistic, frequency and complexity parameters have been used to quantify the EEG signals. However, neural processes are considered nonlinear and nonstationary [4]. Indeed, the application of nonlinear metrics for EEG signals assessment has provided new discoverings that the traditional linear techniques could not report [2]. For instance, quadratic sample entropy (QSE) has shown to be relevant discriminating between different emotional states [5]. However, nonlinear QSE metric requires a high computational burden because it compares arbitrary two vectors for distance calculation to determine whether the paired vectors are similar [6]. This fact is inconsistent with some specific applications like affective computing, where the reaction time from the machine should be short. Consequently, if the processing is not performed fast enough, the computation is forced to be executed off-line, preventing that emotional models can interact in a real-time mode.

To overcome this problem, some methods for accelerating nonlinear computations have been published in the last years. Thus, algorithms based on sliding k-dimensional tree [7], bucket-assisted methodology [8], recursive scheme [9] or the study of symmetrical distance [10] have been proposed. However, recently Yun et al. [11] described an easy-to-implement methodology to speed up the nonlinear computation, by exploiting vectors with dissimilarity (VDS). This algorithm discards the distance calculations of most of vectors because they were dissimilar as predetermined by the VDS decision. Moreover, VDS has shown a good performance with random signals and it has been tested on some of the most frequent nonlinear metrics [11]. However, to the best of our knowledge, it has been never applied using QSE in the context of emotions recognition. Indeed, considering that QSE is a slow and computationally expensive metric, it becomes an ideal candidate to test the performance of VDS technique. Therefore, the main objective of this work consists on testing the performance in terms of computation time, comparing the time required for calculating Conventional QSE (CQSE), and its accelerated version using VDS (QSE-VDS).

This paper is structured as follows. Section 2 describes the database and preprocessing techniques applied to the EEG signals, together with a brief definition of the QSE algorithm and VDS methodology, as well as how to assess the performance of both methods. Section 3 shows the results obtained during this study and finally, Sect. 4 presents a discussion of the results and some conclusions obtained.

2   Materials and Methods

2.1   Dataset Description and Preprocessing of EEG Raw Data

Taking into consideration that we pretend to assess the suitability of VDS in a real scenario, electroencephalographic signals used in this work were chosen from the Database for Emotion Analysis using Physiological Signals (DEAP) [12]. This publicly-available dataset contains a total of 1280 EEG recordings from 32 subjects (50% male, mean age 26.9 years) under different emotional states. EEG signals were acquired with a Biosemi ActiveTwo device at 512 Hz with 32 electrodes located over the scalp according to the international 10–20 system. Although this dataset contains samples from different emotional spaces, only 25 EEG recordings chosen randomly from the database have been used in this study to simplify the results.

Before performing the entropy computation on the recordings, EEG signals need to be filtered and preprocessed, such that only the brain electrical activity is maintained. In this case, signals were downsampled from 512 Hz to 128 Hz [12]. Moreover, two forward/backward high-pass and low-pass filters were applied at 3 and 45 Hz, respectively, to maintain the bands of interest in EEG spectrum [12]. Given those cutoff frequencies, no further actions were needed to subtract baseline and power line interferences.

2.2   Quadratic Sample Entropy and Vectors with Dissimilarity

CQSE metric consists on a regularity-based index that evaluates the rate of appearance of a pattern within a nonlinear and nonstationary time series [6]. Given a time series of N data points, $x(n) = \{x(1), x(2), \ldots, x(N)\}$, vectors $\mathbf{X}_m, \ldots, \mathbf{X}_m(N-m+1)$ of size m are defined by $\mathbf{X}_m(i) = \{x(i), x(i+1), \ldots, x(i+m-1)\}$, for $1 \leq i \leq N - m$. Repetitiveness of vectors $\mathbf{X}_m$, represented as the distance between them, is calculated as the maximum absolute distance between scalar components:

$$d[\mathbf{X}_m(i), \mathbf{X}_m(j)] = \max_{k=0,\ldots,m-1} \left(|x(i + k) - x(j + k)| \right). \tag{1}$$

Hence, if the distance is lower than a threshold r, then vectors $\mathbf{X}_m(i)$ and $\mathbf{X}_m(j)$ are considered similar. After that, self-matches are excluded to compute $B_i^m(r)$, which corresponds to the number of vectors that are similar to $\mathbf{X}_m(i)$.

Then, $B^m(r)$ is quantified as the probability of having two sequences matching for m points:

$$B^m(r) = \frac{1}{N-m} \sum_{i=1}^{N-m} \frac{B_i^m(r)}{N-m-1}. \tag{2}$$

If the same process is repeated for vectors of length $m+1$ samples, the probability of having two sequences matching for $m+1$ points is obtained. Then, CQSE is estimated as

$$CQSE(m,r,N) = -\ln\left(\frac{B^{m+1}(r)}{B^m(r)}\right) + \ln(2r), \tag{3}$$

which determines the probability that two patterns match for m and for $m+1$ points.

In the previous mathematical process, the most time-consuming step is the distance calculation of $d[\mathbf{X}_m(i), \mathbf{X}_m(j)]$. Therefore, optimizing this step is the most important stage for accelerating the overall QSE computation. Thus, VDS focuses on setting a decision method to decide if the two vectors being compared are dissimilar, and can be discarded from the distance computation. Shortly, two arbitrary vectors of dimensions m are considered dissimilar if their distances are greater than r. Hence, a decision expression D to determine if two vectors are dissimilar can be defined as

$$D(i,j) = \left| \sum_{k=0}^{m-1} x_{i+k} - \sum_{k=0}^{m-1} x_{j+k} \right|, \tag{4}$$

such that when D is greater than $m \times r$, the corresponding paired vectors $\mathbf{X}_m(i)$ and $\mathbf{X}_m(j)$ are considered as dissimilar and discarded from the QSE-VDS computation.

2.3 Performance Assessment

Twenty-five EEG signals of 60 s-length (7680 samples) were divided into twelve equal-size non-overlapped windows of $N = 5$ s (640 samples). Considering that we pretend to evaluate the computation of EEG signals, data was treated like a first-in first-out data buffer, emulating the arrival of data within the system. It is worth noting that EEG has a multimodal behavior, so 32 channels were evaluated simultaneously. Figure 1 shows how the EEG data were partitioned and processed for each temporal window. First, a five-second window of EEG recording containing data from every channel was considered. Then, CQSE and QSE-VDS with $m = 2$ and $r = 0.25$ times the standard deviation of the data were computed for each channel within the considered window. This process was repeated for each window and, finally, entropy values computed for each channel and for both methodologies were averaged and stored in data vectors E_{CQSE} and $E_{QSE-VDS}$, respectively.

Similarly, time costs were measured to estimate the computation speed of CQSE and QSE-VDS. To perform this calculation, a stopwatch was started

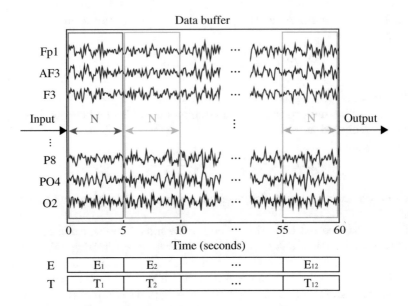

Fig. 1. First-in first-out data buffer emulation.

when each algorithm executed the first iteration at the beginning of first channel (Fp1) and it was stopped when the computation finished processing the QSE value of the last EEG channel (O2). Then, the elapsed time was calculated as the difference between both times, and it was stored in time vectors T_{CQSE} and $T_{QSE-VDS}$, respectively. It is important to remark that all the analyses were performed on a Dell XPS 13 9360 computer with an Intel Core-i7-7560U 2.4 GHz processor, 16 GB of RAM, and a 64-bit Windows 10 Home operating system.

3   Results

The average of entropy values contained in E_{CQSE} and $E_{QSE-VDS}$ was computed, such that a single final value of entropy was provided for each EEG recording, as can be observed in Table 1 (columns 2 and 4). It is worth noting that entropy values obtained for CQSE and QSE-VDS are equal, such that when cross-correlation operations between E_{CQSE} and $E_{QSE-VDS}$ were performed, a correlation index value of 100% was obtained in each case. This analysis was performed to corroborate that both methodologies calculate the same entropy value for each case, and using VDS technique is not altering the entropy results in any case.

Moreover, a global computation time was obtained by performing the sum of the partial times stored in vectors T_{CQSE} and $T_{QSE-VDS}$ for each EEG recording, as can be observed in Table 1 (columns 3 and 5). It is important to note that computation times of all the EEG signals were notably higher in CQSE than

in QSE-VDS, when data was analysed by means of ANOVA ($\rho = 2.47 \times 10^{-59}$).
This fact can also be seen in Fig. 2, which shows the total computation cost values
of CQSE (orange) versus QSE-VDS (green) for the 25 EEG signals under study.
In all the cases, the computation times when computing CQSE were around 25 s
(25.77 ± 0.92), whilst this time decreased around 20 s when computing QSE-VDS
(5.14 ± 0.19). This decrease implies a mean reduction of 5 times the calculation
time when computing CQSE.

Table 1. Entropy values and computational times obtained with CQSE and with
QSE-VDS for the 25 EEG signals

| EEG | Conventional QSE | | QSE-VDS | |
Signal	Entropy	Time (s)	Entropy	Time (s)
1	1.95	25.75	1.95	5.14
2	1.99	25.83	1.99	5.20
3	1.88	25.58	1.88	5.24
4	2.03	25.31	2.03	5.28
5	1.95	26.32	1.95	5.19
6	1.94	24.81	1.94	4.99
7	1.88	24.98	1.88	5.06
8	1.91	25.23	1.91	5.15
9	1.81	25.87	1.81	5.16
10	1.80	25.48	1.80	5.14
11	1.84	25.38	1.84	5.16
12	1.87	26.44	1.87	4.98
13	1.87	25.75	1.87	4.93
14	1.81	25.58	1.81	4.85
15	1.90	24.43	1.90	4.83
16	1.94	25.44	1.94	4.84
17	2.15	24.99	2.15	5.34
18	2.26	24.30	2.26	5.43
19	2.13	24.78	2.13	5.37
20	1.66	27.91	1.66	5.46
21	2.14	27.64	2.14	5.37
22	1.66	27.13	1.66	5.24
23	1.84	26.62	1.84	4.95
24	1.92	26.52	1.92	4.93
25	1.83	26.26	1.83	5.37

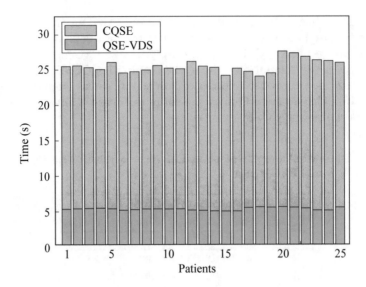

Fig. 2. CQSE and QSE-VDS computation time costs for every EEG recording.

4   Discussion and Conclusions

In this work, a new method proposed recently for accelerating the computation of some entropy metrics has been tested for the first time on the well-known quadratic sample entropy to quantify the regularity of EEG signals. In previous works, CQSE has been applied in the emotional research field, reporting a notable performance discriminating among some emotions [2]. However, CQSE is characterized by its slowness computing the entropy values because it assesses the distance among two arbitrary vectors in a multitude of iterations. This feature make almost impossible its use in real-time applications where the reaction time from the emotional system is critical for an acceptable user experience. Moreover, this drawback is further emphasized if the emotions system is embedded in a limited system with low performance.

To overcome this limitation, a modification of CQSE has been carried out discarding dissimilar vectors to avoid a high amount of calculations, and consequently accelerating the computation [11]. Indeed, as shown in Table 1, computing times obtained when using QSE-VDS are notably lower than when CQSE is utilized. It can also be observed in Fig. 2 that the degree of time reduction is similar for each one of the EEG signals and it is maintained throughout all the analysis. Indeed, although signals with equal duration have been used for testing both methodologies, CQSE experienced a higher variability in the computing time results (std = 0.92), whilst QSE-VDS shows a more stable bahaviour (std = 0.19).

Finally, it is worth noting that both methodologies obtained the same entropy results, regardless of the computation time. Moreover, the EEG signals used for the evaluation have been recorded specifically for assessing different emotional states. In this sense, QSE-VDS has resulted to be very efficient, given that it can be computed in a fifth part of the time required for calculating CQSE. Note that while the time required by CQSE to compute a single window is around 6 s, the time needed by QSE-VDS to process a window is around 0.4 s. Therefore, the proposed modification could help to accelerate affective predictive models based on CQSE calculation opening the door to implement emotions systems working in a real-time fashion.

Acknowledgements. This work was partially supported by Spanish Ministerio de Economía, Industria y Competitividad, Agencia Estatal de Investigación (AEI)/European Regional Development Fund (FEDER, EU) under DPI2016-80894-R and AEI TIN2015-72931-EXP grants, and by the Centro de Investigación Biomédica en Red de Salud Mental (CIBERSAM) of the Instituto de Salud Carlos III. Beatriz García-Martínez holds FPU16/03740 scholarship from Spanish Ministerio de Educación, Cultura y Deporte. Arturo Martínez-Rodrigo holds EPC 2016-2017 research fund from Escuela Politécnica de Cuenca, Universidad de Castilla-La Mancha, Spain.

References

1. Picard, R.W.: Affective Computing. MIT Press, Cambridge (1995)
2. García-Martínez, B., Martínez-Rodrigo, A., Zangróniz, R., Pastor, J.M., Alcaraz, R.: Symbolic analysis of brain dynamics detects negative stress. Entropy **19**(5), 196 (2017)
3. Jenke, R., Peer, A., Buss, M.: Feature extraction and selection for emotion recognition from EEG. IEEE Trans. Affect. Comput. **5**(3), 327–339 (2014)
4. Abásolo, D., Hornero, R., Gómez, C., García, M., López, M.: Analysis of EEG background activity in alzheimer's disease patients with lempel-ziv complexity and central tendency measure. Med. Eng. Phys. **28**(4), 315–22 (2006)
5. García-Martínez, B., Martínez-Rodrigo, A., Zangróniz Cantabrana, R., Pastor García, J., Alcaraz, R.: Application of entropy-based metrics to identify emotional distress from electroencephalographic recordings. Entropy **18**(6), 221 (2016)
6. Richman, J.S., Moorman, J.R.: Physiological time-series analysis using approximate entropy and sample entropy. Am. J. Physiol.-Hear. Circ. Physiol. **278**(6), H2039–49 (2000)
7. Pan, Y.-H., Wang, Y.-H., Liang, S.-F., Lee, K.-T.: Fast computation of sample entropy and approximate entropy in biomedicine. Comput. Methods Programs Biomed. **104**(3), 382–396 (2011)
8. Manis, G.: Fast computation of approximate entropy. Comput. Methods Programs Biomed. **91**(1), 48–54 (2008)
9. Shimizu, S., Sugisaki, K., Ohmori, H.: Recursive sample-entropy method and its application for complexity observation of earth current. In: 2008 International Conference on Control, Automation and Systems, ICCAS 2008, pp. 1250–1253. IEEE (2008)
10. Bo, H., Fusheng, Y., Qingyu, T., Chan, T.-C.: Approximate entropy and its preliminary application in the field of EEG and cognition. In: 1998 IEEE Proceedings of the 20th Annual International Conference of the Engineering in Medicine and Biology Society, vol. 4, pp. 2091–2094. IEEE (1998)

11. Yun, L., Wang, M., Peng, R., Zhang, Q.: Accelerating the computation of entropy measures by exploiting vectors with dissimilarity. Entropy **19**(11), 598 (2017)
12. Koelstra, S., Mühl, C., Soleymani, M., Lee, J.-S., Yazdani, A., Ebrahimi, T., Pun, T., Nijholt, A., Patras, I.: DEAP: a database for emotion analysis using physiological signals. IEEE Trans. Affect. Comput. **3**(1), 18–31 (2012)

Picking Cubes: A Rehabilitation Tool for Improving the Rehabilitation of Gross Manual Dexterity

Miguel A. Teruel[1](✉) ⓘ, Ana de los Reyes-Guzmán[2],
Juan Villanueva[1], Vicente Lozano-Berrio[2],
María Alvarez-Rodríguez[2], Silvia Ceruelo-Abajo[3], Elena Navarro[1] ⓘ,
and Pascual Gonzalez[1] ⓘ

[1] LoUISE Research Group, Computing Systems,
Univ. of Castilla-La Mancha, Albacete, Spain
miguel@dsi.uclm.es, juan.villanueva@alu.uclm.es,
{elena.navarro,pascual.gonzalez}@uclm.es
[2] Biomechanics and Technical Aids Unit,
National Hospital for Paraplegics, Toledo, Spain
adlos@sescam.jccm.es
[3] Rehabilitation Department, National Hospital for Paraplegics, Toledo, Spain

Abstract. After Spinal Cord Injury the loss of motor function, and mainly of the upper limb, is a sharp limitation that difficult people in performing many Activities of Daily Living essential for their autonomy. The Box and Block Test is one the most relevant clinical scale for the rehabilitation of the gross manual dexterity. However, the human-supervision of rehabilitation can introduce errors. For this aim, in this paper we present a tool that simulates this test by using a virtual environment, where patients can grasp a cube with their dominant hand, transport the block and release it into the opposite compartment. For controlling the hand and fingers we use leap motion, a low-cost 3D tracking device that tracks users' palm and fingers. Moreover, in order to increase patients´ motivation, we have included additional feedback so that they become aware of their competence while carrying out their rehabilitation exercises.

Keywords: Upper limb rehabilitation · Box and Block Test · Virtual reality
Leap motion · Motivation

1 Introduction

Loss of motor function is a hallmark after Spinal Cord Injury (SCI). The incidence of SCI varies greatly worldwide from 12.1 to 57.8 SCI cases per million depending on the countries [1]. Among them, the upper limb (UL) is affected in more than 50% of cases [2]. People with UL problems are impaired to perform many Activities of Daily Living (ADL) essential for their autonomy. Therefore, these patients suffer more limitations on their level of activity and participation in social settings, than people who have suffered another central nervous system injury, such as stroke [3].

© Springer Nature Switzerland AG 2019
P. Novais et al. (Eds.): ISAmI 2018, AISC 806, pp. 265–273, 2019.
https://doi.org/10.1007/978-3-030-01746-0_31

In this context, the exploitation of quantitative measures of human movement quality is highly useful to the rehabilitation field. These measures must describe the outcomes of the rehabilitation treatments, discriminating clearly between healthy and pathological conditions [4], in order to provide the decision-making in clinical settings with valuable information that makes the most of the rehabilitation process. With this aim, we present a low-cost rehabilitation tool based on the well-known Box and Block Test [14], designed for people that have suffered spinal cord injury and their upper limbs' movement is affected. The tool can be used without the supervision of a therapist, which eliminates the possibility to introduce errors in measuring how the rehabilitation process is going, and the tool itself encourages the patient to improve by providing them with feedback about their performance in the rehabilitation exercises in real time.

2 Upper Limb Rehabilitation

The upper limb function assessment is mainly based on clinical scales [5]. These scales are standardized measure instruments, valid for their use in clinical settings because they have been previously validated through large patients' samples. However, although these scales are easy to use, the main disadvantage of these quantitative measures is that they have a high subjective component, depending on the observer who visually scores the test. Another aspect to consider is the sensitivity of the scales mainly to gross changes in the health status or UL function, so that subtle changes in the subject may not be detected. Therefore, it is necessary to find measure instruments which, combined with clinical scales, provide objectivity, and at the same time, solve the limitations that such scales have.

Biomechanical studies are examples of objective methods, where several technologies can be used to collect data from the subjects [6]. One of these studies are the kinematic analysis that provide objective information about motor strategies associated with UL goal-oriented tasks. The main disadvantage of this equipment is that it is very expensive, so that their use is constrained to clinical settings while patients are supervised by qualified clinical staff [7]. In recent years, several low-cost 3D tracking systems have become commercial devices exploited for manipulating applications and videogames. Recently, several studies have combined applications of virtual reality with these devices for the treatment of a variety of impairments after injuries of the central nervous system [8, 9]. These studies aim at improving both the motivation and the interaction of the player with the virtual world by facilitating the immersion and imagination within the environment [10]. One of such systems is the Leap Motion Controller (LMC) [11]. Patients with stroke or SCI are a likely target population for such device for its use in treatment and rehabilitation of the UL function. There are previous evidences of its use for stroke patients [12, 13], yet studies for SCI are scarce till now.

The objective of this work is to present the development of *picking cubes*, a virtual serious game designed and developed for simulating the Box and Block Test [14], a clinical scale for the rehabilitation of the gross manual dexterity of spinal cord injured patients. Picking cubes has been designed considering two important features. First, it uses low cost technologies in order to offer rehabilitation to a wider population at reasonable cost. Second, it supports different aspects that look for increasing patients' motivation while they carry out their rehabilitation exercises.

3 Motivation for Appealing Repetitive Rehabilitation Tasks

As Holden [8] stated, repetition is relevant both for motor learning as well as for the cortical changes related to it. However, in order to achieve this motor learning process, the repeated practice of a physical task must go with information about its success regarding previous executions and feedback about performance success, both provided by some stimuli (e.g. visual, proprioceptive or aural). Moreover, motivation of the participants is critical when designing therapies that involve to repeat movements.

This need of motivation is not new as some studies [15] show. Motivation could be defined as "the general desire or willingness to do something" [16]. One of that theories about motivation is Self-Determination Theory [17] that defines three principles: competence, autonomy and relatedness. The motivation of a person is increased whenever he/she feels competent for the task being performed. Autonomy shows that a person increases his/her motivation when he/she feels certain degree of autonomy and, therefore, is able to perceive the cause of success or failure while performing a reha-bilitation task. Lastly, two principles, *relatedness state* defined as the feeling of belonging and *connectedness* as the connection with others, have been acknowledged that influence greatly to increase motivation. These principles are related to intrinsic and extrinsic motivation. The intrinsic motivation reflects "the inherent tendency to seek out novelty and challenges, to extend and exercise one's capacities, to explore, and to learn". On the other hand, extrinsic motivation refers "to the performance of an activity in order to attain some separable outcome". The extrinsic motivation comes from external sources, while intrinsic exists within the individual. In our case, we try to increase intrinsic motivation because intrapersonal events (e.g., rewards, communica-tions, and feedback) that cause feelings of competence during the rehabilitation can, in the end, enhance this type of motivation [18]. Hence, the design of picking cubes was carried out to provide patients with a proper feedback that shows their level of com-petence while carrying out a specific exercise.

As Flores noticed [19], another relevant element to take into account for increasing motivation is the use of games, in particular *serious games* that make use of virtual reality features. Therefore, the proposal presented in this work is a serious game that guides patients while carrying out a repetitive practice, offers an enhanced feedback about performance, and increases patients' motivation to endure practice. Finally, another relevant aspect, which considers our proposal for improving patients' moti-vation is the support for competitive features [20]. Specifically, each patient competes with his/herself as he/she knows and compares his/her current score with the previous ones

Taking into consideration these features and trying to increase the motivation of patients who suffer a spinal cord injury at the cervical level and affects the function of the upper limb, we have created picking cubes that is described in the next section.

4 Picking Cubes

In this section, the Picking Cubes game is presented. Hence, in Sect. 4.1 the game is described as well as the hardware used for its control. Besides, in Sect. 4.2 it is shown a complementary application used to analyze the players' results.

4.1 The Game

In general terms, the game consists in picking up several cubes with the player's dominant hand (left hand in the following examples) and to place them in the opposite part of the screen, i.e. the right side (Fig. 1). Such interaction is performed by using the players' own hands.

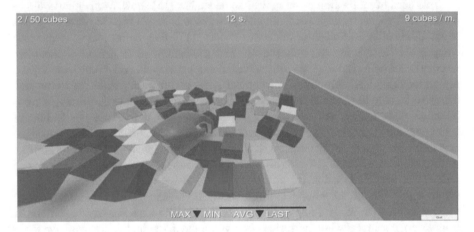

Fig. 1. Screenshot of picking cubes

In order to gather the players' hand movements, a hand tracker is used (Fig. 2). Concretely, we used a Leap Motion Controller as hand-tracking device [11]. This enables us to track players' hands and fingers thanks to its two monochromatic IR cameras and three infrared LEDs. It covers a hemispheric tracking area of around one meter.

The game can be configured depending on patients' needs, that is, it offers facilities to define the dominant hand, the number of cubes that patients will have to pick and the size of such cubes (Fig. 3). After that, the game starts so that the player can interact with the cubes. The game lasts until all the cubes have been put in the non-dominant-hand side area of the screen, or until the patient gives up.

It is worth noting that during the play, patients receive information about the current game and their progression regarding previous plays. As far as the current play is concerned (at the top of Fig. 1), the game shows the current number of moved cubes (score), the number of cubes they have to move to finish the game properly (goal), the elapsed time and the current speed as number of cubes moved per minute. Besides, regarding the current performance in comparison with previous games, it shows

Fig. 2. Player during a game session

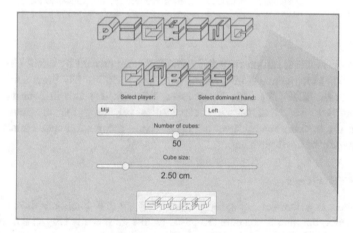

Fig. 3. Start screen

whether patient's current speed is higher than the speed of the player's best and worst, average and last speed of play. In the event of current speed being higher, a green triangle facing up would be shown along with each variable related to previous results (MAX, MIN, AVG and LAST) as shown at the bottom of Fig. 1. Moreover, in order to complete this information, a dynamic status bar is also shown to describe the relationship between patient's current speed and its previous best and worst game's speed. Therefore, this status bar's value changes real time, decreasing if the patient does not move new cubes or increasing when a new cube is moved.

When a play is finished, additional information is presented to the patient in a results screen (Fig. 4). More specifically, this screen shows the player's final score, goal, elapsed time and speed. Furthermore, the up-to-that-game best, worst, average and last speed is also shown. Such speed is shown in green provided the current game's speed was better. Otherwise, it is highlighted in red.

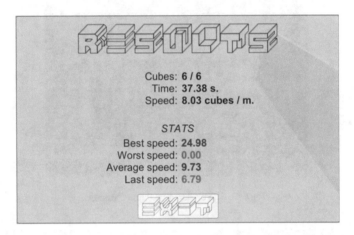

Fig. 4. Results screen

Finally, it is worth noting that this game has been created by using Unity [21] as game engine. Moreover, the integration with Leap Motion has been carried out by using the Orion SDK [22]. This Unity asset, although it is still in beta development stage, improves the hand tracking capabilities regarding the previous SDK by expanding the tracking range, speeding up the hand recognition speed and improving the recognition of hand gestures.

4.2   Results Viewer

Besides, there is a complementary application that not only facilitates the management of the patients, but also it presents the game results graphically (Fig. 5). In this sense, for each patient, it depicts the statistics related to the best, worst, average and last speed. Moreover, the patient's evolution can be graphically seen in order to analyze his/her performance and the relationship between score and goal (top of Fig. 5's chart) as well as that between elapsed time and speed per game (middle and bottom of Fig. 5's chart).

Finally, it is worth noting that the game and its editor have been developed following a distributed architecture. Therefore, several players can be playing the game from different location while their results are analyzed from a different location by a therapist using the Editor. This can be done by using a common database that enables such distributed architecture among players, therapists and game data server.

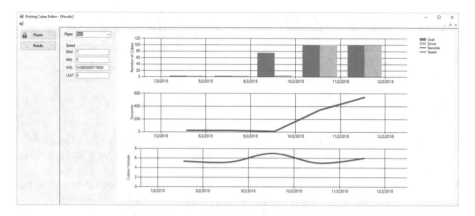

Fig. 5. Picking cubes editor

5   Conclusions and Future Works

As was stated in the introduction, upper limb rehabilitation becomes a serious need because of the number people affected of this problem when suffer a spinal cord injury, is terribly high. Unfortunately, existing measures for its treatment are quite demanding in terms of the therapists needed for its supervision, or very expensive because of the cost of the existing devices.

In this paper we have presented picking cubes, a serious game that exploits a low-cost 3D tracking system called Leap Motion to supervise the therapy execution. This facilitates that the game, in an automatic way, controls the rate of success, speed, etc. of the patient without human-supervision. This was one of the main constraints of its development, as it facilitates that a high number of patients could be treated simultaneously. They could even continue their treatment at home, increasing their performance rate. Moreover, picking cubes also facilitates that therapists can customize the therapy according to the specific needs of the patient, so that patients could make the most of their rehabilitation. This game also exploits an important feature: the introduction of different motivation facilities in order to appeal patients for doing a repetitive task that, otherwise, could be boring.

Several are our future works. First of all, we are designing an experiment to evaluate the acceptability of the presented game in a real setting at the Hospital Nacional de Parapléjicos. Moreover, we are also analysing other options in the field of motivation in order to appeal patients to carry out their therapy in a more fancy and attractive way. This evaluation will also provide feedback about whether the tool's real-time information may be too extensive and could distract the patients. Furthermore, the usability of the tool will also be assessed. Moreover, in future version of the tool, we will try to implement alternative gamification strategies to whether evaluate the level of motivation raised and the rehabilitation progress improve by using each one of the alternatives.

Acknowledgements. This work was partially supported by Spanish Ministerio de Economía, Industria y Competitividad, Agencia Estatal de Investigación (AEI)/European Regional Development Fund (FEDER, UE) under Vi-SMARt (TIN2016-79100-R) and RehabHand (DPI2016-77167-R) grants.

References

1. Van den Berg, M., Castellote, J.M., Mahillo-Fernandez, I., de Pedro-Cuesta, J.: Incidence of spinal cord injury worldwide: a systematic review. NeuroEpidemiology **34**(3), 184–192 (2010)
2. Wyndaele, M., Wyndaele, J.J.: Incidence, prevalence and epidemiology of spinal cord injury: what learns a worldwide literature survey? Spinal Cord **44**(9), 523–529 (2006)
3. Broeks, J.G., Lankhorst, G.J., Rumping, K., Prevo, A.J.H.: The long-term outcome of arm function after stroke: results of a follow-up study. Disabil. Rehabil. **21**(8), 357–364 (1999)
4. Yang, N., Zhang, M., Huang, C., Jin, D.: Motion quality evaluation of upper limb target-reaching movements. Med. Eng. Phys. **24**(2), 115–120 (2002)
5. Van Tuijl, J.H., Janssen-Potten, Y.J., Seelen, H.A.: Evaluation of upper extremity motor function tests in tetraplegics. Spinal Cord **40**(2), 51–64 (2002)
6. Zhou, H., Hu, H.: Human motion tracking for rehabilitation—a survey. Biomed. Signal Process **3**(1), 1–18 (2008)
7. de los Reyes-Guzmán, A., Gil-Agudo, A., Peñasco-Martín, B., Solís-Mozos, M., del Ama-Espinosa, A.J., Pérez-Rizo, E.: Kinematic analysis of the daily activity of drinking from a glass in a population with cervical spinal cord injury. J. Neuroeng. Rehabil. **7**(1), 41 (2010)
8. Holden, M.K.: Virtual environments for motor rehabilitation: review. Cyberpsychology Behav. **8**(3), 187–211 (2005)
9. Rose, F.D., Brooks, B.M., Rizzo, A.A.: Virtual reality in brain damage rehabilitation: review. Cyberpsychology Behav. **8**(3), 241–262 (2005)
10. Burdea, G.C., Coiffet, P.: Virtual Reality Technology, 2nd edn. Wiley, Hoboken (2003)
11. Leap Motion Controller. https://www.leapmotion.com/. Accessed 01 Feb 2018
12. Khademi, M., Mousavi Hondori, H., McKenzie, A., Dodakian, L., Lopes, C.V., Cramer, S.C.: Free-hand interaction with leap motion controller for stroke rehabilitation. In: CHI 2014 Extended Abstracts, pp. 1663–1668. ACM (2014)
13. Iosa, M., Morone, G., Fusco, A., Castagnoli, M., Fusco, F.R., Pratesi, L., Paolucci, S.: Leap motion controlled videogame-based therapy for rehabilitation of elderly patients with subacute stroke: a feasibility pilot study. Top. Stroke Rehabil. **22**(4), 306 (2015)
14. Mathiowetz, V., Volland, G., Kashman, N., Weber, K.: Adult norms for the Box and Block Test of manual dexterity. Am. J. Occup. Ther. **39**(6), 386–391 (1985)
15. López-Jaquero, V., Montero, F., Teruel, M.A.: Influence awareness: considering motivation in computer-assisted rehabilitation. J. Ambient. Intell. Hum. Comput. 1–13 (2017)
16. Cialdini, R.B.: Influence: the Psychology of Persuasion. Morrow, New York (1993)
17. Ryan, R.M., Deci, E.L.: Self-determination theory and the facilitation of intrinsic motivation, social development, and well-being. Am. Psychol. **55**(1), 68 (2000)
18. Ryan, R.M., Deci, E.L.: Intrinsic and extrinsic motivations: classic definitions and new directions. Contemp. Educ. Psychol. **25**(1), 54–67 (2000)

19. Flores, E., Tobon, G., Cavallaro, E., Cavallaro, F.I., Perry, J.C., Keller, T.: Improving patient motivation in game development for motor deficit rehabilitation. In: Proceedings of the 2008 International Conference on Advances in Computer Entertainment Technology, Yokohama, Japan, pp. 381–384 (2008)
20. Goršič, M., Cikajlo, I., Novak, D.: Competitive and cooperative arm rehabilitation games played by a patient and unimpaired person: effects on motivation and exercise intensity. J. Neuroeng. Rehabil. **14**(1), 23 (2017)
21. Menard, M., Wagstaff, B.: Game Development with Unity. Cengage Learning, Boston (2014)
22. Leap Motion Inc.: Orion Beta (2017). https://developer.leapmotion.com/orion

Multilag Extension of Quadratic Sample Entropy for Distress Recognition with EEG Recordings

Beatriz García-Martínez[1], Arturo Martínez-Rodrigo[2(✉)],
Antonio Fernández-Caballero[1,3], and Raúl Alcaraz[2]

[1] Instituto de Investigación en Informática de Albacete,
Universidad de Castilla-La Mancha, 02071 Albacete, Spain
[2] Instituto de Tecnologías Audiovisuales, Universidad de Castilla-La Mancha,
16071 Cuenca, Spain
Arturo.Martinez@uclm.es
[3] Centro de Investigación Biomédica en Red de
Salud Mental (CIBERSAM), Madrid, Spain

Abstract. Distress has become one of the major issues in developed countries because of its negative effects in physical and mental health. In order to control its consequences, a number of researchers have studied distress from an electroencephalographic point of view by means of the use of different nonlinear metrics. However, those studies are only based on non-lag approaches, thus many nonlinear dynamics of brain signals could not be properly assessed. In this sense, this work applies a multilag extension of a nonlinear regularity-based metric called quadratic sample entropy, in order to check the influence of the selection of a time lag for the recognition of distress with electroencephalographic recordings.

Keywords: Distress · Electroencephalography
Quadratic sample entropy · Multilag analysis

1 Introduction

The recognition of emotional states is basic in human communication and inter-action, since they influence in processes related to perception, cognition and rational decision-making [1]. Nevertheless, the lack of emotional intelligence in machines makes them unable to interpret human emotions and react according to those feelings [2]. Given the emerging use of human-machine interfaces (HMIs), it becomes essential to develop new models for automatic emotions recognition, thus humanizing the interactions between people and machines [3,4]. However, creating emotional models is not easy since emotional responses vary accord-ing to social and cultural aspects, personality and previous experience of each subject [5]. Moreover, emotions are highly intercorrelated and there are no gold standards for their definition [1]. In this sense, many affective models of def-inition and classification of emotions can be found in literature [6,7]. In this

© Springer Nature Switzerland AG 2019
P. Novais et al. (Eds.): ISAmI 2018, AISC 806, pp. 274–281, 2019.
https://doi.org/10.1007/978-3-030-01746-0_32

regard, one of the most widely used emotions classification approach is the circumplex model of Russell, which distributes all emotions according to their level of valence (positive/negative) and arousal (activating/deactivating) [8].

Among all those emotional states, negative stress or *distress* is receiving growing attention because of the negative consequences of a long-term exposure to this emotion, severely damaging both physical and mental health. Hence, distress has become one of the major problems in developed societies. In this sense, the automatic recognition of this emotional state would help to improve the quality of life of people suffering from distress by palliating the effects of the aforementioned diseases [9].

In recent years, a number of studies focused on the automatic detection of distress and other emotions have assessed changes in physiological activity, with special interest in electroencephalographic (EEG) recordings. The main reason is that EEG signals represent the first bodily response against any external stimulus, while the rest of physiological variables are secondary effects of brain processes [10]. In addition, neural processes are considered nonlinear and nonstationary at both cellular and global level [11]. Hence, the application of nonlinear metrics for EEG signals assessment has provided new discoverings that the traditional linear techniques could not report [12]. For instance, a nonlinear regularity-based metric called quadratic sample entropy (QSE) has demonstrated to be the first single index able to discern between calm and distress emotional states [13]. Other similar nonlinear approaches have also reported valuable results in distress automatic detection [14–17].

Those nonlinear metrics are based on quantifying similar patterns, typically with consecutive samples in the time series, without the application of any type of time delay or lag between samples in a pattern [18]. Depending on the features of the autocorrelation function of the signal under study, a non-lag analysis may only report information about linear stochastic processes, thus improperly quantifying the nonlinear characteristics of the time series [18]. In those cases, a multi-lag analysis could provide new insights about the nonstationary dynamics of a nonlinear signal [18]. For this reason, in the present manuscript QSE is computed with different time delays to check the influence of the lag on the discrimination between calm and distress emotional states with EEG recordings.

This paper is structured as follows. Section 2 describes the database and preprocessing techniques applied to the EEG signals, together with a brief definition of the multilag QSE algorithm. Section 3 shows the results obtained during this study. Finally, Sect. 4 presents a discussion of the results and some conclusions obtained after this work.

2 Materials and Methods

2.1 Database

Electroencephalographic signals assessed in this work are included in the Database for Emotion Analysis using Physiological Signals (DEAP) [19]. This publicly-available dataset contains a total of 1280 EEG recordings from 32 subjects (50% male, mean age 26.9 years) under different emotional states. For the elicitation of those emotions, 40 one-minute length music videoclips with emotional content were used as stimuli. After each visualization, participants rated their levels of valence, arousal, dominance, liking and familiarity by means of self-assessment manikins (SAM) [20]. EEG signals were acquired by a Biosemi ActiveTwo device at 512 Hz with 32 electrodes located over the scalp according to the international 10–20 system [21]. Other peripheral recordings such as electrocardiogram or respiration rate were also recorded, but their analysis is out of the scope of this manuscript.

Although this dataset contains samples of emotions covering the whole valence-arousal space, only two subsets corresponding to distress and calm emotional states have been selected for further analysis in this work. In this sense, samples from distress group present an arousal higher than 5 and valence lower than 3, whereas calm segments were rated with arousal lower than 4 and valence between 4 and 6. Finally, a total of 259 samples were selected, where 137 belonged to calm group and 122 to distress subset.

2.2 EEG Preprocessing

Raw EEG signals usually contain noise and artifacts that blur the neural information contained in the time series. For this reason, EEG recordings need to be filtered and preprocessed before the application of any kind of analysis, thus only the brain electrical activity is maintained. In this case, signals were downsampled from 512 Hz to 128 Hz [19]. Moreover, two forward/backward high-pass and low-pass filters were applied at 3 and 45 Hz, respectively, to maintain the bands of interest in EEG spectrum [19]. Given those cutoff frequencies, no further actions were needed to subtract baseline and power line interferences.

Some artifacts could not be eliminated by filtering the data. It is the case of physical artifacts like facial and muscular movements, eye blinks, or heart bumps, among others. Technical artifacts, such as electrode-pops, also remained after the filtering process. For those cases, a blind source separation technique called independent component analysis (ICA) was applied to reject those artifacts [22]. On the other hand, noise and other interferences required an interpolation of adjacent channels to be removed [23].

2.3 Quadratic Sample Entropy and Its Multilag Extension

Quadratic sample entropy is a regularity-based index that evaluates the rate of appearance of a pattern within a nonlinear and nonstationary time series [24]. Given a time series of N data points, $x(n) = \{x(1), x(2), \ldots, x(N)\}$, vectors $\mathbf{X}_m, \ldots, \mathbf{X}_m(N-m+1)$ of size m are defined by $\mathbf{X}_m(i) = \{x(i), x(i+1), \ldots, x(i+m-1)\}$, for $1 \leq i \leq N - m$. Repetitiveness of vectors $\mathbf{X}_m$, represented as the distance between them, is calculated as the maximum absolute distance between scalar components:

$$d[\mathbf{X}_m(i), \mathbf{X}_m(j)] = \max_{k=0,\ldots,m-1} \left(|x(i+k) - x(j+k)| \right). \tag{1}$$

Hence, if the distance is lower than a threshold r, then vectors $\mathbf{X}_m(i)$ and $\mathbf{X}_m(j)$ are considered similar. After that, self-matches are excluded to compute $B_i^m(r)$, which corresponds to the number of vectors that are similar to $\mathbf{X}_m(i)$. Then, $B^m(r)$ is quantified as the probability of having two sequencies matching for m points:

$$B^m(r) = \frac{1}{N-m} \sum_{i=1}^{N-m} \frac{B_i^m(r)}{N-m-1}. \tag{2}$$

If the same process is repeated for vectors of length $m+1$ samples, the probability of having two sequences matching for $m+1$ points is obtained. Then, non-lag QSE is estimated as

$$QSE(m, r, N) = -\ln\left(\frac{B^{m+1}(r)}{B^m(r)}\right) + \ln(2r), \tag{3}$$

which determines the probability that two patterns match for m and for $m+1$ points. If a time delay or lag, τ, is considered in the previous expression, then a multilag extension of QSE is obtained:

$$QSE(m, r, N, \tau) = -\ln\left(\frac{B^{m+\tau}(r)}{B^m(r)}\right) + \ln(2r) \tag{4}$$

Hence, instead of choosing consecutive samples in a pattern, only $m+\tau$ samples are selected for QSE calculation. Indeed, non-lag QSE is a particular case of its multilag form ($\tau = 1$).

3   Results

EEG signals of 30 seconds-length (3840 samples) were divided into six equal-size non-overlapped windows of $N = 640$ samples. In each window and for each EEG channel, QSE was computed with $m = 2$, $r = 0.25$ times the standard deviation of the data, and $\tau = 1$ (non-lag), 5, 10, 15 and 20. A final value of QSE for each channel was obtained as the average of the six windows. After that, a one-way analysis of variance (ANOVA) was developed to evaluate the discriminatory power of multilag QSE when discerning between calm and distress samples. In this sense, only values of statistical significance $\rho < 0.05$ were considered significant.

Statistical results of the most relevant channels are shown in Table 1. As can be observed, EEG channels with the best discriminatory power are located in the posterior half of the brain, in parietal and occipital lobes of both left and right hemispheres. It is important to note that the discriminatory power of all channels was notably higher in multilag than in non-lag analysis, while no big differences were found in results with $\tau = 5$, 10, 15 and 20. It can also be seen in Fig. 1, which shows the mean value of QSE for calm (green) and

Table 1. Values of ρ obtained from the most statistically significant EEG channels for different values of τ.

EEG Channel	Statistical significance, ρ				
	$\tau = 1$	$\tau = 5$	$\tau = 10$	$\tau = 15$	$\tau = 20$
P3	2.61×10^{-3}	4.83×10^{-5}	6.50×10^{-5}	5.05×10^{-5}	8.80×10^{-5}
PO3	6.56×10^{-4}	6.99×10^{-6}	7.84×10^{-6}	5.48×10^{-6}	8.01×10^{-6}
O1	1.62×10^{-2}	1.58×10^{-5}	1.04×10^{-5}	5.92×10^{-6}	6.67×10^{-6}
Oz	1.07×10^{-3}	2.19×10^{-6}	2.89×10^{-6}	1.99×10^{-6}	2.38×10^{-6}
Pz	5.95×10^{-5}	1.37×10^{-6}	1.65×10^{-6}	1.67×10^{-6}	1.64×10^{-6}
CP2	2.21×10^{-4}	4.60×10^{-5}	5.98×10^{-5}	2.64×10^{-5}	3.29×10^{-5}
P4	5.03×10^{-4}	7.10×10^{-5}	8.40×10^{-5}	9.15×10^{-5}	1.50×10^{-4}
O2	2.29×10^{-4}	1.45×10^{-6}	2.18×10^{-6}	1.88×10^{-6}	2.46×10^{-6}

distress (red) samples with different time lags for the most relevant channels. As shown, QSE for distress samples is always higher than for calm subjects. Moreover, the statistical difference between both groups is smaller for non-lag analysis ($\tau = 1$) than for the rest of lags, where that statistical difference is consistent for all values of τ. In addition, QSE mean levels in calm and distress from multilag analysis are notably higher than from the non-lag tests. These results were obtained for all the EEG channels analyzed in this study.

4   Discussion and Conclusions

In this work, a multilag extension of the well-known quadratic sample entropy has been computed to discern between calm and distress emotional states with EEG recordings. In previous works, the non-lag form of QSE has provided relevant new insights in this research field [13]. However, results derived from the present study suggest that the application of a time lag for QSE calculation may enhance the possibility of automatically detecting distress conditions. Indeed, as shown in Table 1, statistical significance values of multilag analysis reported an improvement in the discriminatory power of this entropy metric for the recognition of calm and distress with respect to the non-lag test. It can also be observed in Fig. 1 that the statistical difference between QSE mean values from calm and distress is bigger for multilag ($\tau = 5$, 10, 15 and 20) than for non-lag ($\tau = 1$) tests. Thus, multilag QSE could be more suitable than non-lag QSE for the automatic identification of distress emotional state.

In addition, the multilag extension of QSE presents another advantage in comparison with the non-lag approach: computational time is lower for multilag QSE calculation. Non-lag QSE selects consecutive samples ($m + 1$ samples) for patterns creation, hence all points in the time series are used for its calculation. On the contrary, multilag QSE does not require all samples of the signal, but it only selects one sample every τ (which is represented as $m + \tau$ samples). In

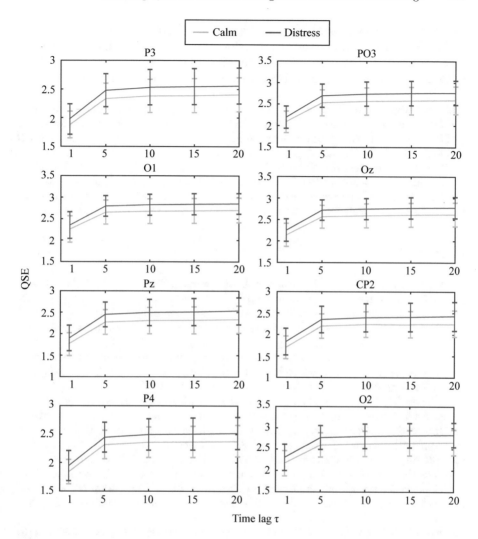

Fig. 1. QSE mean values for calm (green) and distress (red) groups obtained with different values of τ for the most statistically significant EEG channels.

this sense, computational costs are notably reduced since not all the samples in the time series are necessary to be analyzed. This would also help to develop new real-time affective predictive models based on QSE calculation for distress recognition with EEG signals.

With respect to the results of the multilag analyses, no great differences could be found between outcomes reported by different time lags. In this sense, statistical significance values are quite similar for $\tau = 5$, 10, 15 and 20, as shown in Table 1. Furthermore, these similarities can also be seen in Fig. 1, where the statistical difference between calm and distress mean QSE values is consistent for all τ in multilag tests. Thus, all values of τ studied in this work seem to report the same discriminatory power. Hence, it is not clear which τ values are more suitable for calm and distress emotional states identification with EEG recordings. In this sense, further research is still required in this field.

Acknowledgements. This work was partially supported by Spanish Ministerio de Economía, Industria y Competitividad, Agencia Estatal de Investigación (AEI)/European Regional Development Fund (FEDER, EU) under DPI2016-80894-R and AEI TIN2015-72931-EXP grants, and by the Centro de Investigación Biomédica en Red de Salud Mental (CIBERSAM) of the Instituto de Salud Carlos III. Beatriz García-Martínez holds FPU16/03740 scholarship from Spanish Ministerio de Educación, Cultura y Deporte. Arturo Martínez-Rodrigo holds EPC 2016–2017 research fund from Escuela Politécnica de Cuenca, Universidad de Castilla-La Mancha, Spain.

References

1. Coan, J.A., Allen, J.J.B.: Handbook of Emotion Elicitation and Assessment. Oxford University Press, Oxford (2007)
2. Picard, R.W.: Affective Computing. MIT Press, Cambridge (1995)
3. Castillo, J.C., et al.: Software architecture for smart emotion recognition and regulation of the ageing adult. Cogn. Comput. **8**(2), 357–367 (2016)
4. Fernández-Caballero, A., et al.: Human-avatar symbiosis for the treatment of auditory verbal hallucinations in schizophrenia through virtual/augmented reality and brain-computer interfaces. Front. Neuroinform. **11**, 64 (2017)
5. Gomes, M., Oliveira, T., Silva, F., Carneiro, D., Novais, P.: Establishing the relationship between personality traits and stress in an intelligent environment. In: International Conference on Industrial, Engineering and Other Applications of Applied Intelligent Systems, pp. 378–387. Springer (2014)
6. Ekman, P.: An argument for basic emotions. Cogn. Emot. **6**(3–4), 169–200 (1992)
7. Schröder, M., Cowie, R.: Towards emotion-sensitive multimodal interfaces: the challenge of the European network of excellence HUMAINE. Adapting the Interaction Style to Affective Factors Workshop in conjunction with User Modeling (2005)
8. Russell, J.A.: A circumplex model of affect. J. Pers. Soc. Psychol. **39**(6), 1161 (1980)
9. Martínez-Rodrigo, A., Zangróniz, R., Pastor, J.M., Fernández-Caballero, A.: Arousal level classification in the ageing adult by measuring electrodermal skin conductivity. In: Ambient Intelligence for Health. Lecture Notes in Computer Science, vol. 9456, pp. 213–223. Springer (2015)
10. Jenke, R., Peer, A., Buss, M.: Feature extraction and selection for emotion recognition from EEG. IEEE Trans. Affect. Comput. **5**(3), 327–339 (2014)
11. Abásolo, D., Hornero, R., Gómez, C., García, M., López, M.: Analysis of EEG background activity in alzheimer's disease patients with lempel-ziv complexity and central tendency measure. Med. Eng. Phys. **28**(4), 315–22 (2006)
12. García-Martínez, B., Martínez-Rodrigo, A., Alcaraz, R., Fernández-Caballero, A., González, P.: Nonlinear methodologies applied to automatic recognition of emotions: an EEG review. In: International Conference on Ubiquitous Computing and Ambient Intelligence, pp. 754–765. Springer (2017)

13. García-Martínez, B., Martínez-Rodrigo, A., Zangróniz Cantabrana, R., Pastor García, J., Alcaraz, R.: Application of entropy-based metrics to identify emotional distress from electroencephalographic recordings. Entropy 18(6), 221 (2016)

14. García-Martínez, B., Martínez-Rodrigo, A., Zangróniz, R., Pastor, J.M., Alcaraz, R.: Symbolic analysis of brain dynamics detects negative stress. Entropy 19(5), 196 (2017)

15. García-Martínez, B., Martínez-Rodrigo, A., Fernández-Caballero, A., Moncho-Bogani, J., Pastor, J.M., Alcaraz, R.: Nonlinear symbolic assessment of electroencephalographic recordings for negative stress recognition. In: International Work-Conference on the Interplay Between Natural and Artificial Computation, pp. 203–212. Springer (2017)

16. García-Martínez, B., Martínez-Rodrigo, A., Fernández-Caballero, A., González, P., Alcaraz, R.: Conditional entropy estimates for distress detection with EEG signals. In: International Work-Conference on the Interplay Between Natural and Artificial Computation, pp. 193–202. Springer (2017)

17. Hosseini, S.A., Naghibi-Sistani, M.B.: Emotion recognition method using entropy analysis of EEG signals. Int. J. Image Graph. Signal Process. 3(5), 30 (2011)

18. Kaffashi, F., Foglyano, R., Wilson, C.G., Loparo, K.A.: The effect of time delay on approximate & sample entropy calculations. Phys. D Nonlinear Phenom. 237(23), 3069–3074 (2008)

19. Koelstra, S., Mühl, C., Soleymani, M., Lee, J., Yazdani, A., Ebrahimi, T., Pun, T., Nijholt, A., Patras, I.: DEAP: a database for emotion analysis using physiological signals. IEEE Trans. Affect. Comput. 3(1), 18–31 (2012)

20. Morris, J.D.: Observations SAM: the self-assessment manikin - an efficient cross-cultural measurement of emotional response. J. Advert. Res. 35(6), 63–68 (1995)

21. Klem, G.H., Lüders, H.O., Jasper, H., Elger, C.: The ten-twenty electrode system of the international federation. Electroencephalography and Clinical Neurophysiology 52(3) (1999)

22. Jadhav, P., Shanamugan, D., Chourasia, A., Ghole, A., Acharyya, A., Naik, G.: Automated detection and correction of eye blink and muscular artefacts in EEG signal for analysis of autism spectrum disorder. In: Engineering in Medicine and Biology Society (EMBC), 2014 36th Annual International Conference of the IEEE, pp. 1881–1884. IEEE (2014)

23. Reis, P.M.R., Hebenstreit, F., Gabsteiger, F., von Tscharner, V., Lochmann, M.: Methodological aspects of EEG and body dynamics measurements during motion. Front. Hum. Neurosci. 8, 156 (2014)

24. Richman, J.S., Moorman, J.R.: Physiological time-series analysis using approximate entropy and sample entropy. Am. J. Physiol. Hear. Circ. Physiol. 278(6), H2039–49 (2000)

A Customizable Game-Inspired
Application for Memory Stimulation

Rodrigo Rocha[1], Davide Carneiro[1,3(✉)], Ana P. Pinheiro[2], and Paulo Novais[3]

[1] CIICESI, ESTG, Polytechnic Institute of Porto, Felgueiras, Portugal
{8140411,dcarneiro}@estg.ipp.pt
[2] Faculdade de Psicologia, Universidade de Lisboa, Lisbon, Portugal
appinheiro@psicologia.ulisboa.pt
[3] Algoritmi Centre/Department of Informatics,
Universidade do Minho, Braga, Portugal
pjon@di.uminho.pt

Abstract. Demographic changes are leading to a growing older population (>65 years), with repercussions on age-related conditions. From a Computer Science perspective, this also means that there will soon be a significant number of users with changes in perceptual and motor skill capacities. The goal of this work is to develop an environment to support the preservation of memory and functional capacities of the elderly. Health professionals will be able to set up and personalize immersive and realistic scenarios with high ecological validity composed of visual, auditory, and physical stimuli. Patients will navigate through and interact with these scenarios and stimulate memory functions by later recalling distinct aspects of the different exercises of the tool. The long-term goal is to build a behavioral model of how older users interact with technology.

Keywords: Real-time analytics · Big data · NoSQL databases
Behavioral biometrics

1   Introduction

Demographic changes are leading to a growing population of older people, with repercussions on age-related conditions such as Mild Cognitive Impairment (MCI) [1,2]. Pathological aging is related to a noticeable and measurable decline in cognitive abilities, including memory, as well as changes in perceptual and motor skill capacities, considering what would be expected based on age and educational level. Among other consequences, these changes bring important implications for the design of ICT.

The primary aim of this multidisciplinary work is to develop an environment to stimulate and support the preservation of memory and functional capabilities of the elderly, mitigating or delaying negative consequences of ageing and promoting a more active and autonomous ageing process. Indeed, recent results show that memory remains plastic even in an older age: after mnemonic training, the average elderly person can be expected to perform at the 77th percentile

© Springer Nature Switzerland AG 2019
P. Novais et al. (Eds.): ISAmI 2018, AISC 806, pp. 282–289, 2019.
https://doi.org/10.1007/978-3-030-01746-0_33

of the performance distribution of their age group [3,4]. However, the plasticity associated with mnemonic training appears to be largely specific to that training as improvement was found to be higher on tasks allowing for the use of the newly acquired mnemonic than on tasks not allowing for that use [5,6]. Similarly, [7] shows that although the methods used by top "mental athletes" allow them to memorize surprising amounts of information, the acquired abilities are rather irrelevant when performing daily activities. The main drawback in existing approaches is thus that most of them are based on mnemonics or similar mental tasks, that are hard to translate to skills needed in real-life [8,9].

The development of a software environment for such a specific population also constitutes an unprecedented opportunity. Indeed, a population of older technology users will be a reality in a few decades. In this regard, one of the key issues of HCI designs is to understand the impact of older users' abilities and restrictions [5]. However, the inclusion of older people within the design cycle for information technology is until now limited to aspects such as usability or the graphical aspects of user interface.

The long-term goal of this work is thus to build a behavioral model of how old users interact with technological devices. This model will further our understanding of the specific interaction patterns of the elderly, supporting the development of more accessible and inclusive applications and devices. Moreover, it may unveil differences between healthy people and people with different profiles of cognitive impairment, with the potential of acting as an early diagnosis mechanism for age-related conditions such as MCI or dementia.

2   Conceptualization

In the proposed approach, the clinician is able to set up and personalize immersive and realistic scenarios, that will afterwards be navigated by the patients. Each scenario is composed of one or more rooms. Each room is composed of one or more still photos. Each photo may have other superimposed photos in specific coordinates and with a specific size (Fig. 1, left). Each photo may also have one or more auditory stimuli associated, with specific playing patterns, which will be played when the patient visualizes the current photo. Each photo also has information describing its position relatively to other photos, which makes the scenarios navigable. Finally, each photo may also have "passages" to other rooms (Fig. 1, right).

These navigable scenarios may portrait the home of the participant, a grocery store, or virtually any intended environment. Moreover, scenarios can be enriched and fully personalized with additional stimuli (e.g. family pictures, objects, sounds) to stimulate a specific cognitive ability or to recall specific types of memories (e.g. family members).

This is, in fact, one of the most promising features of this project. Recent studies show that the plasticity associated with mnemonic training (which is the most common approach to memory stimulation) appears to be largely specific to that training as improvement was found to be higher on tasks allowing for

Fig. 1. Two still photos of one room of a given scenario. A custom picture has been superimposed by the clinician (the ESTG frame) in the left photo. The photo on the right shows a passage to another room (garden).

the use of the newly acquired mnemonic [3,10]. We believe that if clinicians design realistic scenarios, with realistic stimuli, this will allow patients to reuse the cognitive improvements acquired while using this application in their daily activities (e.g. go shopping, spend time with family members).

Patients use a mobile application to navigate through these scenarios ("Point and Click" style), when and as defined by the researcher. Later, patients recall their experience through specific exercises. These recall tasks may be of four different types: multiple choice (text and/or images), multiple choice single answer (text and/or images), numeric answer or free text.

Moreover, each recall task may reference any of the specific auditory or visual stimuli present in the corresponding scenario (e.g. "In which room was this image located?"). During these recall tasks, memory performance is automatically measured and quantified, providing valuable feedback to the researcher through real-time analytics.

3   Architecture

The architecture that implements the proposed system is composed of three main elements: a mobile application, the server, and tools for administration and real-time analytics (Fig. 2). There are also two different types of users: patients, who interact with the mobile application, and clinicians, who interact with the analytics and administration system. Each of these elements is described in the following sub-sections.

3.1   Server

The server is the central element of this architecture. It is composed of three main components: a NoSQL database, a node.js server, and a Spark instance.

All the main functionalities of the server are implemented by node.js. This includes services for uploading media to the server (e.g. sounds, images) that

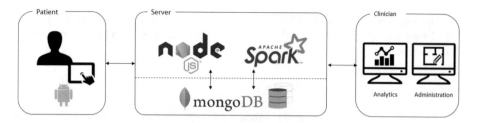

Fig. 2. Architecture of the developed system.

can later be used to construct scenarios, mediating access to the database, and a HTTP web server for communicating with external applications.

The NoSQL database stores all the relevant data, which includes:

- User Data - Identifying and authentication data. For patients this also includes socio-demographic variables;
- Scenarios - This collection contains the structural information of each scenario, including rooms/photos, relative positions, passages, superimposed photos and auditory stimuli;
- Recall Tasks - Collection of all recall tasks created by the clinicians, with information describing type of task, possible answers (in case of multiple choice), correct answer, referenced stimuli (if any) and target patient/scenario;
- Results - Contains all the data collected to measure the patients' performance during recall tasks;
- Behavioural Data - Describes all the data collected from the patients' interaction with the mobile devices, such as touch patterns and time between decisions;
- Administration - Contains data associating scenarios, to patients, to recall tasks. That is, which scenarios/recall tasks should be made available to which patients, at which moment and during how much time.

Finally, the Spark instance is used for data processing and machine learning tasks, namely for computing measures of cognitive performance and for analyzing patients' interaction patterns.

3.2 Mobile Application

The mobile application is developed for Android tablets. It is the patients' point of interaction with the system. It is through this application that the patient navigates in the scenarios that are appointed by the clinician. It is also through this application that the patient participates in the recall tasks.

The application checks, at regular intervals, if there are available scenarios/recall tasks for the logged in patient. Where there are, the application downloads the corresponding media (photos and sounds) as well as the meta-data that describes how photos and sounds should be displayed. When the download is complete, the scenario/recall task becomes available to the patient.

The application also collects two types of data transparently (Sect. 4). This data is temporarily stored locally in the device, and sent to the server when a data connection is available.

3.3   Analytics and Administration

The analytics and administration component implements functionalities targeted at the clinician. The Analytics module provides statistic measures about the different variables collected, describing memory performance and interaction patterns over time, for each user. These variables are very important for clinicians to monitor and evaluate the evolution of the patients' behaviour and cognitive performance over time.

The Administration module is designed for clinicians to design scenarios and recall asks. Clinicians start by uploading the necessary media (e.g. photos and sounds) that they will use. Then, they design the scenario by creating rooms, assigning photos to rooms, placing them relatively to each other, superimposing other photos/sounds (if desired) and specifying passages. Finally, through this module the clinicians can also assign scenarios/recall tasks to specific patients, as well as set the period of availability.

4   Features

The mobile application collects data that are used to produce two different types of features. Operational features describe *what* the patient does in the application, while behavioral features describe *how* the patient does it.

Operational features are extracted from the patients actions while navigating through the scenarios and/or while carrying out recall tasks. The following features are available for the clinicians, all of which can be aggregated by patient/scenario/gender/age group/recall task or any combination of these attributes, allowing different views on the data:

- Correct answers - This is the main feature in assessing the cognitive performance of the patient. It can be provided in percentage or in absolute terms;
- Time between decisions - Quantifies the time, in milliseconds, between each two consecutive decisions of the patient;
- Actions log - This feature quantifies the events and duration of each individual activity of a given participant (e.g. visiting a room, completing a recall task, starting/ending a specific question of a recall task).

These features allow, above all, clinicians to evaluate the cognitive performance of patients with different granularities (e.g. over a long period of time, in a specific recall task). This analysis can also be performed by group, gender, age group, among others. The three features provide significant information for quantifying cognitive performance. The quantification of correct answers is a direct measure. The time between decisions quantifies, in a way, the fitness of this performance, i.e., the faster the decisions the fitter (in principle) the brain is. Finally, the action has similar uses as it quantifies how much time each patient spent in each question/task/room, with longer times being associated to less cognitive performance.

On the other hand, there are the behavioural features, which describe mostly the interaction of the patient with the device. These include:

- Touch duration - The duration of each individual touch on the screen;
- Touch intensity - The average, minimum and maximum values of the intensity exerted by the finger on the screen, for each touch;
- Touch area - The average, minimum and maximum values of the area occupied by the finger on the screen, for each touch;
- Intensity values - The sequence of values of intensity generated during each touch, from the moment the finger first touches the screen to the moment it is lifted;
- Area values - The different areas of finger in contact with the screen, in each touch, from the moment the finger first touches the screen to the moment it is lifted;
- Type of action - The actions on the screen can be further characterized (e.g. touch on an active control vs. touch on a layout inactive element).

These features describe how the patient is interacting with the device. They provide new types of information previously not considered in this kind of applications, as explored in Sect. 5. Indeed, and as put forward in the introductory section, we believe that in the long term and after the collection of extensive interaction data from different groups of patients, we will be able to find differences in interaction patterns due to socio-demographic variables such as gender, age, occupation or health conditions.

5   Validation

This section describes some of the data collected with this application, from patients with different characteristics. Specifically, and to highlight the potential interest of this approach in terms of Human-Computer Interaction, in this section we analyze data collected from four different users: two male (one young and one older) and two female (one young and one older). This is not meant to be a representative data analysis but only the validation of the proposed hypothesis, that is, showing that age- and gender-based differences may exist. We are now collecting and analyzing data from a large group of users, in order to study these hypothesis. Given the scope of the paper, in this section we focus on the behavioral features described in Sect. 4.

Figure 3 highlights the differences according to gender and age of two of the variables: touch duration and touch intensity. It shows that older people tend to have longer and also stronger touches on the touch screen, with differences being more evident in duration. These results are further detailed in Table 1

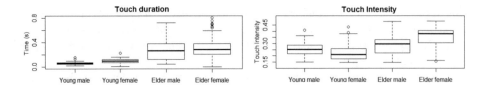

Fig. 3. Distribution of the time between decisions of four different patients: older patients tend to take longer to make decisions when visiting the scenarios or completing recall tasks.

Table 1. General statistics of touch duration and intensity for both genders and age groups.

	Touch duration		Touch intensity	
	$\bar{x}$	σ	$\bar{x}$	σ
Young male	0.061	0.022	0.255	0.052
Young female	0.097	0.041	0.228	0.069
Elder male	0.279	0.174	0.285	0.082
Elder female	0.315	0.178	0.354	0.080

Another interesting feature that can be built from the collected data is the patient's touch pattern, i.e., how touch intensity varies over time during the touch. In this approach, each patient's touch pattern is modelled by fitting a quadratic function to the data. On the one hand, this provides a combined analysis of these two factors. Moreover, the coefficients of the resulting quadratic function can be efficiently used to characterize the "general shape" of a patient's touch. We expect that this kind of data may allow for interesting applications such as user identification. Figure 4 shows the touch patterns of the four different patients, depicted in terms of the intensity values over time and the resulting quadratic function that models it: (a) young male, (b) elder male, (c) young female and (d) elder female.

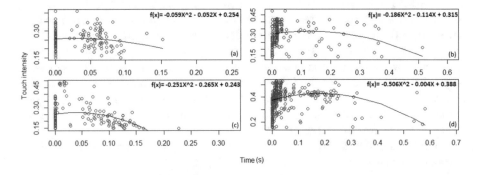

Fig. 4. Touch patterns (intensity over time) of the four patients: (a) young male, (b) elder male, (c) young female and (d) elder female.

6 Conclusions

In this paper we presented a game-inspired mobile application for stimulating memory, especially designed for older people or for people with cognitive disabilities. It has several key innovative aspects. First, it is a tool that can be completely personalized by the clinician, with each patient's specific needs in mind. It allows for the distributed and remote management and administration of interventions, facilitating the whole process. Finally, it acquires, in real-time, a wide range of data describing the patient's memory performance and interaction patterns. In this sense, it provides many new features that were, until now, not considered in this type of intervention. It can thus significantly contribute not only for the study of memory and its improvement but also for a better understanding of Human-computer Interaction, especially in these slices of the population.

References

1. Petersen, R.C., Caracciolo, B., Brayne, C., Gauthier, S., Jelic, V., Fratiglioni, L.: Mild cognitive impairment: a concept in evolution. J. Intern. Med. **275**(3), 214–228 (2014)
2. DeSA, U.: World population prospects: the 2012 revision. Population division of the department of economic and social affairs of the United Nations Secretariat, New York (2013)
3. Verhaeghen, P., Marcoen, A., Goossens, L.: Improving memory performance in the aged through mnemonic training: a meta-analytic study (1992)
4. Park, D.C., Bischof, G.N.: The aging mind: neuroplasticity in response to cognitive training. Dialogues Clin. Neurosci. **15**(1), 109 (2013)
5. Jochems, N.: Designing tablet computers for the elderly a user-centered design approach. In: International Conference on Human Aspects of IT for the Aged Population, pp. 42–51. Springer (2016)
6. Verhaeghen, P.: Memory training and mnemonics. In: The Encyclopedia of Adulthood and Aging (2016)
7. Foer, J.: Moonwalking with Einstein: The Art and Science of Remembering Everything, Penguin (2012)
8. Pereira, G., Ninaus, M., Prada, R., Wood, G., Neuper, C., Paiva, A.: Free your brain a working memory training game. In: International Conference on Games and Learning Alliance, pp. 132–141. Springer (2014)
9. Jensen, K., Valente, A.: Development of a memory training game. In: Technologies of Inclusive Well-Being, pp. 25–38. Springer (2014)
10. Deveau, J., Jaeggi, S.M., Zordan, V., Phung, C., Seitz, A.R.: How to build better memory training games. Front. Syst. Neurosci. **8**, 243 (2015)

An Innovative Tool to Get Better at Expressing Facial Emotions

Arturo S. García[1]([⊠]), Pascual González[1,2], Antonio Fernández-Caballero[1,2], and Elena Navarro[1,2]

[1] Instituto de Investigación en Informática de Albacete,
Universidad de Castilla-La Mancha, 02071 Albacete, Spain
{ArturoSimon.Garcia,Elena.Navarro}@uclm.es
[2] Centro de Investigación Biomédica en Red de Salud Mental (CIBERSAM),
28029 Madrid, Spain

Abstract. Social cognition disorders cause important social and communicative impairments in people suffering from autism or schizophrenia, as they have difficulty recognising emotions in others and expressing their own. This paper introduces the prototype of an innovative computer-based training tool as an attempt to improve the expressiveness of facial emotions in people with socio-cognitive deficits. This tool is designed so that the therapist plans training and evaluation sessions, which are composed of several stages and adapted to each patient. In some of them, the patient learns about facial expressions using pictures while in other stages he/she proves what he/she has learnt, and the system automatically evaluates the performance. The system provides the therapist with post-therapy tools to analyse the results and the performance of the patient.

Keywords: Social cognition · Facial emotional expressiveness

1 Introduction

Emotions play a relevant role in human relationship and enable people to respond adaptively to threats and opportunities found in the environment. Emotional faces communicate both feelings and behavioural intentions of an individual. They also activate conduct tendencies in the perceiver, namely approach or avoidance [15]. In fact, emotional facial expressions are rich and powerful means of communicating information about one's affective states, as well as about the environment in which we live in [11,13].

Thus, the inability of certain persons to recognise or express these emotions may be critical, and could contribute to social dysfunction. There are several collectives where this problem represents a serious challenge. For example, people with psychiatric disorder, such as schizophrenia [10] or autism spectrum [1], or those that suffer acquired brain injury (ABI) [14] after an accident, a stroke, and so on, are among the most relevant groups affected by this problem.

© Springer Nature Switzerland AG 2019
P. Novais et al. (Eds.): ISAmI 2018, AISC 806, pp. 290–297, 2019.
https://doi.org/10.1007/978-3-030-01746-0_34

One main strategy to recognise an emotion is by considering some physiological signal that individuals display as a reaction to an emotion. In this case, the human face plays a fundamental role in emotion recognition. As recently stated [4], facial expressions have both a physical component related to morphological changes of a face and an affective component conveying information about the expresser's internal feelings. This relevance of facial expression in emotion recognition has been used in several studies that recognise a human emotion by analysing an individual's face [5,7,8]. Other studies try to define learning strategies to avoid the inability to recognise emotions of certain individuals [4].

The training of emotion recognition in patients diagnosed with autism spectrum disorder or Asperger syndrome has been proven useful by previous studies [2,3]. The relevance of facial expression in emotion recognition has led to use static photos showing a specific emotion as one of the most frequently used therapies [9]. However, computer-based tools are becoming more and more common. Some examples are Emotion Trainer [12] and Let's Face It! [16]. These tools show a set of predefined pictures of people expressing an emotion and ask the patient to identify the emotion expressed. However, it is not only emotion recognition that is an important handicap for people with social cognition disorders but also the emotional expression [17]. In this sense, there has been less effort in developing tools to assist in this kind of training.

Therefore, this paper explores the nature of a tool to assist therapists in training emotional expression for people affected by social cognition deficit. Besides, a prototype of this tool is introduced, which also allows the therapist to design a series of training sessions composed of different stages, including learning and evaluating facial expressions.

2   Requirements for a Training Tool for Facial Emotion Expression

The main idea behind the training tool presented in this paper is to help people with social cognition deficits to express emotions through facial expressions. For this, based on the existing literature and helped by expert therapists, we defined a training session as a series of stages for the patients to go through:

- Initial stage:
  - Stage 0: The system will evaluate the patient's performance by asking him/her to express an emotion using his/her face. The patient will be able to see his/her own face on the screen using a web cam. The expressed facial emotion will be automatically evaluated by the system and receive a score. The rating obtained is not visible to the patient, but it will be used by the therapist to compare the results before and after the session. This is the manner to know whether there is an improvement in performance. The facial expression will also be recorded for latter study.
- Learning stages:
  - Stage 1: The system will show a set of images depicting facial expressions associated to several emotions.

- Stage 2: During this stage, the system will ask the patient to try to mimic the facial expression shown on the screen. Both expressions will be shown on the display side-to-side to allow comparison and offer the possibility to adapt the patient's own facial expression to the requested one. The patient's facial expression will be automatically evaluated and will receive a score to help the patient to improve.

– Evaluation stage:

- Stage 3: Finally, the system will ask the patient to express an emotion without any helpful visual clue apart from his/her own face. The patient should just replicate what he/she learnt from the set of images used during training. An automatic evaluation of the patient's performance will be visible to him/her and recorded for a latter study. The facial expression will also be recorded for later study.

After this, the therapist will have the capability to compare a patient's performance before (stage 0) and after (stage 3) having gone through the therapy. This includes recalling all facial expressions that the system asked the patient to mimic, as well as the score received. This way, the progression of the patient's performance can be easily followed in a latter study. Figure 1 shows a state chart depicting the structure of a session.

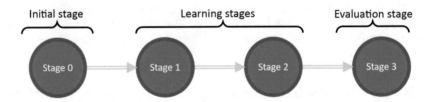

Fig. 1. Transition of stages for one session during the use of the tool.

A therapist will be able to design his/her own therapies adapted to the particularities of each patient. Several parameters allow the configuration of each session, which includes pictures used to depict facial expressions of people of different age, race and gender. The difficulty level of each session should also be configurable by using pictures ranging from exaggerated to subtle facial expressions, including micro-expressions, and reducing the time available to mimic and offer the facial expression asked by the system.

To sum up, the proposed system will allow the therapist to design and adapt therapies. The interpretation of the results will enable progression of the patient in getting better at expressing facial emotions.

2.1 Design of the Prototype for the Training Environment

According to the description provided in the previous section, the following requirements were extracted. The system should have a library of pictures of people's faces, including different race, age and gender, and expressing distinct emotions. These moods should be based on the seven universal emotions (neutral, joy, anger, fear, disgust, sadness and surprise) [6]. The system should allow the therapist to configure the type of images shown to the patient and select the difficulty level of the session.

After that, the system should be able to randomly display pictures from the selected set of images to either present how a facial expression of an emotion looks like in someone else's face or request the patient to mimic that facial expression. Moreover, the system should be able to recognise the facial expression of the patient by means of a web cam and provide a score according to how good he/she is at mimicking the requested expression. Finally, the system should be able to request a specific facial expression of an emotion by just using its name and evaluate how similar it is to what was previously exhibited.

Apart from all this, the system should record the patient's face every time he/she is requested to express a facial emotion. After the session is over, the therapist should be able to see a summary of the results, including the score given by the system in stages 0 and 3, so that the therapist can decide whether the improvement is enough to increase the difficulty level, or the patient should keep practising at the same level. This summary can be extended by the exploration of all the screen shots taken during the session.

In such a system, the emotion recognition module becomes paramount. For that reason, different alternatives for emotion detection engines were studied. Of all of them, we chose Affectiva (https://www.affectiva.com/), mainly because it provides emotion recognition in real-time using a standard web cam. In addition, Affectiva works off-line while all the calculations are done in the local machine (such as desktop computer or smart phone). This contrasts with other alternatives such as IBM Watson (https://www.ibm.com/watson/), Microsoft Emotion API (https://docs.microsoft.com/en-us/azure/cognitive-services/emotion/home) and Google Cloud Vision API (https://cloud.google.com/vision/) that make use of web services or cloud computing for emotion recognition. Also, Affectiva can be integrated in HTML5, Android, iOS or Unity 3D.

In our implementation, Unity 3D (https://unity3d.com/) was used to develop the two-dimensional interface of the emotion training tool. The four stages designed for the patients to go through are depicted in the screen shots shown in Fig. 2. The four screens share the same structure. There are two buttons on the lower section of the screen, one to exit the therapy and the other one to move towards the next image. The latter is disabled until the system considers that the patient has spent a certain amount of time (configured by the therapist) looking at an image or trying to mimic a facial expression.

The middle section shows the live stream of the patient's web cam so that he/she watches his/her performance (top left of Fig. 2 for stage 0), the facial expression picture that the patient has to learn (top right of Fig. 2 for stage 1), the picture to mimic side-by-side with the live stream of the web cam so he/she can compare them (bottom left for stage 2), and an empty image side-by-side with the web cam image whenever the system is asking the patient to show

Fig. 2. Screen shots of the prototype tool for stage 0 (top left), stage 1 (top right), stage 2 (bottom left) and stage 3 (bottom right).

a facial expression by just naming it (bottom-right part of Fig. 2 for stage 3). Finally, the upper section of the screen shows the name of the facial expression that is being displayed or requested.

Apart from this, the description of stages 2 and 3 includes that the system should provide the patient with some feedback about his/her performance as a score (from 0 to 100). This is implemented using a bar in both screen shots of the bottom row of Fig. 2 that also shows the score using a numeric value. This bar is updated in real-time according to how close the expression shown by the patient is to the wanted one. The value is provided by the emotion recognition engine.

Once the patient has gone through all the stages, and the session is over, the therapist compares the results of pre- and post-therapy evaluations, and examine the screen shots taken when the system asked the patients to mimic or show a specific facial expression. Then, this information is used to plan the next session. Figure 3 shows a screen shot of the therapist's tool for the exploration of the outcomes. It depicts the results of patient '01' showing 'anger' during session '1'. The left picture shows the image captured during the initial stage, together with the score given by the emotion recognition engine, and it is compared side-by-side to the one taken during the evaluation stage.

If several pictures have been taken during those stages, the therapist goes through all of them by using the red arrows located under each picture. The therapist also examines all the facial expressions evaluated by using the arrows located in the upper section of the screen shot. Finally, there is an exit button on the bottom-left that allows the therapist to go back to another screen that enables choosing the next session to examine.

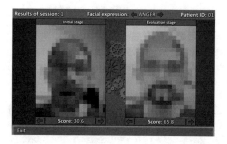

Fig. 3. Screen shots of the therapist's tool to explore the results of a session.

3 Conclusions

This paper has introduced an innovative tool to better expressing facial emotions for patients with socio-cognitive deficits. Emotional processing, as part of social cognition, refers to the ability to perceive, recognise and manage emotions, normally by means of non-verbal behaviour including facial expression and body movements. Different mental disorders associated with cognition deficits have an impact on how people recognise and express emotions, causing difficulties to socially interact with other people.

Moreover, the present paper has described the prototype of a tool to teach them how to recognise and express facial expressions associated to different moods (six plus one international emotions [6]). The process covers several sessions with different stages in which a patient is first shown how facial expressions look like, by means of standard pictures, and is then encouraged to mimic those facial expressions (learning stages). In the last stage of a session, the patient is asked to exhibit facial expressions just by naming them without providing any other visual feedback than his/her own face, so that he/she manifests what he/she has assimilated during the learning stages.

The system is continuously monitoring a patient while he/she is learning to follow his/her progress in a latter analysis of the results. This enables comparing the performance of a patient before and after going through all the stages. The system has been developed using Unity 3D, but it is planned to deploy it on mobile phones and as a web application so that it can be used by more patients. Finally, we are planning to use the system with real patients affected by social cognition disorders for its clinical validation.

Acknowledgements. This work was partially supported by Spanish Ministerio de Economía, Industria y Competitividad, Agencia Estatal de Investigación (AEI)/European Regional Development Fund under EmoBioFeedback (DPI2016-80894-R), HA-SYMBIOSIS (TIN2015-72931-EXP) and Vi-SMARt (TIN2016-79100-R) grants, and by the Centro de Investigación Biomédica en Red de Salud Mental (CIBER-SAM) of the Spanish Instituto de Salud Carlos III.

References

1. Baron-Cohen, S., Wheelwright, S., Skinner, R., Martin, J., Clubley, E.: The autism-spectrum quotient (AQ): evidence from Asperger syndrome/high-functioning autism, malesand females, scientists and mathematicians. J. Autism Dev. Disord. **31**(1), 5–17 (2001)
2. Berggren, S., Fletcher-Watson, S., Milenkovic, N., Marschik, P.B., Bölte, S., Jonsson, U.: Emotion recognition training in autism spectrum disorder: a systematic review of challenges related to generalizability. Dev. Neurorehabilitation 1–14 (2017)
3. Bölte, S., Feineis-Matthews, S., Leber, S., Dierks, T., Hubl, D., Poustka, F.: The development and evaluation of a computer-based program to test and to teach the recognition of facial affect. Int. J. Circumpolar Health **61**, 61–68 (2002)
4. Calvo, M.G., Nummenmaa, L.: Perceptual and affective mechanisms in facial expression recognition: an integrative review. Cogn. Emot. **30**(6), 1081–1106 (2016)
5. Castillo, J.C., Castro-González, Á., Fernández-Caballero, A., Latorre, J.M., Pastor, J.M., Fernández-Sotos, A., Salichs, M.A.: Software architecture for smart emotion recognition and regulation of the ageing adult. Cogn. Comput. **8**(2), 357–367 (2016)
6. Ekman, P.: Universals and cultural differences in facial expressions of emotion. In: Nebraska Symposium on Motivation, pp. 207—282. University of Nebraska Press (1972)
7. Fernández-Caballero, A., Latorre, J.M., Pastor, J.M., Fernández-Sotos, A.: Improvement of the elderly quality of life and care through smart emotion regulation. In: Pecchia, L., Chen, L.L., Nugent, C., Bravo, J. (eds.) Ambient Assisted Living and Daily Activities, pp. 348–355. Springer (2014)
8. Fernández-Caballero, A., Martínez-Rodrigo, A., Pastor, J.M., Castillo, J.C., Lozano-Monasor, E., López, M.T., Zangr'oniz, R., Latorre, J.M., Fernández-Sotos, A.: Smart environment architecture for emotion detection and regulation. J. Biomed. Inform. **64**, 55–73 (2016)
9. Ginani, G.E., Pradella-Hallinan, M., Pompéia, S.: Identification of emotional expressiveness in facial photographs over 36 h of extended vigilance in healthy young men - a preliminary study. Cogn. Emot. **31**(2), 339–348 (2017)
10. Gottesman, I.I.: Schizophrenia Genesis: The Origins of Madness. WH Freeman/Times Books/Henry Holt & Co (1991)
11. Kaiser, J., Crespo-Llado, M.M., Turati, C., Geangu, E.: The development of spontaneous facial responses to others' emotions in infancy: An EMG study. Scientific Reports 7, 17500 (2017)
12. LaCava, P.G., Golan, O., Baron-Cohen, S., Myles, B.S.: Using assistive technology to teach emotion recognition to students with Asperger syndrome a pilot study. Remedial Spec. Educ. **28**(3), 174–181 (2007)
13. Lozano-Monasor, E., López, M.T., Vigo-Bustos, F., Fernández-Caballero, A.: Facial expression recognition in ageing adults: from lab to ambient assisted living. J. Ambient. Intell. Humanized Comput. **8**(4), 567–578 (2017)
14. Rosenberg, H., McDonald, S., Dethier, M., Kessels, R.P., Westbrook, R.F.: Facial emotion recognition deficits following moderate-severe traumatic brain injury (TBI): re-examining the valence effect and the role of emotion intensity. J. Int. Neuropsychol. Soc. **20**(10), 994–1003 (2014)
15. Seidel, E.M., Habel, U., Kirschner, M., Gur, R.C., Derntl, B.: The impact of facial emotional expressions on behavioral tendencies in females and males. J. Exp. Psychology. Hum. Percept. Perform. **36**(2), 500–507 (2010)

16. Silver, M., Oakes, P.: Evaluation of a new computer intervention to teach people with autism or Asperger syndrome to recognize and predict emotions in others. Autism **5**(3), 299–316 (2001)
17. Walsh, J.A., Creighton, S.E., Rutherford, M.D.: Emotion perception or social cognitive complexity: What drives face processing deficits in autism spectrum disorder? J. Autism Dev. Disord. **46**(2), 615–623 (2016)

Situational-Context for Virtually Modeling the Elderly

Jose Garcia-Alonso[1]([✉]), Javier Berrocal[1], Juan M. Murillo[1], David Mendes[2], Cesar Fonseca[3], and Manuel Lopes[3]

[1] University of Extremadura, Cáceres, Spain
{jgaralo,jberolm,juanmamu}@unex.es
[2] DECSIS SA, Évora, Portugal
david.m.mendes@decsis.pt
[3] University of Evora, Évora, Portugal
{cfonseca,mjl}@uevora.pt

Abstract. The generalized aging of the population is incrementing the pressure over, frequently overextended, healthcare systems. This situations is even worse in underdeveloped, sparsely populated regions like Extremadura in Spain or Alentejo in Portugal. In this paper we propose an initial approach to use the Situational-Context, a technique to seamlessly adapt Internet of Things systems to the needs and preferences of their users, for virtually modeling the elderly. These models could be used to enhance the elderly experience when using those kind of systems without raising the need for technical skills. The proposed virtual models will also be the basis for further eldercare innovations in sparsely populated regions.

Keywords: Situational-context · Internet of Things
Eldercare · Virtual profiles

1 Introduction

The aging of the population is a confirmed fact in most developed countries. Over 20% of people in developed countries are elderly (65 or more years old), and the growth of this age group means that it is likely to reach some 26% of the population of these countries in 2030 [16].

This trend is even more acute on rural European regions like Extremadura in Spain or Alentejo in Portugal. These regions have a lower population density than the average of the European Union (EU), ranging from 14,8 people per km^2 on Baixo Alentejo and 31,9 people per km^2 on Badajoz compared to the 116,6 people per km^2 averaged on the EU. And even lower that the averaged population density of the rural regions of the EU, averaged on 48,4 people per km^2. Additionally, these regions keep losing its young population, due to migration to more socioeconomically developed regions, which is increasing the aging ratio.

P. Novais et al. (Eds.): ISAmI 2018, AISC 806, pp. 298–305, 2019.
https://doi.org/10.1007/978-3-030-01746-0_35

As a result of these conditions, these regions have a higher than average aged population, which can be understood as a signal of development. However, the reality is that these are economically disadvantaged regions with an especially fragile cultural and socioeconomic context. Literacy index are lower than average in these regions and, due to the low population density and migration of the young population to richer regions, the elders frequently live alone. Therefore, public health policies, that are already suffering to cover the aging population in more developed regions, are stretched to their limits in these regions.

In recent years, healthcare companies and researchers have made significant efforts to improve the quality of life of the elders. Particularly, from a technological point of view, areas like Ambient Intelligence and e-Healthcare are trying to bring technological advances to eldercare. Works like [9, 15] propose different solutions that will result in an improved quality of life of elders with some specific diseases. However, most of these works do not take into account the particular conditions of rural sparsely populated areas and their socioeconomic context.

In this paper, we propose the creation of comprehensive virtual profiles of the elderly. The proposal is built around the Situational-Context [1], a model to analyze the conditions that exist at places where Internet of Things (IoT) systems are present. We propose to use this model to gather all the information about the elderly that is available from the different devices and systems they interact with in their normal lives. By using the Situational-Context the information gathering is transparent for the elderly and, therefore, no technical skills are needed. Then, these profiles can be used to automatically adapt the behavior of IoT systems to the specific needs and preferences of each elderly. The proposed system can be applied in regions like Alentejo or Extremadura not only to try to improve the quality of life of the elderly and their formal and informal caregivers but also to promote further innovations adapted to the socioeconomic context of those regions.

To describe the proposed system for gathering comprehensive virtual profiles of the elderly, the rest of this paper is organized as follow. Section 2 presents the motivations of this work and briefly present the Situational-Context, the technical framework in which this work is based. Section 3 details the proposed virtual profiles of the elderly for rural sparsely populated regions and describe the possible innovations that can follow from the information contained on such profiles. In Sect. 4 the more relevant related works are analyzed. And finally, Sect. 5 presents the conclusion of this work.

2   Motivations

The computational capabilities of embedded and mobile devices have kept increasing over the last years. Thanks to this trend, the relevance of the IoT in different domains has also increased. However, the use and configuration of these systems require that users have certain technical skills, which is not always possible, particularly in domains like eldercare. Therefore, solutions are needed to transparently and effortlessly integrate the users' preferences and needs into

IoT system. In that regard, works like [5,7] focus on gathering and processing the contextual information of users in order to create virtual profiles. These virtual profiles can then be used to adapt other systems to the users needs and preferences.

In this context, some of the authors of this paper proposed the Internet of People (IoP) [14]. This proposal aims to use the information that can be gathered from IoT sensors to better adapt IoT systems to their users' needs. To this end, smartphones are made the architectural core of the systems. These devices are in charge of gathering all the available information about their owners and use it to generate a sociological profile with their preferences and needs. Then, this profile is offered as a service to third parties using a technique called People as a Services (PeaaS) [6].

By using this approach, eldercare applications can be built taking into account the preferences and needs of the elders and their caregivers while keeping their privacy. In [2], some of the authors of this paper presented a mobile application for monitoring cognitively impaired elderly. The proposed application can monitor the users' daily activities to detect their routines and to alert their caregivers in case of need.

However, as the relevance of the IoT increases so does the number of devices and systems in which the users are involved. This make the development of ad-hoc solutions, like the previous one, that adapt IoT systems to theirs users' preferences unsustainable. The Situational-Context [1] is an effort to bring advances in areas like Context-Aware [7], Ubicomp [3], User Modelling [8] and Ambient Intelligence [11] to analyze the conditions that exist at a particular time and place. This analysis is then used to predict, at run-time the expected behaviour of IoT systems.

To achieve this goal, the Situational-Context is defined as the composition of the virtual profiles of all the entities involved in a situation. Understanding by entities not only the users physically present in a place and represented by their smartphones, but also the different sensors and actuators of IoT systems. The virtual profiles of these entities contain, at least, the following information:

- A Basic Profile with raw information about the entity's status, its relationship with other devices and its historic information. This Basic Profile is represented as a timeline with the contextual changes and interactions that happened to the entity.
- A Social Profile with the information obtained as a result of high level inferences performed over the Basic Profile.
- The Goals that detail the status of the environment desired by the entity.
- The Skills or capabilities that an entity has to make decisions and perform changes in the environment.

By combining the profiles of all the entities involved in a given situation the Situational-Context obtains the combined history of the entities ordered in a single timeline, high level inferences performed over the combination of the virtual profiles and the set of Goals and Skills present in the context. From all this information strategies to maximize the achievement of Goals given the present Skills are identified. These strategies guide the interactions between the present devices.

In this paper we propose the use of the Situational-Context in an eldercare environment were access to several sources of information about the elderly are available. With all this information, the Situational-Context will build a comprehensive virtual representation of the elderly. These virtual models of the elderly will then be the basis for further innovations in eldercare. The proposed solution is meant to be deployed in sparsely populated underdeveloped regions where access to technological advanced healthcare is limited and the technical skills of the elderly and their informal caregivers is also reduced.

3   Elderly Virtual Profiles

Providing healthcare to the elderly in sparsely populated regions like Extremadura or Alentejo present its own challenges. The elderly in this regions, as in many others, can be characterized by comorbidity (presence of two or more illness in the same person) and a progressive reduction of their functional capacity (defined as the potential of the elderly to take care of themselves). Additionally, these regions have low literacy, high loneliness due to low population density and emigration, and economic vulnerability.

Traditionally, health and social care in these regions are provided by a network of hospitals in the biggest cities, often located more than one hour away of the living places of the elderly, and a network of primary health centers. However, these services are rendering insufficient for the everyday older population with their associated health problems. They are also very limited in terms of their capacity to bring technological advances to these regions.

In this work, we propose the use of the Situational-Context technology to create virtual models of the elderly living in these regions. In order to do that, the proposed system will transparently gather information from several sources and combine it to form these profiles. Then, these comprehensive profiles of the elderly can be used to improve the coordination between different healthcare solutions and to generate an scenario that favors the implementation of new technological advances in less developed regions. Figure 1 shows the basic architecture of the proposed solution. Each of the different sources of information for the creation of the virtual profiles are described below.

- **Personal Electronic Health Record.** As in most healthcare systems, the access to comprehensive health information of the elderly is key. In this approach, the PEHR of each elder is gathered and kept in a companion device, usually a smartphone. By using the Situational-Context, this device constantly updates the PEHR with information obtained by its own sensors and also with information obtained through any other entities that interact with the elder, either at home or in public places such as health centers.

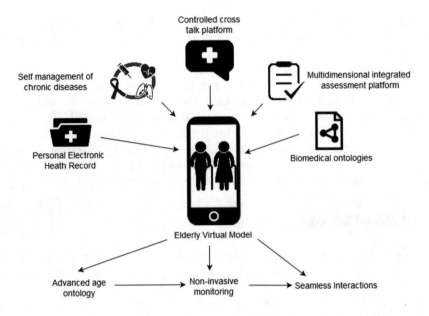

Fig. 1. Elderly virtual model.

- **Self management of chronic diseases app.** This application, also present in the elderly companion device, records the day to day activities and any incident regarding different chronic diseases managed by the elderly. The study of the behavior of self-care in the elderly can result in a predictive indicator of care needs for that age group in different contexts [4]. By providing this information to other entities in the system through the Situational-Context caregivers and health professional can provide better attention to the elderly and optimize the use of resources.
- **Controlled cross talk platform.** The goal of this system is to serve as the interface between the elderly and their professional caregivers. To simplify these communications this platform translate the medical language of the professionals to the natural language of their users. In order to do that, a clinical controlled language is used that appears perfectly natural, but it is in fact a formal language that is computer-processable and can be unambiguously translated [13]. By accessing the information of the previous applications through the Situational-Context, the platform can provide richer communications between the elderly and their caregivers.
- **Multidimensional integrated assessment platform.** This platform is designed to be used by health professionals. By using all the information originated from the rest of the applications in the system, caregivers can apply a model of care based on a multidimensional assessment of the elderly functionality that can be constantly adapted as new information is gathered [10]. Once the functionality and care needs of the elderly are established, they

are integrated, by means of the Situational-Context, into the virtual profile so they can be later consulted by other devices.

– **Representation of knowledge guided by biomedical ontologies.** Finally, all the health care information interchanged between the different application is represented by biomedical ontologies. This allow the Situational-Context to have the information semantically controlled with all the technical and scientific validation assured [12].

The comprehensive virtual profile of the elderly created by combining the information of these sources will simplify the elderly and caregivers interaction among themselves and also with IoT systems. They will also create a breeding ground for innovation in eldercare in sparsely populated areas.

In that regard, the authors are working on several aspects emerging from the described virtual profiles. First, the creation of an advanced age ontology. An extension of the existing biomedical ontologies that takes into account all the details and specificities of the elderly living in lowly populated regions.

Second, the profiles can be used to improve non-invasive monitoring applications, like the one presented in [2]. By having a common representation of the elderly information through the Situational-Context, monitoring applications can gather richer information from IoT systems.

And third, with more advanced monitoring applications more seamless interactions with other systems can be orchestrated transparently for the users. As stated above, elderly living in sparsely populated regions like Extremadura or Alentejo usually do not have the technical skills needed to interact with complex technological systems. Therefore, the use of the richer information gathered can be used to decrease the burden of interacting with healthcare solutions.

Summarizing, we believe that the use of the Situational-Context paradigm to create virtual profiles of the elderly by integrating information from a set of different health care solutions will generate an integrated environment that would simplify the daily usage of those systems for both the elderly and their caregivers. It will also help further innovation in eldercare, specially for sparsely populated regions with an aged population.

4   Related Works

The generalized aging of the population has made eldercare a growing subject of study. Numerous works have analyzed different aspects of bringing technological advances to this area.

From a technical perspective, works like [9,15] focus on monitoring the elderly with cognitive disorders. The obtained information is then used to help the elderly in their daily activities or to raise an alarm if needed. As demonstrated in [2], a richer contextual information helps to enhance the result of this type of applications.

Other works try to improve the quality of life of the cognitively impaired elderly. In [17], Vuong et al. review some of the most relevant works in this area.

Most of the studies they review share features with the present work. Moreover, in the present work we have tried to address some of the shortcomings that Vuong et al. identify in their review. In particular, monitoring as much contextual information as possible to obtain richer virtual profiles of the elderly.

The virtual profiles of the elderly proposed in this work can further improve the results of monitoring applications. Additionally, works focusing on specific diseases are not prepared to deal with the comorbidity usually found in the elderly and the lack of resource for healthcare in sparsely populated regions.

From a healthcare perspective, works like [4,10] propose interesting approaches to enhance eldercare in these regions. Their inclusion in a techno-logical proposal like the one presented here would facilitate their application in resources constrained healthcare budgets without raising the need for technical skills by the users.

Finally, to the best of the authors' knowledge, there has been no work focusing on virtually modeling the elderly in aged, sparsely populated regions.

5 Conclusions and Future Work

In this work we have presented an early stage proposal to use the Situational-Context to combine information from different sources in order to create com-prehensive virtual profiles of the elderly. These profiles can be used to integrate the needs and preferences of the elderly into different IoT systems.

These profiles are the first step in an ambitious project to further innovation in eldercare in aged, sparsely populated regions like Extremadura or Alentejo. However, there is still a significant amount of work required. The different sources of information have to be extended to not only provide information to the virtual profiles but also to take advantages of them to improve their interactions with the elderly. We are also currently working on improving the complex information that can be obtained from processing the virtual profiles. As more information is available more complex inferences can be made, which opens the door to more advanced seamless interactions between the elderly and IoT systems.

Acknowledgments. This work was supported by the Spanish Ministry of Economy, Industry and Competitiveness (TIN2015-69957-R (MINECO/FEDER)), by 4IE project (0045-4IE-4-P) funded by the Interreg V-A España-Portugal (POCTEP) 2014-2020 program, by the Department of Economy and Infrastructure of the Government of Extremadura (GR15098), and by the European Regional Development Fund.

References

1. Berrocal, J., García-Alonso, J., Canal, C., Murillo, J.M.: Situational-context: a unified view of everything involved at a particular situation. In: Bozzon, A., Cudré-Mauroux, P., Pautasso, C. (eds.) Proceedings of the 16th International Conference on Web Engineering, ICWE 2016. Lecture Notes in Computer Science, vol. 9671, Lugano, Switzerland, 6–9 June 2016, pp. 476–483. Springer, Cham (2016). https://doi.org/10.1007/978-3-319-38791-8_34

2. Berrocal, J., García-Alonso, J., Murillo, J.M., Canal, C.: Rich contextual information for monitoring the elderly in an early stage of cognitive impairment. Pervasive Mob. Comput. **34**, 106–125 (2017). https://doi.org/10.1016/j.pmcj.2016.05.001

3. Caceres, R., Friday, A.: Ubicomp systems at 20: progress, opportunities, and challenges. IEEE Pervasive Comput. **1**, 14–21 (2011)

4. Fonseca, C., Lopes, M., Fonseca, C., Lopes, M.: Modelo de autocuidado para pessoas com 65 e mais anos de idade, necessidade de cuidados de enfermagem. Universidade de Lisboa, p. 195 (2013). http://hdl.handle.net/10451/12196

5. Gronli, T.M., Ghinea, G., Younas, M.: Context-aware and automatic configuration of mobile devices in cloud-enabled ubiquitous computing. Pers. Ubiquit. Comput. **18**(4), 883–894 (2014)

6. Guillén, J., Miranda, J., Berrocal, J., García-Alonso, J., Murillo, J.M., Canal, C.: People as a service: a mobile-centric model for providing collective sociological profiles. IEEE Softw. **31**(2), 48–53 (2014). https://doi.org/10.1109/MS.2013.140

7. Hong, J.Y., Suh, E.H., Kim, S.J.: Context-aware systems: a literature review and classification. Exp. Syst. App. **36**(4), 8509–8522 (2009)

8. Kobsa, A.: Generic user modeling systems. User Model. User-adapted Interact. **11**(1–2), 49–63 (2001)

9. Lin, C.C., Lin, P.Y., Lu, P.K., Hsieh, G.Y., Lee, W.L., Lee, R.G.: A healthcare integration system for disease assessment and safety monitoring of dementia patients. IEEE Trans. Inf. Technol. Biomed. **12**, 579–586 (2008)

10. Lopes, M.J., Escoval, A., Pereira, D.G., Pereira, C.S., Carvalho, C., Fonseca, C.: Evaluation of elderly persons' functionality and care needs. Rev. Lat. Am. Enferm. **21**, 52–60 (2013)

11. Marzano, S.: The New Everyday: Views on Ambient Intelligence. 010 Publishers, Rotterdam (2003)

12. Mendes, D., Rodrigues, I., Rodriguez-Solano, C., Baeta, C.: Enrichment/population of customized CPR (computer-based patient record) ontology from free-text reports for CSI (computer semantic interoperability). J. Inf. Technol. Res. **7**(1), 1–11 (2014)

13. Mendes, D.: Clinical Practice Knowledge Acquisition and Interrogation using Natural Language. Ph.D. thesis (2014). https://www.rdpc.uevora.pt/bitstream/10174/12553/1/dissertation_Dm_PhDi.pdf

14. Miranda, J., Mäkitalo, N., García-Alonso, J., Berrocal, J., Mikkonen, T., Canal, C., Murillo, J.M.: From the internet of things to the internet of people. IEEE Internet Comput. **19**(2), 40–47 (2015). https://doi.org/10.1109/MIC.2015.24

15. Mulvenna, M., Nugent, C., Moelaert, F., Craig, D., Draes, R.M., Bengtsson, J.: Supporting people with dementia using pervasive healthcare technologies. In: Mulvenna, M.D., Nugent, C.D. (eds.) Supporting People with Dementia Using Pervasive Health Technologies. Advanced Information and Knowledge Processing, pp. 3–14. Springer, London (2010)

16. Rodrigues, R., Huber, M., Lamura, G.: Facts and Figures on Healthy ageing and long-term care. European Centre for Social Welfare Policy and Research, Vienna (2012)

17. Vuong, N., Chan, S., Lau, C.: mHealth sensors, techniques, and applications for managing wandering behavior of people with dementia: a review. In: Adibi, S. (ed.) Mobile Health. Springer Series in Bio-/Neuroinformatics, vol. 5, pp. 11–42. Springer, Cham (2015)

Towards the Design of Avatar-Based Therapies for Enhancing Facial Affect Recognition

Arturo S. García[1(✉)], Elena Navarro[1,2], Antonio Fernández-Caballero[1,2], and Pascual González[1,2]

[1] Instituto de Investigación en Informática de Albacete,
Universidad de Castilla-La Mancha, 02071 Albacete, Spain
{arturoSimon.Garcia,Pascual.Gonzalez}@uclm.es
[2] Centro de Investigación Biomédica en Red de Salud Mental (CIBERSAM),
28029 Madrid, Spain

Abstract. This paper focuses on improving facial affect recognition, which is a fundamental aspect of social cognition. This is an important issue, as deficits in social cognition influence the functionality and quality of life of patients with brain disorders. In this regard, a proposal for designing avatar-based therapies dedicated to impaired patients is introduced. In addition, the validation of the tool by expert therapists is presented.

Keywords: Social cognition · Facial affect recognition · Therapy
Virtual reality · Avatar

1 Introduction

Social cognition refers to cognitive processes involved in social interactions, which incorporates processes of perceiving, interpreting and generating responses to the intentions, dispositions, and behaviours of others [4,13]. Moreover, it is fundamental to the proper functioning and development of humans as social beings. Social cognition greatly influences the functionality and quality of life of patients with brain disorders [12]. Thus it seems essential to accurately explain social stimuli during interactions with others. In fact, social cognition impairments affect social relations and work environments. Several diseases such as schizophrenia [10], autism spectrum disorders [2], and acquired brain injury (ABI) [16] have been associated with social cognition deficits.

Social cognition is usually described as multidimensional, incorporating four principal domains: emotional processing, social perception, attributional bias and theory of mind [14]. One of them, emotional processing, is related to the capacity to perceive, recognise and manage emotions [6,11]. Some non-verbal behaviours play an important role in emotional processing, including facial expressions and body movements [17]. One of the most frequent therapies to

© Springer Nature Switzerland AG 2019
P. Novais et al. (Eds.): ISAmI 2018, AISC 806, pp. 306–313, 2019.
https://doi.org/10.1007/978-3-030-01746-0_36

improve emotion recognition is by means of photographs showing a specific affect. However, the main criticism to the use of images is associated to its low ecological validity [9,15]. Indeed, they are static stimuli, whilst temporal aspects of facial motion are also relevant to the recognition of expressions. Thus, it is paramount to include not only static images but also dynamic ones to improve such process [9]. Because patients and their specific disabilities are so diverse, an important drawback is that a huge database of expressions is necessary to include different race, age and gender. The growth is even higher when images or videos of faces are considered from different angles.

Virtual humans can be used to overcome these limitations. They are depicted as human representations inside a virtual world and are either computer-controlled agents or avatars controlled by a real person. Virtual humans are modelled and animated to match any combination of race, age and gender, being observed from any angle, under any lighting condition and in any social context (an office, a pub, etc.). Moreover, they can have a conversation with the observer and even react to that dialogue. Up to date, there have been some promising approaches that employ agents for facial affect recognition, mainly tested with people affected by autism [1,3]. However, there are still few papers about the use of remotely-controlled avatars [7].

Therefore, this paper investigates the requirements and important functional characteristics of a software framework aimed at supporting therapists to conduct social cognition therapies through the use of remotely-controlled avatars. Moreover, a preliminary evaluation of the framework with expert therapists is also outlined.

2   Towards Avatar-Based Therapies to Interpret Facial Affect

The main motivation for the design of future therapies based on remotely-controlled avatars is that they provide the therapist with a practically unlimited set of avatars, which can even speak accordingly to their physical representation. Thus, they can be used to have a conversation with the patient and express any emotion at any time. Selecting the right type of person to interact with the patient also results beneficial to engage and encourage them. Moreover, the fact of having patients inside a controlled virtual environment opens the door to monitor them. For example, it is possible to keep track of eye gaze to identify whether their problem recognising facial affect is a result of an inability to carry out a correct face exploration.

In addition, we believe that a therapist who puppeteers an avatar would react better than computer-controlled agents to unpredictable behaviours of patients affected by dissimilar deficits, mainly due to all the possible variations in their symptoms.

2.1 Requirements of Avatar-Based Therapies for Facial Affect Recognition

After analysing the characteristics of future avatar-based therapies and through co-creation workshops with expert therapists, the following list of system requirements was extracted for the tool:

– Avatar selection: The patient should be able to select the avatar that makes them feel more comfortable with.
– Facial affect recognition: The system should recognise facial expressions of the therapist to identify the emotion that they want to transmit to the patient. A set of seven different emotions (neutral, joy, anger, fear, disgust, sadness and surprise), based on universal emotions [5], with two levels of intensity (strong and mild emotion), were selected for their use during the therapy.
– Ability to modify the facial expression of the avatar: The system should allow the modification of the avatar's facial expression. Ultimately, this facial expression will be selected by the therapist to reflect it on the patient's user interface.
– Voice chat: Therapist and patient should be able to speak to each other.
– Voice modulation: The therapist should be able to modify his/her voice so that it matches the age and gender of the employed avatar.
– Ability to observe the patient's reactions: The therapist should be able to see the patient's reactions so that he/she can adapt the therapy to each particular patient.
– Ability to see where the patient is looking into the scene: The therapist should be able to see where the patient is looking inside the virtual world as he/she may be distracted or not paying attention to the avatar.
– Transmission of data: The system should support information exchange between therapist and patient systems, as they may be geographically distributed and connected through the Internet or a local network.
– Real time interaction: The system should allow for real time interplay between therapist and patient.

Once these functional requirements were extracted, we explored the use of our recently described tool [8] as an approach to implement this type of therapies. This section also provides a short description of the tool and its system architecture as well as the procedure followed during the execution of the therapy.

2.2 Description of the Distributed Therapies Supporting Tool

This tool was designed as a generic test-bed for the implementation of social cognitive deficit therapies by using avatars. Its main goal is to convey voice, facial expressions and body movements of a user (therapist) by means of an avatar so that he/she interacts remotely with another user (patient) to run several therapies without meeting face-to-face. It is based on a distributed layer-based architecture whose main components are:

- Capture Hardware: This refers to the equipment used on the therapist's side of the distributed application that gathers body movements, face and voice.
- Software Processing: Once the previous stimuli have been gathered, a software layer processes them to obtain the therapist's facial expression and modulated voice (aimed at adapting to the avatar's age and gender).
- Networking Software: The previous processed stimuli are sent through a networking layer to the patient's side application.
- Avatar Representation Software: The patient's application implements the final avatar characterisation by combining the information received into avatar's expressions, lip movements and voice.

Figure 1 depicts the tool's architecture used as an extension of a previous proposal [8]. The main differences are (1) references to body movement acquisition, transformation, transmission and representation have been removed, as they are not needed here, and (2) a division into therapist and the patient systems (left and right of the figure, respectively) is provided.

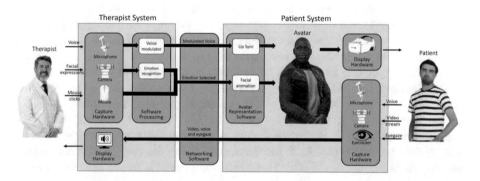

Fig. 1. System architecture to convey voice and facial expression.

This architecture fulfills the requirements for avatar-based therapies by delivering different modules. From the therapist perspective, Capture Hardware provides audio and video streams, and mouse clicks to the Software Processing. Within this stage, the Voice Modulator module takes the therapist's voice stream as input and supplies a modulated voice matching age and gender of the avatar selected. The Emotion Recognition module takes the video feed as input and issues one of the thirteen different emotions identified as output. The same emotion output can be generated by directly selecting one affect on the user interface. These two outputs are sent through the network and received by the Patient System. Within the Avatar Representation Software, the LipSync module is used to animate the avatar mouth and the Facial Animation module to change the facial expression of the avatar. This avatar is visualised using the patient's Display Hardware. For patient-generated data, Capture Hardware provides the voice signal, a video stream of the camera and the eye gaze data that are sent through the network to the Therapist System where it is displayed and reproduced.

This architecture is implemented under Unity3d (https://unity3d.com) by using a set of different assets for acquisition, processing, transmission and avatar representation [8].

3   Therapy Execution

The execution of this kind of therapies can be divided in two main steps for both the therapist and the patient: pre-therapy and therapy. From the therapist point of view, pre-therapy consists of emotion calibration and session configuration. The former allows the association of the therapist's facial expression to an emotion, while the latter is used to select an avatar and tweak the voice modulation (if it is needed). As already mentioned, the recognition system makes use of seven different emotions with two levels of intensity. Thus, the therapist needs to train the system for each of them before starting the therapy.

During the therapy, the therapist can select the facial expression of the avatar at any time by means of a button panel on the screen, or by using his/her own facial expression. This has been considered to be convenient as the therapist can choose to act and impersonate the character represented by the avatar, or select the emotion by using the mouse, accordingly to the needs of a particular stage of the therapy. The left side of Fig. 2 shows the main screen of the therapist. As can be seen, a live video stream of the patient is included in the user interface of the therapist, as well as the point where the patient is looking inside the virtual environment. It is represented by a semi-transparent red sphere that is only visible to the therapist. A panel informs about the state of the connection and provides the therapist with three buttons to conclude the session, sending different feedback to the patient screen.

From the patient point of view, the pre-therapy stage includes the connection to the session created by the therapist. In this prototype, it can be done either by introducing the IP address of the therapist computer, or by automatic discovery

Fig. 2. Screen shots taken during the therapy. Left: the therapist screen. Right: The patient screen.

of the session inside the same local network. During the therapy, the patient can experience the virtual environment through a Head Mounted Display (HMD) or using a standard screen. The HMD is preferable as it provides higher levels of immersion and allows an easy change of the viewpoint (as the perspective is associated with the patient's head position and orientation), as can be seen in Fig. 2. At this stage, the interaction with the therapist is reduced to the live video stream and the audio link, as the patient is not aware of the eye-tracking capabilities of the system. Finally, when the therapist concludes the session, a summary screen provides feedback about the performance of the patient, using always positive messages to encourage them to improve and avoid demotivation.

4   Tool Validation

The main objective of this functional validation was to get feedback from experts in the field. Thus, in these validation, five expert therapists tested the system to assess if all the functionality and actions described could be performed using the system developed. Moreover, another objective was to get some valuable ideas about new features that could help to improve the system.

Since this was the first experience of the therapists with avatar-based virtual reality technology, a demonstration session showed them the capabilities of the system before they tested it by themselves. This was followed by the calibration of the facial emotion recognition software so that this feature could be used during the tests. After that, they created a session by selecting a predefined avatar and voice, and then the patient joined. They were able to speak to each other and the avatars' mouth was animated accordingly. During these talks, the therapist was able to modify the facial expression of his/her avatar, and that was reflected on the patient screen.

These early tests did not consider the use of real patients, as we needed a functional validation prior to its real application. Therefore, the participants of the evaluation played the role of the therapist, while a member of the development team played the role of the patient. To help the participants to understand the results of their actions, both therapist and patient setups were based on the same room. This allowed them to quickly compare the therapist and the patient screens. However, in a real scenario, they would need to be in different rooms. Both the therapist and the patient setups consisted of a high-end desktop computer, a web cam and headphones. The only difference was that the patient could choose to use a head mounted display with eye-tracking embedded capabilities (https://www.getfove.com).

We used different techniques to get feedback from the therapists. We encouraged them to think aloud during the validation, observed how they coped with the system and interviewed them afterwards. The validation of the tool demonstrated that the system performed properly, supporting the interaction between the patient and the therapist. The participants did not have problems to control the facial emotions of the avatar, neither using the user interface nor the video-based facial expression recognition. Regarding the participants' expectations, all of them mentioned that the system opened new possibilities of interaction that may improve the treatment of social cognitive deficits. However, it would be interesting to test it with real patients so that they could see their reactions and the acceptance level of the tool.

We received some valuable ideas about future improvements of the system. Some participants stated that it would be interesting to have a wider range of avatars to choose from, covering all sorts of ages and races, as the prototype they tested only had a small number of them. Ideally, the system would include an avatar customisation that would allow the therapists to adapt the avatar to any patient who will use it during a therapy. They requested this because therapists spend time becoming familiar with their patients, and they know if they are more receptive to interact with a specific type of person. They also suggested that it would be interesting to give the patients the possibility to reply to questions using a set of predefined answers. In their experience, this would help to focus the conversation in a particular topic, making the talk more dynamic and reducing digression. Finally, even though the sessions are recorded so that they can be recalled later on, they missed some summary information that could help them to validate the therapy's efficiency. One suggestion was to include a heat map of the eye gaze of the patient, as this information is not available on face-to-face meetings and would be interesting to study its application in real therapies.

5   Conclusions

This paper has introduced a proposal for designing avatar-based therapies. The approach bases on a recently described tool aimed at enhancing social cognition in impaired patients. Firstly, the requirements of avatar-based therapies for facial affect recognition have been established. Then, a brief description of the therapies supporting tool has been introduced. Afterwards, the execution of therapies has been described. The implementation has been subdivided into pre-therapy and therapy, both for therapist and patient.

In addition, the tool has been validated by a series of expert therapists. Their opinion has been quite positive and encouraging. Moreover, the validation has helped improving some unforeseen aspects on therapies oriented towards enhancing social cognition abilities.

Acknowledgments. This work was partially supported by Spanish Ministerio de Economía, Industria y Competitividad, Agencia Estatal de Investigación (AEI)/European Regional Development Fund under EmoBioFeedback (DPI2016-80894-R), HA-SYMBIOSIS (TIN2015-72931-EXP) and Vi-SMARt (TIN2016-79100-R) grants, and by the Centro de Investigación Biomédica en Red de Salud Mental (CIBERSAM) of the Instituto de Salud Carlos III.

References

1. Abirached, B., Zhang, Y., Aggarwal, J.K., Fernandes, T., Carlos, J., Orvalho, V.: Improving communication skills of children with ASDs through interaction with virtual characters. In: IEEE 1st International Conference on Serious Games and Applications for Health, pp. 3493–3502. IEEE (2011)

2. Baron-Cohen, S., Wheelwright, S., Skinner, R., Martin, J., Clubley, E.: The autism-spectrum quotient (AQ): evidence from Asperger syndrome/high-functioning autism, malesand females, scientists and mathematicians. J. Autism Dev. Disord. **31**(1), 5–17 (2001)
3. Beer, J.M., Fisk, A.D., Rogers, W.A.: Emotion recognition of virtual agents facial expressions: the effects of age and emotion intensity. In: Proceedings of the Human Factors and Ergonomics Society, vol. 53, No. 2, pp. 131–135 (2009)
4. Brothers, L.: The neural basis of primate social communication. Motiv. Emot. **14**(2), 81–91 (1990)
5. Ekman, P.: Universals and cultural differences in facial expressions of emotion. In: Nebraska Symposium on Motivation, pp. 207–282. University of Nebraska Press, Lincoln (1972)
6. Fernández-Caballero, A., Martínez-Rodrigo, A., Pastor, J.M., Castillo, J.C., Lozano-Monasor, E., López, M.T., Zangr'oniz, R., Latorre, J.M., Fernández-Sotos, A.: Smart environment architecture for emotion detection and regulation. J. Biomed. Inf. **64**, 55–73 (2016)
7. Fernández-Caballero, A., Navarro, E., Fernández-Sotos, P., González, P., Ricarte, J.J., Latorre, J.M., Rodriguez-Jimenez, R.: Human-avatar symbiosis for the treatment of auditory verbal hallucinations in schizophrenia through virtual/augmented reality and brain-computer interfaces. Front. Neuroinformatics **11**, 64 (2017)
8. García-Sánchez, M., Teruel, M.A., Navarro, E., González, P., Fernández-Caballero, A.: A distributed tool to perform dynamic therapies for social cognitive deficit through avatars. In: Ubiquitous Computing and Ambient Intelligence, pp. 731–741. Springer, Cham (2017)
9. Garrido, M.V., Lopes, D., Prada, M., Rodrigues, D., Jerónimo, R., Mourão, R.P.: The many faces of a face: comparing stills and videos of facial expressions in eight dimensions (SAVE database). Behav. Res. Methods **49**(4), 1343–1360 (2016)
10. Gottesman, I.I.: Schizophrenia Genesis: The Origins of Madness. WH Freeman/Times Books/Henry Holt & Co., New York City (1991)
11. Green, M.F., Penn, D.L., Bentall, R., Carpenter, W.T., Gaebel, W., Gur, R.C., Kring, A.M., Park, S., Silverstein, S.M., Heinssen, R.: Social cognition in schizophrenia: an NIMH workshop on definitions, assessment, and research opportunities. Schizophr. Bull. **34**(6), 1211–1220 (2008)
12. Lahera, G., Ruiz-Murugarren, S., Iglesias, P., Ruiz-Bennasar, C., Herrería, E., Montes, J.M., Fernández-Liria, A.: Social cognition and global functioning in bipolar disorder. J. Nerv. Ment. Dis. **200**(2), 135–141 (2012)
13. Pinkham, A.E., Penn, D.L., Green, M.F., Buck, B., Healey, K., Harvey, P.D.: The social cognition psychometric evaluation study: results of the expert survey and RAND panel. Schizophr. Bull. **40**(4), 813–823 (2014)
14. Pinkham, A.E., Penn, D.L., Green, M.F., Harvey, P.D.: Social cognition psychometric evaluation: results of the initial psychometric study. Schizophr. Bull. **42**(2), 494–504 (2016)
15. Roark, D.A., Barrett, S.E., Spence, M.J., Abdi, H., O'Toole, A.J.: Psychological and neural perspectives on the role of motion in face recognition. Behav. Cogn. Neurosci. Rev. **2**(1), 15–46 (2003)
16. Rosenberg, H., McDonald, S., Dethier, M., Kessels, R.P., Westbrook, R.F.: Facial emotion recognition deficits following moderatesevere traumatic brain injury (TBI): Re-examining the valence effect and the role of emotion intensity. J. Int. Neuropsychol. Soc. **20**(10), 994–1003 (2014)
17. Tracy, J.L., Randles, D., Steckler, C.M.: The nonverbal communication of emotions. Curr. Opin. Behav. Sci. **3**, 25–30 (2015)

On the Feasibility of Blockchain
for Online Surveys with Reputation
and Informed Consent Support

Hélder Ribeiro de Sousa[1(✉)] and António Pinto[1,2]

[1] CIICESI, ESTG, Politécnico do Porto, Porto, Portugal
8100151@estg.ipp.pt
[2] CRACS & INESC TEC, Porto, Portugal
apinto@inesctec.pt

Abstract. Economical benefits obtained by large Internet corporations from gathering and processing user information at a global scale led the European Union to legislate on behalf of individual rights and the privacy of personal information. Data collectors, in particular, must now obtain proof of the user's consent for every single operation comprising their data. Considering the conflicting interests of all involved parties, we propose that consent should be stored in a blockchain. By being a distributed, immutable and verifiable ledger, the blockchain presents itself as an almost tailor-made solution to harmonize conflicting interests while enabling the regulators' supervision.

Keywords: Blockchain · Surveys · Informed consent · GDPR
Reputation

1 Introduction

The Internet and social media applications have clearly shown that there is economical benefit in gathering and processing individual user information at a global scale. Large Internet corporations, and their avidity for profit, may hamper the protection of individual persons' rights. Moreover, recent cases have also shown that big corporations may not be prepared to maintain the required confidentiality of the personal information of their users. Example of such, is the recent data leak of Equifax that could potentially affect 143 million US citizens [1]. Legislation aiming at the protection of the rights of individual persons and international data flows is starting to appear [2,3]. European Union's recent General Data Protection Regulation (GDPR) is being identified by the market as a game changer, mostly due to its commitment to protect the rights of individual persons and the high values that foreseen penalties can reach. Particularly, the

This work is partially funded by the ERDF through the COMPETE 2020 Programme within project POCI-01-0145-FEDER-006961, and by National Funds through the FCT as part of project UID/EEA/50014/2013.

© Springer Nature Switzerland AG 2019
P. Novais et al. (Eds.): ISAmI 2018, AISC 806, pp. 314–322, 2019.
https://doi.org/10.1007/978-3-030-01746-0_37

GDPR states that the person is the sole owner of all information regarding himself and the only one with control over it.

Online surveys, online forms or similar technologies, are a common way of information gathering where users are compelled to fill these forms with personal information or with information regarding their opinion or habits. In this situations, the GPDR imposes new requirements, such as the informed consent, the right to be forgotten, among others. The informed consent, in particular, requires the online survey platform to inform the user about the purpose of the information being collected and for how long said information will be stored while using plain language. During this interaction the entity behind the survey platform must also obtain proof of the user's consent.

Blockchain, initially described in [4], may be used as the underlying, distributed, verifiable and immutable ledger for proof of consent persistence. The work herein consists in evaluating the use of blockchain for this purpose.

The paper is organized in sections. Section 2 describes the blockchain technologies, while Sect. 3 describes reputation systems. The proposed solution is presented in Sect. 4. Section 5 concludes the work and proposed some future work.

2   Blockchain

Blockchain and digital money have gained recent attention from both the academia and from society in general. This is mainly due to the appearance of Bitcoin [5] and its use of blockchain as a way of solving the double spend problem of digital money. Currently, has seen in any digital money market, multiple *crypto-currencies* are available. Blockchain solutions have evolved from being an immutable ledger of Bitcoin transactions, to enable more versatile and robust interactions between peers, namely smart-contracts [6–8].

The different flavors of blockchain solutions all base their operations on being a distributed, immutable and verifiable ledger. Such ledger is achieved by means of cryptographic operations over sets of transactions, in sequence, that are verified by the peers. A block, an element in the chain, comprises: a non-repeating random number (nonce); a list of transactions; the cryptographic hash value of the previous block; and the cryptographic hash value of itself. This way, a future block will include a reference to its previous block, hence the name blockchain. This means that changing one block will have impact if all the blocks that follow it and, thus, the bigger the blockchain, the harder it is to change a past block. The blockchain is stored in a public and distributed manner. All peers that wish to do so, can store a complete copy of the blockchain on their machine. Adding a new block to this distributed storage requires some form of agreement between peers. The agreement protocol used in Bitcoin was Hashcash and consists in finding a nonce that, in combination with the rest of the block information, generates a hash value with a leading number of zeros (eighteen zeros, currently) [9].

Solutions, other than Hashcash, can be used and are widely named as proof-of-work or mining. The rationale is that if you present a valid hash value, it is assumed that you have performed the required computations to discover it. Due to its distributed nature, and by assigning value to these proof-of-works, one is generating involvement of the peers in making the ledger continuously secure. Mining is a computational demanding operation that can be maintained as such by increasing the number of leading zeros in the hash that serves as proof-of-work. Differences in the blockchain may occasionally appear, as distinct peers may try to add different blocks to the end of the blockchain at the same time. In such situations, peers must always adopt the longest blockchain.

While blockchain fundamentals guarantee the integrity of the stored transactions, they do not enforce any business logic or requirements. In Bitcoin, business logic consists solely in verifying that when one person wants to send money to another there are enough credits in his digital wallet. Such operation is performed by verifying all previous transactions associated with that specific wallet, thus, making sure there is enough credit to perform the new transaction and attesting the legitimacy of the money ownership. After which, a new transaction is created, assigning the money to the second person. This form of operation was initially named as smart contracts by Nick Szabo [10]. Despite its more restricted role in Bitcoin, smart contracts must be seen as a broader form of business logic description. In the Ethereum platform blockchain [11] a more generalized form of transactions is foreseen, enabling its use in a multitude of scenarios. The authors proposed the adoption of Solidity, a Turing-complete language for transaction specification, meaning that the Ethereum blockchain can be used to implement virtually any business logic and requirements.

3 Reputation Systems

The key purpose of a reputation system is to allow people to assign a reputation indicator of a third party's trust-ability. This reputation indicator is confined to a specific domain and opinions are expressed by assigning values to the other party's behavior in a particular transaction or interaction. The origin of reputation systems goes back to real life interactions, when people exchange opinions about services or other people. Professional recommendation letters are another example of such [12].

The advent of e-commerce and the massive worldwide adoption of platforms, such as those from eBay and Amazon, for instance, in the last decade of the XX century, triggered the need for online reputation systems [12]. The more recent, global adoption and use of smartphones and the ubiquity of Internet access in these devices led to a second wave in the adoption and use of reputation systems. Moreover, such an availability of Internet connected smartphones also triggered the ubiquity of reputation systems to an unthinkable degree. In its most recent versions, reputation systems are provided by the platforms themselves and are seen as a key module for the community, aiming at the experience exchange between users and thus fostering trust [13]. In some cases, the instigation of the use of reputation systems goes to such lengths that discounts, loyalty points or similar benefits are offered to users that rate their interactions with other members of the community.

The way a user shares its experience may differ from one platform to another. Some adopt a classification scale while others expect users to write small reviews, whereas some others combine both approaches. The most common experience representation setups are: binary, numerical, symbolical, textual or audiovisual. In binary input systems, the user expresses if he liked the experience or not. In numerical, the user inputs a level from 1 to five stars, for instance. In symbolical, the user leaves feedback by clicking on a symbol that represents his experience, such as a thumbs up icon or a like button. In textual, the user inputs a short review in the form of a text. In audiovisual, the user provides a video or audio record of his spoken experience [13].

4   Proposed Solution

The challenges introduced by EU's new data protection regulation are multiple. Among them is the requirement to obtain the user's consent to collect and process his data. Moreover, such consent must be verifiable by a data protection authority. The proposed solution addresses such challenges by leveraging a distributed ledger.

The proposed solution is expected to work independently of the underlying blockchain technology and the collected data structure. Nonetheless, the initial specification was written in Solidity, one of the Ethereum Platform smart contracting languages, and considers the case of data collection through online forms. Ethereum's platform was adopted, in detriment of others, due to its extensibility [11].

The proposed solution's architecture is depicted in Fig. 1 and comprises four macro components: (1) the data subject's device; (2) the data protection regulator's infrastructure; (3) the data controller's infrastructure; (4) the blockchain instance. The data subject's device is the one used by the data subject to interact with the system, both for filling surveys and to interact with the mobile application. The data protection regulator's infrastructure is the one that will be interacting with the blockchain instance to validate the data processing operation's legitimacy. Since its internal architecture is unknown, it will integrate with the blockchain instance through interfaces relying on common web communication protocols (e.g. REST). The data controller's infrastructure is the one where the regulator's services and the collected data will be hosted. Finally, the blockchain instance is the component used to persist the system's tamper-proof data.

The adopted use case starts with the data subject filling a survey through the data controller's survey platform. Once that initial step is finished, the data subject will then interact with the proposed solution using a mobile application that is paired with the data controller's back-end. The pairing consists in the exchange of previously generated RSA-2048 public keys through a secure channel, such as a secure web connection, between the smart-phone and the back-end (both generate a key-pair). From then on, this cryptographic material will be used to secure all communication channels; to record the data subject's consent; and to make proof that the data subject's consent was, in fact, given. The

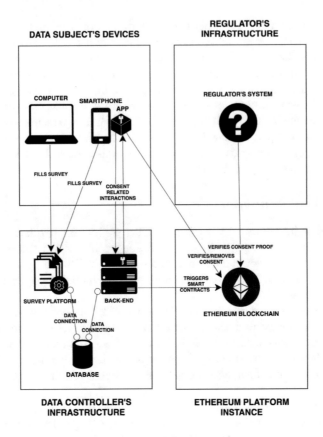

Fig. 1. Proposed solution's architecture

data controller will also be able to request the data subject's consent by sending a push notification to his smart-phone. Upon action, the data subject will be presented with the all GDPR mandatory information and given the option on whether to give his consent or not. The data controller's back-end component will store evidence of the data subject's consent in the blockchain. Finally, the regulator will be able to traverse the blockchain whenever a consent needs to be verified.

The selection of the RSA cryptosystem was based on similar adoptions of this mechanism by different European states, such as Austria, Belgium, Estonia, Finland, Germany, Italy, Liechtenstein, Lithuania, Portugal and Spain, for their citizens' ID cards. Moreover, the 2048-bit key-pair size is secure enough while maintaining a reasonable computation/power cost on mobile devices [14].

```
1   contract Consent {
2
3       address owner;
4       string public uid;
5       uint public timestamp;
6       string public dataDigest;
7       string public consentTerms;
8       string public ctrllerIdentifier;
9       string public ctrllerPubKey;
10      string public ctrllerDataDigestAndUidSig;
11      string public ctrllerConsentTermsAndUidSig;
12      string public subjectIdentifier;
13      string public subjectPubKey;
14      string public subjectDataDigestAndUidSig;
15      string public subjectConsentTermsAndUidSig;
16
17  }
```

Listing 1. Smart contract for consent

The data subject's consent will be stored in the blockchain by leveraging Ethereum Platform's smart contracting feature. For that purpose, three different smart contracts named "Consent", "RemConsent" and "Reputation" were written in Solidity. These smart contracts will originate blocks of three different types. The "Consent" smart contract will originate blocks that store data regarding a given consent. The "RemConsent" smart contract will originate blocks that store data regarding a consent removal. Finally, The "Reputation" smart contract will originate blocks that store the data regarding each subjects view on an interaction with a data controller.

The "Consent" smart contract (as shown in Listing 1) comprises, among other data, identifiers, a timestamp, cryptographic material and the terms of the consent. The *uid* uniquely identifies the consent operation. The *dataDigest* variable contains the digest of all collected information, while *ctrllerDataDigestAndUidSig* and *subjectDataDigestAndUidSig* contain the result of signing *dataDigest+uid* with the controller's and the subject's private keys, respectively. The *consentTerms* variable contains the text presented to the data subject in order to explain him what is being collected, by whom, for how long and with what purpose. Analogously, *ctrllerConsentTermsAndUidSig* and *subjectConsentTermsAndUidSig* contain the result of signing *consentTerms+uid* with the controller's and the subject's private keys, respectively. This behaviour guarantees that both the controller and the subject have agreed with the same consent terms for the same data. The inclusion of the Consent's *uid* variable in the values to be cryptographically signed prevents attackers from forging new consent blocks for the same controller, subject, data and terms.

The mobile application, installed on the data subject's smart-phone, can also be used: to share the subject's view on data controller's conduct (reputation); to check the data controllers' reputation; to review consented terms; to request clarifications about the usage of his data; and to request the removal of a previously given consent.

The support for the removal of a previously given consent, an additional smart contract, named "RemConsent", was written. This smart contract will be triggered whenever the data subject requests, through the use of the mobile application, the removal of a previously given consent. This smart contract creates a new block that stores, among other data, a reference to the consent block it supersedes. The *subjectRemTokenAndConsentUidSig* variable stores a cryptographic evidence of the data subject's will to remove his consent.

```
1  contract Reputation{
2      enum ReputationIndicator {NEGATIVE,NEUTRAL,
3          POSITIVE}
4      uint public timestamp;
5      string public uid;
6      address public consentBlock;
7      ReputationIndicator public repInd;
8      string public repText;
9      string public subject reputationDataSig;
10 }
```

Listing 2. Reputation Smart Contract

A data subject can share his view on an interaction with a data controller by associating a reputation indicator and a text. The reputation indicator can be: **Positive** if the interaction was positive; **Neutral** if the interaction was neither positive nor negative; or **Negative** if the interaction was negative. The text is an optional small textual description of the subject's view on the interaction. The smart contract shown in Listing 2 was written to store the reputation entry on the blockchain. Other stored data includes a reference to the block where the associated consent is stored (*consentBlock*) and the cryptographic proof that the entry was indeed created using the consent block of the data subject (*reputation-DataSig*). The reputation of a data controller's is calculated as the percentage of **Positive** entries of all valid **Positive/Negative** reputation entries. **Neutral** entries are excluded from the calculation.

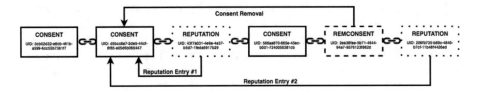

Fig. 2. Block disposition throughout the blockchain

Figure 2 depicts an example of the resulting blockchain, where the proof of a subject's consent will be stored in a block that will be referenced by all further related operations of consent removal and reputation.

5   Conclusions and Future Work

Economical benefits obtained by large Internet corporations from gathering and processing user information at a global scale led the European Union to legislate on behalf of individual rights and the privacy of personal information. The GDPR institutes the individual as the sole owner of all information regarding himself and the only one with control over it. The informed consent, in particular, requires online survey platforms to inform its users about the purpose of the information being collected and for how long said information will be stored while using plain language. During this interaction, the data collector must also obtain proof of the user's consent.

Due to the conflicting interests of all parties involved, the authors propose that consent should be stored in a blockchain instance. By being a distributed, immutable and verifiable ledger, the blockchain presents itself as an almost tailormade solution to harmonize conflicting interests while enabling the regulators' supervision. Given their very specific characteristics, smart contracts are an appropriate and fitting choice for organizations who must comply with different regulatory requirements. In summary, while exploiting blockchain technology, the authors came to conclude that the proposed solution fulfills the most immediate challenges faced by data subjects, data controllers and regulators once the GDPR starts to be enforced.

References

1. CNN: Giant Equifax data breach: 143 million people could be affected (2017). http://money.cnn.com/2017/09/07/technology/business/equifax-data-breach/index.html. Accessed 31 Jan 2018
2. European Union: Regulation (EU) 2016/679 of the European Parliament and of the Council (2016). http://eur-lex.europa.eu/legal-content/EN/TXT/?uri=CELEX:32016R0679. Accessed 31 Jan 2018
3. Sater, S.: Blockchain and the european union's general data protection regulation: a chance to harmonize international data flows. SSRN (2017)
4. Diffie, W., Hellman, M.: New directions in cryptography. IEEE Trans. Inf. Theory **22**(6), 644–654 (1976)
5. Nakamoto, S.: Bitcoin: a peer-to-peer electronic cash system (2008). http://bitcoin.org/bitcoin.pdf
6. Raval, S.: Decentralized Applications: Harnessing Bitcoin's Blockchain Technology. O'Reilly Media Inc., Sebastopol (2016)
7. Swan, M.: Blockchain: Blueprint for a New Economy. O'Reilly Media Inc., Sebastopol (2015)
8. Mougayar, W.: The Business Blockchain: Promise, Practice, and Application of the Next Internet Technology. Wiley, Hoboken (2016)
9. Back, A., et al.: Hashcash-a denial of service counter-measure (2002)
10. Szabo, N.: Formalizing and securing relationships on public networks. First Monday 2(9) (1997)
11. Wood, G.: Ethereum: a secure decentralised generalised transaction ledger. Ethereum Proj. Yellow Pap. **151**, 1–32 (2014)

12. Resnick, P., Kuwabara, K., Zeckhauser, R., Friedman, E.: Reputation systems. Commun. ACM **43**, 45–48 (2000)
13. von Reischach, F., Michahelles, F., Schmidt, A.: The design space of ubiquitous product recommendation systems. In: Proceedings of the 8th International Conference on Mobile and Ubiquitous Multimedia, MUM 2009, pp. 2:1–2:10. ACM, New York (2009)
14. Barker, E., (NIST): Nist special publication 800-57 part 1 revision 4: Recommendation for key management, technical report, National Institue of Standards and Technology (2016)

On the Use of the Blockchain Technology in Electronic Voting Systems

João Alves[1(✉)] and António Pinto[1,2]

[1] CIICESI, ESTG, Politécnico do Porto, Porto, Portugal
8000055@estg.ipp.pt
[2] CRACS & INESC TEC, Porto, Portugal
apinto@inesctec.pt

Abstract. The benefits of blockchain go beyond its applicability in finance. Electronic Voting Systems (EVS) are considered as a way to achieve a more effective act of voting. EVS are expected to be verifiable and tamper resistant. The blockchain partially fulfills this requirements of EVS by being an immutable, verifiable and distributed record of transactions. The adoption of EVS has been hampered mainly by cultural and political issues rather than technological ones. The authors believe that blockchain is the technology that, due to the overall attention it has been receiving, is capable of fostering the adoption of EVS. In the current work we compare blockchain-based EVS, identifying their strengths and shortcomings.

Keywords: Electronic Voting System · Blockchain

1 Introduction

Blockchain [1], as the solution for the double spend problem of digital money in use in the Bitcoin *cryptocurrency* [2], has also received attention by the both the market and the academic community. The benefits of blockchain go beyond its applicability in the area of economics [3]. By being an immutable, distributed and verifiable ledger, the concept of blockchain becomes directly applicable to any public record system. Its independent and decentralized nature also compels its use in other areas such as social, human and political [4].

Electronic Voting Systems (EVS) are considered as a way to achieve a stronger and more effective act of voting in democratic societies [5]. High abstention rates, as well as the less fraud resistance and delay in the calculation of results of the traditional voting processes propel the use of EVS. Moreover, EVS may significantly reduce the costs of electoral acts by requiring less human resources per act. However, its application as not seen a tremendous success as

This work is partially funded by the ERDF through the COMPETE 2020 Programme within project POCI-01-0145-FEDER-006961, and by National Funds through the FCT as part of project UID/EEA/50014/2013.

P. Novais et al. (Eds.): ISAmI 2018, AISC 806, pp. 323–330, 2019.
https://doi.org/10.1007/978-3-030-01746-0_38

the digitalization of other processes. While being evermore adopted, it still has a long way to go when compared to the number of more traditional election acts.

While EVS requirements may differ, they are generically expected to be verifiable and tamper resistant. The blockchain seems to fulfill, at least, some of the key requirements of EVS by, in particular, being an immutable, verifiable and distributed record of transactions. Other requirements, such as voter anonymity, inability to know which vote was cast by whom, may not be viable with blockchain. Nonetheless, efforts have been made in implementing EVS with the help of blockchain. The National Settlement Depository is using blockchain to tally votes in its shareholder meetings [3]. The US startup *Follow My Vote* is designing an EVS that aims to combat electoral fraud and protect user credentials by developing a convenient, secure and auditable end-to-end open-source voting system [6]. In the work herein we compare the existing EVS solutions that make use of blokchain.

The paper is organized in sections. Section 2 introduces the blockchain technology. Section 3 describes electronic voting systems and presents it main characteristics and requirements. Section 4 makes an overview of blockchain EVS solutions and compares related work on EVS supported by a blockchain backend. Section 5 concludes the work.

2   Blockchain

Blockchain, and similar concepts, can be seen as a distributed, verifiable and immutable ledger, where records are grouped together in blocks. These blocks are then appended to a list of blocks and replicated throughout the distributed peer-to-peer network it operates on. Each block includes a set of references to itself and to the previous block in the list, hence the name of blockchain (see Fig. 1). The references result from applying cryptographic hash functions to each block and, due to its one-way operation, guarantees integrity of the block. The original version of blockchain uses SHA-256 hash functions [7]. These hash functions are also used has an agreement protocol between peers, also referred to as mining, that makes the overall system secure but very inefficient. The use of specialized hardware, known as mining platforms, created to handle the work of processing transactions is a reality. These mining platforms accelerate the multiple hash computation required to confirm each block, also known as proof of work (PoW), in hope of being the first do do so and of collecting the respective reward.

Anyone can change its copy of the blockchain, but convincing others to work on it is very complex. The rule is to always adopt the longest chain, the one where more participants have worked on and continue to do so. In order to alter the distributed blockchain, one needs to have more computational power than the remaining nodes. At least more than 50% of the mining computational power must be under control of the attacker in order to, eventually, produce a longer chain with altered blocks.

More recently, aside from just including a list of transactions, some blockchain implementations introduced smart contracts. These can be seen as distributed

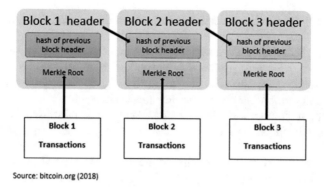

Source: bitcoin.org (2018)

Fig. 1. Simplistic blockchain example

programs that are executed on a network of nodes. The nodes do not trust each other, nor they trust a central authority. Trust is attained from the immutable nature of the blockchain, making smart contracts highly resilient to tampering. The use of smart contracts is more adequate to situations where parties agree to exchange money if certain conditions can be verified at the moment of the exchange [4].

In a more architectural approach, a blockchain is comprised by a network of cooperating nodes, that store transacted data, and by processors, which are capable of verifying the introduction of new information. It uses a trustless proof mechanism, in which all transactions of the blockchain are *a priori* verified [8] before being stored worldwide on multiple different decentralized nodes. Thus becoming a trusted public ledger system. This feature eliminates any third-party trust authority or any intermediary, like banks or brokers in an financial appliance, involving only two entities in each transaction.

3   Electronic Voting Systems

Electronic voting solutions is yet another step in the ongoing path of technological evolution and processes dematerialization. Governments and public administrations that promote the use of an EVS expect efficiency and economic gains. The EVS must maintain some key characteristics of more traditional ballot systems. Vote secrecy is such a key characteristic, in most cases. In other words, independently of the voting systems being electronic or not, some requirements must exist. Some requirements are even imposed by law, such as constitutional principles present in several countries. In [5], five generic principal for EVS are identified. These being: Generality, Freedom, Equality, Secrecy, Directness, Democracy.

- **Generality** means that every voter has the right to participate in an electoral process.
- **Freedom** means that the election process must ensure that it occurs without any violence, coercion, pressure or any other manipulative intervention that may be inflicted on the voter by a third party. EVS assumes an important role here, guaranteeing the voters are incoercible [9].
- **Equality**, a democracy pillar, means that all candidates, or parties, are equally qualified to be elected and that voting rights are equal among all voters.
- **Secrecy** means that no one should be able to know the vote of another person. This requirement, in particular, may not be present in all types of elections. There are scenarios were elections are made by a show of hands.
- **Directness**, in traditional voting systems, means that there shouldn't be intermediaries in the voting process.
- **Democracy** means that an EVS must respect the requirements of traditional electoral systems. However, additional requirements must also be met due to the nature of electronic voting. These requirements relate to the preservation of attributes and properties, such as transparency, accountability, security, accuracy and legitimacy of the system. Voters should be able to understand how elections are conducted.

Based on the above principles, multiple requirements of EVS emerged [5]. In particular, a reliable certification procedure for hardware and software must exist. The entire infrastructure, as well as any system functionality, must be identified and registered. All operations (authentication, recording of votes, etc.) must be monitored, while preserving the confidentiality of voters. The infrastructure must be open for inspection by authorized bodies. Voters, parties and candidates must be assured that there has been no malpractice. Adequate system security must be ensured, simple and easy to use.

Moreover, from the user standpoint, additional requirements can also be identified. The participation in the voting process should be confirmed. Uncoercibility should be ensured. Ability for consciously non-valid vote should be provided for. Only eligible voters should be able to vote. Each eligible voter should be able to vote only once. No voter should be able to duplicate or change his or another person's vote. The voter should be able to verify that his vote is considered in the final tally. Voters should be able to have indiscriminate access to the voting infrastructure. Registration, authentication and voting procedures should be visibly distinct. Votes should be validated separately and independently from voter authentication. No intermediaries should be involved in the voting process (i.e. no person can be authorized to vote for another person). Each and every ballot should be recorded and counted correctly.

4   Blockchain Based EVS

Agora is a company working on voting technology that commercializes a blockchain and bitcoin based EVS for national elections in any country [10]. Their goal is to increase transparency and promote fair elections worldwide. They provide a customized blockchain with two main features, the *Skipchain* and the *Catena*. *Skipchain* is the adopted consensus mechanism with a highly efficient transaction validation. It consists of a proactive Byzantine consensus permitting software clients to navigate arbitrarily in long blockchain timelines. Hashes are used as backward navigation pointers, whereas signatures collected by a group of witnesses are used as forward navigation pointers. The Catena [11] consists of a logging mechanism built on top of the Bitcoin blockchain where criptographic proofs are stored. Agora only uses Bitcoin block headers and small Merkle proofs. This solution introduced new methods of participation, fulfilling modern voters requirements such as voting in their own digital device. Common requirements such as transparency, privacy, integrity, affordability and accessibility are also guarantied. The use of this solution in a real world scenario has been accomplished [12]. They claimed to have registered 70 % of votes in blockchain, increasing voter participation while reducing the costs of the electoral act. Their results are publicly available online (https://sierraleone2018.agora.vote/).

Ayed proposed a blockchain based EVS [13]. He identified four main requirements for an EVS, these being: authentication, anonymity, accuracy and verifiability. Accuracy and verifiability can easily be extracted from the blockchain. Authentication, to assure that only registered voters can cast votes, is a parallel process not supported by blockchain. The solution proposed by Ayed, will hash the voter's name, number, vote and the hash of the previous block. Of this information, only the vote is *a priori* unknown and, by being a limited set of options, it should be feasible to break as one would only need to run has many hash operations as the number of voting possibilities. Moreover, anonymity of the voter's identity is said to be achieved by not allowing any links between voters and ballots.

Bitcoin transactions are considered pseudo-anonymous as only wallet addresses are used in transactions and not the identifiers of the people behind each transaction. However, upon discovery of who owns the wallet, anyone can view all transactions with that address. To introduce anonymity, solutions like Zerocoin [14] started to appear. Zerocoin is an extension to the bitcoin protocol that enables fully anonymous currency transactions. These services are also known as bitcoin laundry services [15]. Takabatake et al. in [16] proposed the use of Zerocoin as a basis for an EVS. Doing so, the transparency, integrity and anonymity requirements are resolved. The transparency and integrity due to the distributed and public nature of the blockchain. The anonymity with the use of Zerocoin. This solution, however, is not completely anonymous because the IP address of public nodes can be known. Moreover, it is not clear if registration, authentication and voting procedures are separated.

BlockVotes, proposed in [17], is yet another blockchain based EVS. BlockVotes uses a ring signature algorithm to generate values that are stored in the blockchain. While satisfying requirements as ballot privacy, individual verifiability, eligibility completeness, uniqueness, robustness, coercion-resistant, it does not satisfy fairness nor receipt-freeness requirements.

Bistarelli et al. proposed an EVS based on Bitcoin [18]. It comprises 3 phases: pre-voting, voting and post-voting. The resulting implementation is completely decentralized as you can vote without any intermediaries. All votes can be verified by anyone who reads this public ledger. This solution can only be applied when voter anonymity is not a requirement. Uncoercibility, confidentiality and neutrality are also not satisfied. Once a vote is cast, it is broadcasted to the peer-to-peer network and, by not being confidential, it can influence future voters.

Jason and Yuichi proposed another EVS based on Bitcoin [19]. It consists in using the Bitcoin Protocol in conjunction with a blind signature scheme [20]. Blind signatures enable that a third party can attest for the contents of data (a vote) without having access to the data. This solutions enables anonymity in voting but requires the usage of *cryptocurrency* to cast a vote. The authors suggest the use of prepaid bitcoin cards and proposed three distinct entities: the voter, an administrator and a counter. The administrator is the entity that makes use of blind signatures.

Tasarov et al. proposed another blockchain EVS [21] that, instead of bitcoin, uses Zcash [22] as the underlying payment system. The proposed EVS achieves transaction anonymity using Zcash, unmodified. This anonymity relied on zero-knowledge proofs as a substitution of the PoW scheme present in bitcoin. It supports both anonymous and transparent transactions and has two types of addresses, which differs from the bitcoin's single address and uses four distinct steps: registration, notification, voting and counting. For voters identification and verification it uses X.509 certificates and a certificate authorities (CA).

Table 1. Comparison of blockchain based EVS

EVS characteristics	Takabatake (1)	Ben Ayed (2)	Wu (3)	Bistarelli (4)	Jason (5)	Tarasov (6)	Agora (7)
Confirmation	Y	Y	Y	Y	Y	Y	Y
Eligibility	Y	Y	Y	Y	Y	Y	Y
Verifiability	Y	Y	Y	Y	Y	Y	Y
Unreusability	Y	Y	Y	Y	Y	Y	–
Anonimity/Secreacy	–	–	Y	–	Y	Y	Y
Uncoercibility	Y	Y	–	–	Y	–	Y
Integrity	Y	Y	–	Y	Y	–	Y
Uniqueness	Y	Y	Y	–	Y	Y	Y
Separation	–	–	Y	Y	–	Y	–
Validation	–	Y	Y	Y	Y	Y	Y
Ballot counting	Y	–	Y	–	Y	Y	Y
Currency usage	Y	–	Y	Y	Y	Y	Y

Table 1 compares the described voting systems based on blockchain. Requirements such as the confirmation of the participation in the voting process, or only allowing eligible voters to vote, or permitting voters to verify that their vote is considered in the result, or that votes are not reused are generally satisfied by all solutions. Vote integrity, secrecy and anonymity are requirements that are not enforced by all solutions. Having the registration, authentication and voting procedures as separated procedures is not satisfied by solutions 1, 2 and 5. The separation of the vote validation and voter authentication procedures is also not present in solution 1. Solutions 1, 3, 5 and 6 perform the ballot counting only at end of the voting process, as required.

5 Conclusions

Cultural and political issues have been hampering the adoption of EVS. Bitcoin, blockchain and similar technologies are attracting attention. Blockchain, by being an immutable, verifiable and distributed ledger, at first, seems to be adaptable to a multitude of scenarios. Electronic voting is such a scenario. The authors believe that blockchain will foster a wider adoption of EVS.

Voting based on show of hands, where there is no need for vote secrecy or anonymity, appears to be the best match between EVS and blockchain. Some solutions support vote anonymity at the expense of scalability. Other solutions support vote anonymity but, by being based on *cryptocurrencies*, require that voters must spend digital money to cast their vote or pay a fee to miners to obtain a proof-of-work.

Nonetheless, upon comparing existing proposals, we can clearly conclude that there is no one blockchain-based EVS solution that fits all scenarios. In particular, we didn't identify a solution that fulfills the common requirements of EVS, plus voter anonymity, while being scalable and without requiring the spending of digital currency. As future research work, we will seek to develop such an EVS.

References

1. Diffie, W., Hellman, M.: New directions in cryptography. IEEE Trans. Inf. Theor. **22**(6), 644–654 (1976)
2. Nakamoto, S.: Bitcoin: A peer-to-peer electronic cash system (2008)
3. William, J.: Blockchain the simple guide to everything (2016)
4. Bartoletti, M., Pompianu, L.: An empirical analysis of smart contracts: platforms, applications, and design patterns. In: International Conference on Financial Cryptography and Data Security, pp. 494–509. Springer (2017)
5. Gritzalis, D.A.: Principles and requirements for a secure e-voting system. Comput. Secur. **21**(6), 539–556 (2002)
6. Koven, J.B.: Block The Vote: Could Blockchain Technology Cybersecure Elections? (2016)
7. Norton, J.: Blockchain Easiest Ultimate Guide To Understand Blockchain (2016)
8. Swan, M.: Blockchain: Blueprint for a New Economy. O'Reilly Media Inc., Sebastopol (2015)
9. Wang, K.-H., Mondal, S.K., Chan, K., Xie, X.: A review of contemporary e-voting: requirements, technology, systems and usability. Data Sci. Pattern Recognit. **1**(1), 31–47 (2017)
10. Agora, Bringing our voting systems into the 21st century in Whitepaper version 0.1 (2018)

11. Tomescu, A., Devadas, S.: Catena: efficient non-equivocation via bitcoin. In: 2017 IEEE Symposium on Security and Privacy (SP), pp. 393–409. IEEE (2017)
12. Sierra Leone just ran the first blockchain-based election | TechCrunch (2018)
13. Ayed, A.B.: A conceptual secure blockchain based electronic voting system. Int. J. Netw. Secur. Appl. **9**(3), 01–09 (2017)
14. Miers, I., Garman, C., Green, M., Rubin, A.D.: Zerocoin: anonymous distributed e-cash from bitcoin. In: 2013 IEEE Symposium on Security and Privacy (SP), pp. 397–411. IEEE (2013)
15. de Balthasar, T., Hernandez-Castro, J.: An analysis of bitcoin laundry services. In: Nordic Conference on Secure IT Systems, pp. 297–312. Springer (2017)
16. Takabatake, Y., Kotani, D., Okabe, Y.: An anonymous distributed electronic voting system using zerocoin. IEICE technical report (2016)
17. Wu, Y.: An e-voting system based on blockchain and ring signature, Master. University of Birmingham (2017)
18. Bistarelli, S., Mantilacci, M., Santancini, P., Santini, F.: An end-to-end voting-system based on bitcoin. In: Proceedings of the Symposium on Applied Computing - SAC 2017, pp. 1836–1841 (2017)
19. Jason, P.C., Yuichi, K.: E-voting system based on the bitcoin protocol and blind signatures. TOM **10**(1), 14–22 (2017)
20. Chaum, D.: Blind signature system. In: Advances in Cryptology, pp. 153–153. Springer (1984)
21. Tarasov, P., Tewari, H.: Internet voting using zcash (2017)
22. Hopwood, D., Bowe, S., Hornby, T., Wilcox, N.: Zcash protocol specification, technical report, 2016-1.10. Zerocoin Electric Coin Company (2016)

Industry 4.0 Multi-agent System Based Knowledge Representation Through Blockchain

Pedro Pinheiro[1,2], Ricardo Santos[1,2(✉)], and Ricardo Barbosa[2(✉)]

[1] ESTG.IPP, School of Management and Technology, Institute Polytechnic of Porto, Porto, Portugal
{8140403,rjs}@estg.ipp.pt
[2] CIICESI, Center for Research and Innovation in Business Sciences and Information Systems, Felgueiras, Porto, Portugal
rmb@estg.ipp.pt

Abstract. In a world where technology evolves at an exponential rate and society tends to be ever more connected with technological tools, consumers tend to have a more well-built opinion on product manufacturing and personalization. This associated with the introduction of Internet of Things powered devices into everyday lives, creates a demand for an evolution in industry, moving it into its fourth industrial revolution. This revolution, often called industry 4.0 is the next big change in industry and in the whole manufacturing process. At the same time, blockchain tends to be a powerful technology, that coupled with the concept of industry 4.0, can create endless possibilities for companies to share value and knowledge, in a decentralized environment, giving them advantage in a competitive world. In this paper, an overview of blockchain will be presented, as well a presentation of existing blockchain-based models. It is also introduced our approach to mitigate the problem of dependencies between companies, using a multi-agent system.

Keywords: Multi-agent system · Blockchain · Industry 4.0

1 Introduction

The manufacturing environment has been facing big changes in the recent years [11]. These changes are due to the ever evolving and emerging technologies such as Internet of Things and smart data analysis. Some governments have made a great effort of pushing these technologies into the manufacturing processes of several industries, to improve the performance, quality and controllability of this processes [6].

The steps that have been taken, have started to guide industry into its fourth industrial revolution, often called industry 4.0, which aims to create a system, oriented to production, that integrates production facilities, multiple warehousing and logistics systems and even social requirements to create a global value

P. Novais et al. (Eds.): ISAmI 2018, AISC 806, pp. 331–337, 2019.
https://doi.org/10.1007/978-3-030-01746-0_39

network [14]. In an industrial environment, where entities have dependencies between them, there is a need for a solution that helps to mitigate the overload of dependencies and relationships established between entities. With that great amount of dependencies becomes hard to know who an entity should rely on to help in its business processes.

Our proposal consists in using blockchain and a multi-agent system (MAS) to represent an industry, as an entity, in an industrial environment where there is both collaborative and competitive contexts. With this model we believe to be possible to improve the decision-making processes, helping an entity to make a better choice concerning which other entities should establish connections with. This model will represent the interactions, reasoning and knowledge of an entity.

This paper is structured as follow. In Sect. 2 is made an explanation about blockchain, and also contains some state of the art solutions that use this technology. In Sect. 3, our proposal is presented. Section 4 is dedicated the conclusion and the future work.

2   Blockchain as the Engine for Industry 4.0

The Blockchain technology was popularised by the introduction of bitcoin, as a solution to the problem of making a database both secure and widely distributed [12]. Blockchain is a technology that combines peer-to-peer networks, cryptographic algorithms and a decentralization mechanism [15], enabling fully trusting interactions between individuals without the need for a trusted intermediary [2]. For this to be possible blockchain guarantees that data can't be tampered with and uses key concepts such as smart contracts, data distribution and proof-of-work.

Smart contracts are digital contracts that make sure that contractual conditions are meet before any value is exchanged, reducing the amount of human involvement required to create, execute, and enforce a contract [1]. This type of contracts allows rules to be created in a simpler, faster way, since the contract is represented in the source code of the block [15]. Proof-of-Work is a requirement that defines that an advanced mathematical computational problem must be solved, also called *mining* [5], to prevent attacks and other abuses that deteriorate the network [15], when performing transactions on the blockchain. In each block, is stored data regarding all recent transactions made in the network, keeping in the blockchain records for all transactions [12].

In a blockchain anyone can add data, review this data, but no one can change it, making this a secure and immutable network of activities [1]. The fact that blockchain operates on a decentralized network and in this network, it must exist consensus regarding the state of data, creates an environment where is easy for two or more entities, that may not know each other, to exchange value safely [1].

There are several types of blockchains, namely public, private and hybrid, where the main differences are in the way they use permissions [10]. The public blockchain enables a free participation from anyone, establishing the number of blocks that are going to be added to the chain, defining their current state. A private blockchain works based on permissions, establishing who can write and read. The hybrid blockchain consists of public and private state of network, where is ensured that every transaction is private but still verifiable by immutable record on the public state of blockchain [10].

Industry 4.0 will create new demands in the supply chain, making it smarter, more transparent and efficient, enabling industries to have a clearer vision to the suppliers, while answering more easily consumers needs [3]. With the application of blockchain is possible to keep transparent and immutable records from the supply chain, sharing data between vendors and the corporations becomes simpler, and helps setting a registry of products and track their progression through different points of the supply process [2]. In industry 4.0, information regarding new products and intellectual property, for instance, will have to be protected against abuse [3]. Storing data on blockchain, takes advantage of all its properties and guaranties a level of security necessary since data stored on the blockchain is immutable [1]. The decentralized and distributed structure of blockchain enables the creation of a peer-to-peer network where the risk of failure is mitigated, since there is no central point where a failure might happen [1].

With all these elements, this technology can be used to power the creation of decentralized organizations, that can be autonomous and still share value between organization's departments and with exterior entities [15], with transparency, trust and integrity [12].

2.1   A State of the Art of Blockchain-Based Systems

The blockchain technology is developing a new blockchain-based society [12] and this technology will revolutionise many existing traditional systems, transforming them into a more secure, distributed, transparent and collaborative systems [1]. There is also a potential for this technology to have a much bigger impact, changing the ways for firms to exchange value without relying in a central infrastructure [12].

Some real-world applications have been studied. One of these applications was the Blockchain Platform for Industrial Internet of Things (BPIIoT) [2]. The objective of this platform is to improve the functionality of Cloud-based Manufacturing (CBM) platforms. CBM is a service-oriented model, that enables consumers to select and manage manufacturing resources [2]. BPIIoT is based on a blockchain network where a smart contract is built to set an agreement between the service consumers and the manufacturing resources [2]. This model would improve on-demand manufacturing, machine maintenance and product traceability [2].

Another model, that consists on a multi-participation Blockchain-based collaborative manufacturing model, called E-Chain [17], aims to be able to solve problems like the lack of liquidity and cooperation and how to fulfil value shifting between people and its organization. This system is formed by several, smaller and independent, blockchains that together form a production network. In this network, each blockchain maintains its own history records, and it is where the digitalization of assets is going to happen. Furthermore, the way this network is structured will allow for participants to do several possible transactions [17]. Other component of this model is the value delivery network. This component is responsible for maintaining and managing the resource and value exchange between blockchains and is open to the real world, allowing for transactions to happen between E-chain and other blockchain systems [17].

Another application of blockchain technology is OriginChain which aims to use this technology to improve product traceability [7]. This platform is based on a geographically distributed private blockchain, and the objective is to be a trustworthy traceability platform that covers multiple organizations. Origin-Chain offers transparent tamper-proof traceability data, improves data's availability and automates regulatory-compliance cheeking [7].

3 Blockchain as a Solution to Knowledge Representation from Multi-agent Systems

In every industrial environment, organizations have established between them a set of business processes, that typically operate in a supply chain. With the introduction of industry 4.0, consumers can have a more demanding opinion on the products they want to buy, and this increases the level of complexity in the response that must be given to the costumer, from the organizations. It will also bring a bigger demand to the supply chain and a bigger need for smart collaboration between entities [3]. In the supply chain, there are a set of dependencies established between one, or more, entities exchanging value between them, becoming hard to have an overall idea of all the transactions made [1].

Despite the existing work on supply chains analysis, our approach aims to solve the problem of dependencies and how to improve the decision-making process in an environment formed by a set of multiple entities, with different objectives. Our approach, presented in Fig. 1, begins by assuming the existence of a network of entities, with multiple relationships established between them, where the set of these relationships create a recurrent process that occurs in the network of entities, generating interactions, reasoning and knowledge. Therefore, our solution is formed by two components, a MAS and a blockchain.

An agent is anything that can be viewed as perceiving its environment through sensors and acting upon that environment through effectors [4]. As result, a MAS is defined as a network, of agents, that are problem solving entities, working together to solve problems hard to solve only by one agent [4]. These are a particular type of intelligent systems, where autonomous agents reside in a world with no control, or persistent knowledge [9], making MAS mostly suited for modelling distributed problems involving multiple agents [8], therefore, one of our components relies in a MAS and in this component is where the core processes of an entity will be yield. This component is formed by two layers, the interaction layer and the reasoning layer.

The interaction layer represents the procedures that are established between the multiple agents [13], which can be communications, dependencies, connections or other types of relations. This layer works as the entry point of activities

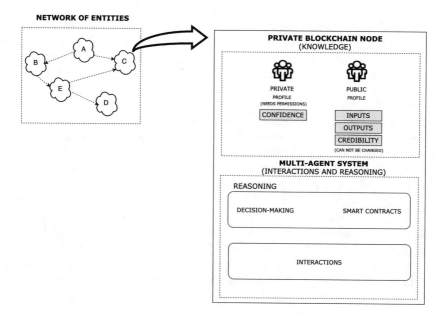

Fig. 1. Multi-agent system and blockchain solution

for our model and can create a historic of interactions to be used as a reference point in the future.

The reasoning layer is formed by two different components: the decision-making and the smart contracts. The decision-making component is responsible for supporting the entity in the process of choice [13]. The second component, will rely on the smart contract concept, to define a set of rules to make sure that both parts are following the agreed terms, when making a value exchange.

The other component, a blockchain, is where the representation of the knowledge, is going to be made. We use a private blockchain, so it becomes possible to attribute a set of permissions to write and read certain data. With knowledge representation being made in the blockchain its assured the authenticity and immutability of data, as well security in its access. The blockchain will be formed by a private profile of the entity and by a public profile.

The private profile, which needs permissions to be accessed, contains data about the confidence of the entity in the network and its entities, establishing a level of trustworthiness. This profile uses a permission-based access to ensure that only the entity which the profile belongs to can read and write data. The public profile contains the inputs, the outputs, and information about its credibility.

The inputs represent the needs of the entity, namely what it needs from the rest of the network, which can refer, for instance, to raw materials. The outputs represent what the entity has to offer to the network and its participants and, in the end, is what other entities might need as an input.

The credibility module represents how credible we, as an entity, are to the network and its entities. The credibility is represented in the public profile, to allow for other entities to see how credible we are, but this component cannot be changed by the entity to whom the value refers, despite being stored in the public part. This can be managed by setting permissions to only allow reading of this variable. The way confidence and credibility are represented in the two profiles allows a representation of a scenario where, for instance, assuming two entities A and B, where B has a low level of credibility in the network, but because previous interactions with entity A were successful, entity A has a high level of confidence in B which enables it to rely on B for future relations.

In this model, the multi-agent system interacts with the blockchain, starting with the interaction layer, which is the entry point in this model, where all the data here generated will be used to create or update the knowledge in the blockchain. On the other hand, the reasoning layer from the MAS will use the knowledge stored in the blockchain to improve decision-making. With this solution, we believe in the possibility to solve inter-entities dependencies, allowing for a faster assessment of the network, by each entity, enabling any entity to make a secure choice when selecting who its inputs supplier should be.

4   Conclusion and Future Work

In industry 4.0, smart supply chains will become the new standard for a network of corporations, because of developments in the industrial environment creating a more demanding costumer. We present an initial approach, to tackle the dependency problem, established in the future of supply chain. Associating blockchain technology with a multi-agent system, to represent an entity and provide a solution to gather knowledge from the connections made between companies in the supply chain, between interactions.

In the future, our work will move towards solving some identified problems such as how to handle the limited computing and energy resources of some industrial devices, that becomes a critical problem when blockchain is applied to industrial systems, because of the mining process [16]. This is a barrier to overcome with our model. Some important work, to be made in the future, is to develop knowledge representation and the decision-making process, where both need to be specified.

References

1. Abeyratne, S.A., Monfared, R.P.: Blockchain ready manufacturing supply chain using distributed ledger. Int. J. Res. Eng. Technol. **05**(09), 1–10 (2016). https://doi.org/10.15623/ijret.2016.0509001. http://esatjournals.net/ijret/2016v05/i09/IJRET20160509001.pdf
2. Bahga, A., Madisetti, V.K.: Blockchain platform for industrial internet of things. J. Softw. Eng. Appl. **9**, 533–546 (2016). https://doi.org/10.4236/jsea.2016.910036. http://www.scirp.org/journal/jsea

3. Deloitte: Industry 4.0. Challenges and solutions for the digital transformation and use of exponential technologies, pp. 1–30. Deloitte (2015)
4. Glavic, M.: Agents and multi-agent systems: a short introduction for power engineers, pp. 1–21 (2006)
5. Hovland, G., Kucera, J.: Nonlinear feedback control and stability analysis of a proof-of-work blockchain. Model. Identif. Control. Nor. Res. Bull. **38**(4), 157–168 (2017). https://doi.org/10.4173/mic.2017.4.1. http://www.mic-journal.no/ABS/MIC-2017-4-1.asp
6. Kang, H.S., Lee, J.Y., Choi, S., Kim, H., Park, J.H., Son, J.Y., Kim, B.H., Noh, S.D.: Smart manufacturing: past research, present findings, and future directions. Int. J. Precis. Eng. Manuf. Green Technol. **3**(1), 111–128 (2016). https://doi.org/10.1007/s40684-016-0015-5
7. Lu, Q., Xu, X.: Adaptable blockchain-based systems: a case study for product traceability. IEEE Softw. **34**(6), 21–27 (2017). https://doi.org/10.1109/MS.2017.4121227
8. Marreiros, G., Santos, R., Ramos, C., Neves, J.: Context-aware emotion-based model for group decision making. IEEE Intell. Syst. Mag. **25**(2), 31–39 (2010)
9. Oprea, M.: Applications of Multi-agent Systems. https://link.springer.com/content/pdf/10.1007%2F1-4020-8159-6_9.pdf
10. Pilkington, M.: blockchain technology: principles and applications. In: Research Handbook on Digital Transformations, pp. 1–39 (2015). https://doi.org/10.4337/9781784717766.00019. http://papers.ssrn.com/abstract=2662660
11. Qin, J., Liu, Y., Grosvenor, R.: A categorical framework of manufacturing for industry 4.0 and beyond. Procedia CIRP **52**, 173–178 (2016). https://doi.org/10.1016/J.PROCIR.2016.08.005. http://www.sciencedirect.com/science/article/pii/S221282711630854X?via%3Dihub
12. Rabah, K.: Overview of blockchain as the engine of the 4th industrial revolution. Mara Res. J. Bus. Manag. **1**(1), 125–135 (2016). The Africa Premier Research Publishing Hub www.mrjournals.org
13. Santos, R., Marreiros, G., Ramos, C., Bulas-Cruz, J.: Argumentative agents for ambient intelligence ubiquitous environments. In: Proceedings of Artificial Intelligence Techniques for Ambient Intelligence, ECAI 2008 – 18th European Conference on Artificial Intelligence (2008)
14. Wang, S., Wan, J., Li, D., Zhang, C.: Implementing smart factory of industrie 4.0: an outlook. Int. J. Distrib. Sens. Netw. **12**(1), 3159805 (2016). https://doi.org/10.1155/2016/3159805. http://journals.sagepub.com/
15. Wright, A., De Filippi, P.: Decentralized Blockchain Technology and the Rise of Lex Cryptographia. SSRN Electron. J. (2015) . http://www.ssrn.com/abstract=2580664
16. Xiong, Z., Zhang, Y., Niyato, D., Wang, P., Han, Z.: When mobile blockchain meets edge computing: challenges and applications, pp. 1–17 (2017). http://arxiv.org/abs/1711.05938
17. Zhang, F., Liu, M., Shen, W.: Operation modes of smart factory for high-end equipment manufacturing in the internet and big data era. Smc2017.Org (2017). http://www.smc2017.org/SMC2017_Papers/media/files/0642.pdf

On the Interoperability of European National Identity Cards

Abubakar-sadiq Shehu[1,2]($\boxtimes$), António Pinto[3,4], and Manuel E. Correia[1,4]

[1] Department of Computer Science, Faculty of Science,
University of Porto, Porto, Portugal
msabubakar-sadiq.it@buk.edu.ng, mcc@dcc.fc.up.pt
[2] Department of Information Technology, Bayero University Kano, Kano, Nigeria
[3] CIICESI, ESTG, Politécnico do Porto, Porto, Portugal
apinto@inesctec.pt
[4] CRACS & INESC TEC, Porto, Portugal

Abstract. Electronic identity (eID) schemes are key enablers of secure digital services. eIDs have been adopted in several European countries using smart-cards for secure authentication and authorization. Towards achieving a European digital single market where European citizens can seamlessly access cross-border public services using their national eIDs, the European Union (EU) developed the electronic IDentification, Authentication and trust Services (eIDAS) regulation. eIDAS creates an interoperable framework that integrates the eIDs adopted in the EU Member States (MS). It is also an enabler of a cross-border operation, harmonized with the General Data Protection Regulation (GDPR) regulation by protecting the privacy of personal data. If one can use the same procedure for authentication and authorization abroad, one can better understand new services that use eIDs. This paper provides a comparative analysis of eID cards adopted in EU MS and their privacy features in preparedness for eIDs cross-border interoperation.

Keywords: eID · Interoperability · STORK · eIDAS

1 Introduction

The Internet enables access to e-Government and e-Business services. eID guarantees unambiguous identification and authentication of persons using digital services. eID schemes were introduced as a form of physical and digital identification of citizens when accessing digital services of Governments, but can also be accepted and recognized by public and private organizations. These trends occurred separately, at different speeds, in several MS without EU coordination and without cross-border considerations.

This work is partially funded by the ERDF through the COMPETE 2020 Programme within project POCI-01-0145-FEDER-006961, and by National Funds through the FCT as part of project UID/EEA/50014/2013.

© Springer Nature Switzerland AG 2019
P. Novais et al. (Eds.): ISAmI 2018, AISC 806, pp. 338–348, 2019.
https://doi.org/10.1007/978-3-030-01746-0_40

An eID card bears a chip that uses public key infrastructure to embed digital certificates containing user attributes for definitive proof of identity and electronic signatures. The communication interface of an eID card can either be contact, contact-less, dual or hybrid. An eID card with contact interface requires physical contact using a card reader for identification or authentication, while contact-less interface uses electromagnetic induction technologies such as RFID and NFC. Dual interface cards have both contact and contact-less connection to a single chip, hybrid interface cards have two chips, one connected via a contact and other via a contact-less interface [11].

eID card projects began in Europe in the late 1990s with a few countries adopting its use as an identity document with electronic functions [2].

Its use has reduced the use of user-created logins with improved digital safety for identity claim [6], for example, contact eID card requires the physical presence of the card and a secret PIN to activate the digital certificate. With a single eID card, a user can access multiple e-Government and e-Business services, instead of using multiple login credentials.

European citizens and businesses while trying to access cross-border online services face multiple problems due to the nonexistent interoperation of eID. Users need to maintain multiple sets of credentials (username and password), typically, one per each service. The non-recognition of the Portuguese eIDs in other MS makes its use of cross-border online services impossible. For example, a Spanish, Portuguese or Belgian eID card would not be accepted as a means of identification in a German online service, and vice versa. Some research projects to tackle this problem have appeared, examples being the epSOS (cross-border exchange of health data), the e-CODEX (cross-border legal services), the STORK (secure identity across borders linked) and the eIDAS (regulation on electronic identification and trust services in e-transactions) projects [15].

This paper is divided into four sections. Section 1 provides a general overview of the paper. Section 2 presents an overview of eID schemes, using Belgium, Portugal and Germany as representative scenarios. Section 3 discusses the interoperability status of MS eIDs. Section 4 concludes the paper.

2   eID Overview

Prior to the penetration of Internet and e-Government, paper-based identification was the adopted practice to identify one's self. Government being the primary provider of identity documents, issues documents to its citizens in the form of drivers license, citizen cards, passports, tax papers and birth certificates. For example, a passport identifies the holder and verifies the holders issuing country, while a drivers license confirms the holder's eligibility to drive. Other attributes such as name, blood group or home address are also usually included in these documents. In 1999, Finland became the pioneer European country to adopt eID card as a national identity document [26], a decade later Germany also mandated its use [10]. Currently, most European countries have adopted ID cards (some with electronic functions) as a means of national identification, with exception of a few countries like Slovenia, Denmark, Ireland, Norway and United Kingdom (discontinued in 2011) [12,24]. A recent study conducted by the EC showed that eID cards have become a dominant solution with 41% usage over mobile-ID and others as a means of identification and authentication in e-Government services [28]. In Estonia, 98% of the citizens possess an eID card activated to be used for digital signatures, online voting and as a public transport card [21]. Depending on the country, these cards can be mandatory. Some eID card implementations have grown out of the Governments traditional functions of issuing travel and identity documents, by having as a primary goal the means of strong e-identification and authentication with trustworthy identifiers [25]. These and other purposes are gradually converging and giving rise to interoperability challenges, affecting the emergence of a European digital single market [4].

Secure log-ins are still used in some countries such as United Kingdom, Ireland, Sweden, Spain, Denmark and Netherlands. Citizens are assigned a combination of a username and password to be used for identification on public e-services. In Britain, citizens online identity management is managed by *GOV.UK Verify*, where citizens login with personal credentials, that assures their claimed identity to a third party requested service.

Mobile ID is another eID solution adopted by some MS including Austria, Finland, Estonia, Luxembourg, Sweden, Portugal and Netherlands [8]. In most cases, its adoption is to complement or simplify the use of eID cards. eID cards are the predominant solution with exception of Austria, which has more mobile ID users than eID cards [16].

Existing eID infrastructures and legal frameworks differ in each MS. For example, in Austria, Finland and Portugal, eID cards are issued to citizens based on a national register that contains citizens information and unique identifiers. In Germany, the country's law prohibits the issuance and use of unique identifiers, therefore eID cards are issued based on citizen's records at the local municipal offices [19]. The UK's identity management system adopts a federated model while the model in Estonia, Belgium, Portugal, and Germany is centralized. The identity management in Austria uses a mixture of the centralized and distributed model [18]. These differences contribute with barriers to the use of eIDs in cross-border online services. Therefore, the eIDAS regulation does not opt to harmonize existing infrastructures and legal frameworks, instead, it creates an interoperable platform based on the STORK framework.

2.1   Summary of eID Schemes

A summary of eID schemes is given in Table 1. While eID cards in Greece, France, Cyprus, Romania and Bulgaria are still in a development stage, other countries with active eID cards adopt distinct implementation models with different roles for each stakeholder. For example in Austria eID cards are issued by both private and public authorities. The scheme is based on a specific ID model that relies on sector-specific pins (ssPINs) generated for every Government service, which prevents the linkability of citizens information across agencies [1]. In Portugal, Belgium, Estonia and Germany eID cards are issued by Government agencies [9].

Table 1. eID schemes in EU states

Country	eID Means	Status	Type/Interface
Austria (AT)	Smart-card (Bürgerkarte), other cards, mobile-ID	In-use	Contact, contact-less
Belgium (BE)	Smart-card Belgian Personal Identity Card (BELPIC)	In-use	Contact
Bulgaria (BG)	Smart-card (Lichna Karta)	In-development	-
Croatia (HR)	Smart-card (Osobna iskaznica)	In-use	Contact-less
Cyprus (CY)	Certificates(ARIADNI) national ID	In-use In-development	Digital certificate
Czech Republic (CZ)	National identity card	In-use	Contact
Denmark (DK)	NemID	In-use	Secure login
Estonia (ES)	EsteID, DigitID Mobil-ID	In-use	Contact
Finland (FI)	Smart-card (FINeID)	In-use	Contact
France (FR)	Certificates	In-use	Digital certificates
Germany (DE)	Smart-card (der personalausweis)	In-use	Contact-less
Greece (GR)	ERMIS National ID	In-use	Digital certificates
Hungry (HU)	Smart-card (eSzemelyi)	In-use	Contact-less
Ireland (IE)	MyGovID	In-use	Secure login
Italy (IT)	Smart-card (Carta di identita electronica) national service card	In-use	Contact-less digital certificate
Latvia (LV)	Smart card (eParaksts)	In-use	Contact
Lithuania (LT)	Smart-card	In-use	Contact
Luxembourg (LU)	Smart-card (La carte d'identité nationale)	In-use	Contact
Malta (MT)	Smart-card (Karti tal-e-ID)	In-use	Contact
Netherlands (NL)	DigID eHerkenning	In-use	Secure login
Poland (PL)	National ID card	In-use	-
Portugal (PT)	Smart-card (Cartão de Cidadão) Digital mobile key	In-use	Contact
Romania (RO)	National ID card	In-development	-
Slovakia (SK)	Smart-card (elektronické identifikačné karty)	In-use	Contact
Slovenia (SI)	Certificates (eUprava)	In-use	Digital certificates
Spain (ES)	Smart-card (Documento Nacional de Identidad), certificates& Cl@ve	In-use	Contact, digital certificates, secure login
Sweden (SE)	Bank ID	In-use	Contact, contact-less cards, secure login, Mobile ID
United Kingdom (UK)	GOV.UK Verify	In-use	Secure identity assurance

eID cards are mandatory in Belgium, Estonia, Germany and Portugal, but not in Austria or Finland. In countries like Poland, despite having eID cards, its electronic functions are yet to be implemented [7].

2.2   eID Card Issuance and Citizens Privacy

The emergence of eID schemes in Europe occurred independently based on heterogeneous infrastructures [19,25]. Irrespective of the scheme, citizen identification, authentication and digital signatures are common electronic functions [20]. An eID card is embedded with more data than the one physically printed on the card. For example, in Portugal, Germany and Finland the citizen's address is embedded on the chip but omitted on the physical card. In Poland and Germany, eID cards are embedded with citizens' height and eye colour despite being omitted on the physical card. Before the introduction of EU's GDPR [22] and eIDAS regulations [23], most eID card implementations adopted some privacy control and security features at national levels. Examples of such are card issuance control procedures, PIN protection and digital certificates that protect against threats like falsification of consent and eavesdropping. The privacy control mechanisms adopted by MS [9,19] in securing eIDs is encapsulated into eIDAS regulations and used to determine each eID's assurance level.

For example, in Belgium, a citizen needs to make a formal request at the municipal office of their residence in order to obtain the eID card. After which an address verification letter is dispatched to the citizen's address, inviting the citizen to commence the registration process. At the municipal office, the citizen is requested to deposit a recent passport photograph, a valid Government identity document (birth certificate or passport) and sign a formal eID request form. Citizen information is collected and sent to the Government certified card personalizer (CP) and card initializer (CI). CP prints the eID card, while CI provides on-card key pair generation. CP then releases part of the eID card activation code (PIN unblocking key) PUK1 to National Register (RRN), and request new certificates for the card from a (certificate authority) CA. In turn, the CA issues two new certificates which are stored on the card by CI. An update to the certificate revocation list (CRL) is also performed. CI also writes printable citizens data on the new card and deactivates it before sending an invitation letter containing a default (personal identification number) PIN and PUK2. Once the letter is received and the card activated, eID automatically computes a signature with each private key and the CA removes its certificates from the CRL.

In Portugal, citizens card are issued by the Portuguese Government under a partnership of three Government entities. The card bears it all, as it accumulates citizens identity information like citizen's civil, health, tax and social security number, thus enabling citizens to carry one single card instead of five different ones. The Registers and Notaries Institute (IRN, Instituto dos Registos e do Notariado) conducts civil registrations of citizens, maintains the Government database for such registrations and issues citizens cards. The Electronic Certification Entity of the State (ECEE) serves as the certificate authority that manages service providers (SP). Gemalto provides the smart-cards while the Portuguese Mint and Official Printing Office (INCM) prints and personalizes the cards. To acquire a card, a citizen must present a formal application by filling a form at any IRN office, or other authorized public institutions. An identification document will be requested (birth certificate, civil identity, taxpayer, social security, health insurance or voters card [9]) to complete the citizen's application. After the identity of a citizen has been established, his data is accumulated including biometrics (fingerprint and face capture) and forwarded to INCM for card personalization. The Portuguese eID card bears two certificates that contain electronic data for authentication and digital signatures [5] both secured with secret PINs stored on the card.

To obtain an eID card in Germany, citizen needs to apply at the identity card authority office of their local Government office. The applicant needs to prove his identity and German nationality by providing a valid identity document or an old ID card (if available). In rare cases of data mismatch on population register or for a citizen who never acquired an ID card, they may be requested to provide their birth certificate. Finally applicant checks and confirm the filled details, signs and pays an application fee for the eID card. Service office captures the necessary applicant's personal data and transmits this data to Federal Printing Office (FPO) for production. FPO inscribes all data on the card and stores similar and more on the chip. It then sends a letter to the applicant containing an initial randomly generated activation PIN, a PUK and a revocation password all protected by a PIN in a tamper-evident scratch code. After which, applicants pick up their cards at the application service or from an authorized representative. Before the first use of a new eID card, citizens must activate the card by changing the initial PIN. Activation can be performed at the issuing authority or locally by the cardholder. Besides using the German card as a physical identity document, information contained on the contact-less chip also provides a means for mutual electronic proof of identity and authentication, a travel document (by accessing the biometric data) and serves as a signature creation token.

The German scheme is considered to have emerged with improved privacy features when compared to eID cards issued in Finland, Estonia, Belgium, Austria and Portugal [10,13,19]. In compliance with [22], the privacy features of this card protect citizen's information and ensure minimal information disclosure through encrypted data transmission. A strong and secure mutual authentication between SP and citizens is achieved through Password Authenticated Connection Establishment (PACE) and Extended Access Control (EAC), ensuring trust between both parties. Citizen consent on information release and secure storage of personal data is also achieved [14][17]. For example, an online service can verify a citizen's age without knowing the date of birth or verify citizen's address without disclosing the full address.

3 Interoperability

Existing eID schemes emerged in isolation and based on different legal frameworks and infrastructures. Activities of MS on their eIDs and online

Table 2. Summary of eID cards

Country	eID card issuer	eID status	Age required (years)	Validity (years)	Accessible data	PIN length	Certificate based	Separate certificates
AT	Source PIN Registry-Authority	non-mandatory	non-specific	5	ID link & source pin	4	No	No
BE	Local residence office	Mandatory	12	10	some data	4-6	Yes	Yes
CZ	Ministry of Interior	Mandatory	15	15 < 5, 15>10& 70 > 35	Basic	4-10	Yes	-
ES	Police and border guard board	Mandatory	15	3	-	4-5	Yes	Yes
FI	Police department	non-mandatory	18	5	-	4	Yes	Yes
DE	Federal ministry of Interior	Mandatory	16	10	Sector-specific	6	Yes	Yes
HU	Ministry of Interior	Mandatory	18	18<3 18>6	-	5-6	-	-
IT	Municipalities issues CIE cards	Mandatory	18	3	Whole data	-	Yes	Yes
LV	Citizenship and Migration office	Mandatory	15	5	-	-	Yes	Yes
LT	Ministry of Interior	Mandatory	16	16<5, 16>10	-	8	Yes	Yes
LU	Municipal authorities	Mandatory	15	4<2, 4>5	-	8	Yes	Yes
MT	National identity management office	Mandatory	14	10	whole data excluding bio-metrics	6	Yes	Yes
PL	Office of civic affairs	Mandatory	18	10	-	-	-	-
PT	Institute of registers & notaries	Mandatory	6	25<5 & 25>10	Whole data excluding bio-metrics & symmetric key	4	Yes	Yes
SK	Police department	Mandatory	15	10	-	-	Yes	Yes
ES	Ministry of interior	Mandatory	14	10 2 for minors 13< 5	Basic	-	Yes	Yes
SE	Swedish police Private partnership	non-mandatory	18	5	Basic	-	Yes	Yes

administrative services are guided by respective national regulations. However, eIDAS regulation kicks in to create an interoperable framework for cross-border integration of eID schemes for the identification of citizens while accessing administrative services across MS where these services support electronic identification [4], provided that the assurance level of the requesting eID is the same or higher than MS offering the service. Although the regulation has already kicked off, the established deadline for mandatory recognition of notified eIDs by all MS is set as 28 September of 2018. eIDAS interoperability framework is based on the STORK architecture, which supports MS infrastructures to operate on the same protocol irrespective of their model. Citizen's attributes might differ on MS eID schemes. For instance, in Portugal besides using a unique civil identity number, citizens card contains other information like gender, height, nationality, date of birth, passport photograph and marital status. While in Germany other attributes like eye colour and academic qualification are included. eIDAS regulation repels the existing EU electronic signature directive 1999 and sets out an interoperable minimum data-set for unique identification of citizens that comprises the citizen name, the family name, the name at birth, the date of birth, the address and the identification number of the citizen. MS can include additional information (Table 2). The regulation and its infrastructure are considered a practical solution for eID integration that gives MS and verifying authorities some level of trust on the data retrieved from each other. The regulation is based on four key principles, which are (1) eID notification, (2) mutual recognition of eIDs, (3) interoperability nodes and (4) eID assurance levels (low, substantial and high). Inclusion in the eIDAS interoperable framework starts with a MS eID notification, where a MS informs the other states of an adopted eID scheme. As of November 2017, only Germany had officially notified others of their eID cards. Also, Austria, Germany and Netherlands have an eIDAS compliant node connected to their identification and authentication systems [24]. With the notification of the German eID scheme, a German eID card can be used to access cross-border administrative services across the 28 MS on ePSOS, STORK or any other public service that supports electronic identification and authentication. The decision to participate is voluntary, but afterword's, the mutual recognition and acceptance of a notified eID scheme for online administrative services by all MS is mandatory.

In the interoperable framework, the citizen's attributes are expected to travel across different nodes, therefore assurance levels of eID schemes is designed to provide the required confidence on the MS that submitted the attributes. Likewise, citizens can be assured of the security and privacy of their information across nodes. The assurance level of a notified eID scheme is categorized based on their security, covering the complete life-cycle of the scheme, from enrolment to the revocation procedures and all activities involved in this processes [27]. The factors considered in calculating the assurance level are multiple. An authoritative source has to be established by all MS, which should be a nationally trusted source that provides valid data to confirm citizens attributes. In Finland, a national register is maintained by the population information commission that stores citizens information, likewise in Belgium, Estonia and Portugal. A definition of the number and categories of authentication methods available must exist. Examples factors being knowledge, possession or inherent based. Knowledge-based considers something known to the owner and is adopted in Belgium, Portugal and Germany where citizens need to know a PIN number [9]. Possession-based requires citizens to demonstrate possession of the eID card. Inherent based is based on a physical attribute of the natural person (fingerprint, hair colour, eye colour or height). Lastly, processes and procedures are designed to an acceptable level of risk with respect to information security. The assurance level of the eID cards of a MS and its current eIDAS status is shown in Table 3.

Based on the STORK infrastructure, models for cross-border integration of notified eIDs and their recognition can either assume a centralized or decentralized architecture [3]. In a centralized framework, a proxy is used to authenticate citizens from other states. In the decentralized framework, a receiving MS uses a middleware to authenticate citizens of other MS. A recognizing MS using the centralized model receives attributes of citizens requesting national service in their country and forwards it to respective SP, while a decentralized model involves the SP directly handling and receiving service requests. With seven months to the eIDAS deadline, some encouraging achievements have been recorded from some countries that have notified their eID schemes. Although its general success presents some challenges as most eIDAS nodes are still in the pre-notification stage.

Table 3. Minimum eIDAS requirements

Country	eID card notified	Established eIDAS node	Node integrated with eID	e-Government connection to node	eID assurance level
AT	No	Yes	Yes	No	High
BE	No	No	No	No	High
CZ	No	No	No	No	Low/substantial
ES	No	No	No	No	High
FI	No	No	No	No	High
DE	Yes	Yes	Yes	No	High
HU	No	No	No	No	-
IT	No	No	No	No	-
LV	No	No	No	No	-
LT	No	No	No	No	Substantial
LU	No	No	No	No	High
MT	No	No	No	No	-
PL	No	No	No	No	-
PT	No	No	No	No	High
SK	No	No	No	No	High
ES	No	No	No	No	High
SE	No	No	No	No	High/substantial

4   Conclusion

eID card remains a dominant form of secure digital identification and authentication to public services in European states. Adoption of these cards has addressed the many challenges faced by citizens trying to access e-Government and e-Business services at national level. Implementations of privacy-preserving methods in MS have in a way reduced possible threats to citizens data, although some vulnerability still exists in the use of certificates which does not really address privacy threats.

The large-scale projects (STORK and STORK 2.0) subscribed by nineteen MS for cross-border services in eLearning and Academic Qualifications, eBanking, Public Services for Business and eHealth has set the path for the eIDAS framework and it perceived success. Extending all MS online public services to accept and take advantage of foreign eID cards will benefit EU citizens in ease of business and movement. It will encourage seamless sharing of information across states. The framework will as-well provide the possibility to safely enroll users and quickly validate their identity and mitigate the potential risks attributed to unverified users. Although, the framework set minimum data-set and assurance level of eIDs, protecting citizens attribute and privacy across MS still lays on national infrastructures and full implementation of GDPR regulation.

References

1. Aichholzer, G., Strauss, S.: The citizen's role in national electronic identity management- A case-study on austria. In: 2009 Second International Conference on Advances in Human-oriented and Personalized Mechanisms, Technologies, and Services, CENTRIC 2009, pp. 45–50. IEEE (2009)
2. Arora, S.: National e-ID card schemes: a European overview. Inf. Secur. Tech. Rep. **13**(2), 46–53 (2008)
3. Bender, J.: eIDAS regulations: eID-Oppurtunities and risks. In: SmartCard Workshop, pp. 1–11, September 2015
4. Bowyer, K.W.: Biometrics research (2016)
5. Campos, L.F.G.: IFPortal: uma plataforma Web para a caracterização e comparação de Frameworks nacionais de interoperabilidade. PhD thesis (2013)
6. connectis December 2016
7. Domanski, T.: Ger ready for another exchange of document. e-Proof will be available in two years, May 2017. https://www.spidersweb.pl/2017/05/e-dowod.html
8. EC: Study on the use of Electronic Identification (eID) for the European Citizens' Initiative". In: Final Assessment Report, pp. 1–210, September 2017
9. Bud Bruegger, F.H.G., et al.: Survey and Analysis of Existing eID and Credential Systems (2013)
10. Fumy, W., Paeschke, M.: Handbook of EID Security: Concepts, Practical Experiences, Technologies. Wiley, New York (2010)
11. Datacard Group: Durability of Smart Cards for Government eID. In: Datacard Group: Secure ID and card personalization solution (), pp. 1–15
12. Hansteen, K., Ølnes, J., Alvik, T.: Nordic digital identification (eID). Nordic Council of Ministers (2016)
13. Harbach, M., et al.: On the acceptance of privacy-preserving authentication technology: the curious case of national identity cards. In: International Symposium on Privacy Enhancing Technologies Symposium, pp. 245–264. Springer (2013)
14. Federal Office for Informtion Security. German eID based on Extended Access Control v2. Online. Federal Office for Information Security Post Box 20 03 63 D-53133 Bonn, February 2017
15. Ivan, P.: The Juncker Commission past midterm: Does the new setup work? EPC Discussion Paper, 18 October 2017
16. Kubach, M., et al.: SSEDIC. 2020 on Mobile eID. In: Open Identity Summit 2015 (2015)

17. Kügler, D.: Advanced security mechanisms for machine readable travel documents. Technical Report, Federal Office for Information Security (BSI), Germany (2005)
18. Leitold, H., Tauber, A.: A systematic approach to legal identity management-best practice Austria. In: ISSE 2011-Securing Electronic Business Processes, Wiesbaden (2011)
19. Naumann, I., Hogben, G.: Privacy features of European eID card specifications. Netw. Secur. **2008**(8), 9–13 (2008)
20. OECD: OECD WORKSHOP Improving the measurement of digital security incidents and risk management. In: Working Party on Security and Privacy in the Digital Economy. Ed. by Alice Weber eLettra Ronchi, OECD. OECD Headquarters, Paris, France, pp. 1–36, October 2017
21. Pappel, I., et al.: Systematic Digital Signing in Estonian e-Government Processes. In: Transactions on Large-Scale Data-and Knowledge-Centered Systems XXXVI. Springer, pp. 31–51 (2017)
22. EU Regulation: 2016/679 Of the European parliament and of the council of 27 April 2016 on the protection of natural persons with regard to the processing of personal data and on the free movement of such data, and repealing Directive 95/46/EC (General Data Protection Regulation). In: European Union, pp. 1–88 (2016)
23. EU Regulation: No 910/2014 Of the European parliament and of the council of 23 July 2014 on electronic identification and trust services for electronic transactions in the internal market and repealing Directive 1999/93/EC. In: European Union, pp. 1–42 July 2014
24. EU Regulation: No 910/2014 of the European Parliament and of the Council of 23 July 2014 on electronic identification and trust services for electronic transactions in the internal market and repealing Directive 1999/93/EC (eIDAS Regulation). In: European Union, pp. 44–59 (2014)
25. Ribeiro, C., et al.: STORK: a real, heterogeneous, large-scale eID management system. Int. J. Inf. Secur. pp. 1–17 (2017)
26. Rissanen, T.: Electronic identity in Finland: ID cards vs. bank IDs. Identity Inf. Soc. **3**(1), 175–194 (2010)
27. De Soete, M.: eIDAS Regulation – eID and assurance levels – Outcome of eIAS study, June 2015. https://docbox.etsi.org/workshop/2015/201506_ecurityweek/eidas_thread/s03_eid/security4biz_de_soete.pdf
28. Vasilescu, A.: eID under eIDAS: MS Status and timeline: Where are we now, and what are the future plans? pp. 1–7, October 2017

The Usage of an Intelligent Virtual Sensor as a Form of Approximation to the Final Consumer

Mário Macedo[1,2], Ricardo Barbosa[2(✉)], and Ricardo Santos[1,2(✉)]

[1] ESTG.IPP, School of Management and Technology, Institute Polytechnic of Porto, Porto, Portugal
{8140183,rjs}@estg.ipp.pt
[2] CIICESI, Center for Research and Innovation in Business Sciences and Information Systems, Felgueiras, Porto, Portugal
rmb@estg.ipp.pt

Abstract. Industry 4.0 is revolutionizing the processes and the products of an organisation. With an increase of incidence on their consumers interests and preferences, organisations are facing the need to understand their needs and preferences, based on the data available online, which can contribute to an improvement of the manufacturing process including the personalisation factor. This work contains a model proposal for an Intelligent Virtual Sensor that is intended to create a knowledge layer for organisations by gathering and processing data regarding their consumers interests and preferences. By looking at industry as a Smart Environment, is possible to enhance the smart manufacturing process by providing data and knowledge gathered from numerous sensors, including the proposed sensor model. With that knowledge layer provided by the proposed sensor, the organisational decision making processes can include the consumers needs and preferences.

Keywords: Industry 4.0 · Smart environments
Intelligent virtual sensor · Decision making

1 Introduction

In a world marked for eras and revolutions, the innovation in technology that we face every day contribute to numerous fields of study. The new industrial revolution, entitled Industry 4.0, is focused on the restructure of the value chain process, and is a result of the integration of computation, networking, and physical processes. With the focus shifting towards their consumer's needs and preferences, organisations are facing a new challenge: understand the interests and preferences of their consumers, while being able to respond to the personalisation of their products. The personalisation of products and the insights about their consumers interests and preferences is dependable of data, that usually is not directly available to organisations. By looking at organisations as a smart

© Springer Nature Switzerland AG 2019
P. Novais et al. (Eds.): ISAmI 2018, AISC 806, pp. 349–356, 2019.
https://doi.org/10.1007/978-3-030-01746-0_41

environment, is possible to enhance the intelligence behind their smart manufacturing processes, by providing knowledge that can aid the decision making
processes. Typically, on a smart environment context, the gathering of data is
done trough the usage of physical sensors, however, in this work we present
an intelligent virtual sensor model, intended to act as an 'external window' for
organisations, by providing a knowledge layer based on data present in multiple
external data sources. This knowledge layer can help organisations to respond
to their consumers needs and interests while understanding and improving their
relationship to their consumers.

This work is divided into the following sections: the second section contains a
state of the art of industry 4.0 and emphasis the focus on the consumer; The third
section introduces the concept of ambient intelligence and smart environments,
and how they can be implemented on an industry scenario trough the usage of
sensors; The Intelligent Virtual Sensor model description and integration with
industry 4.0 is present on section four; And this work ends with a conclusion
and a small description of future work.

2   Industry 4.0

The eighteen century marks the start of industry revolutions. From the mechani-
sation and water/steam power that marked the first revolution, to the inclusion
of electricity and mass production assembly line at the beginning of the twentieth
century, to automation via computer systems in the early seventies, the industrial world has seen three revolutions already. Currently the industry is facing a
new revolution, known as industry 4.0, focused on the restructure of the value
chain process involved in manufacturing industry [12], trough the adoption of
Cyber-Physical Systems (CPS). CPS are integrations of computation, networking, and physical processes, and enable real-time information extraction, data
analysis, decision making, and data transmission. The usage of CPS technology
drives smart manufacturing and leads to efficiency gains, real-time logistics, and
improved demand response [8], that give industrial organisations the control and
knowledge in processes, that is required in order to be able to produce products
with levels of personalisation with low costs [12,13].

This revolution, with origin on the German manufacturing sector circa 2010
[9], was driven by the advances of technology such wireless sensor networks,
Internet of Things, and cloud computing [13]. These changes resulted on the
necessity for organisations to adapt themselves to this evolution and the new
market challenges. Under the Industry 4.0, there has been an astounding growth
in the advancement and adoption of information technology that has increasingly influenced consumers perception on product innovation, quality, variety
and speed of delivery.

2.1   The Focus on the Consumer

One of the lead factors on industry 4.0 is the increase focus on the consumers, in
particularly their interests. With consumers taking an important role in indus-

try production, organisations aim to be able to offer the personalisation of their products to consumers. In order to achieve this goal is imperative for organisations to understand their preferences and interests, which can be obtained by external sources of data that are not directly available to the organisation.

The focus on the consumer by organisations is justified by the desire to increase their satisfaction, while seeking their opinion and feedback. The index of satisfaction of consumers can be represented in three categories [7]: perceived value; consumer satisfaction; and consumer Loyalty. Related to this categories are four variables, namely: image; expectations; perceived quality of 'hard ware'; and perceived quality of 'human ware'. Additionally, the consumers opinion is a very important measurement, and represents numerous possibilities of products and/or services changes. As example, organisations can use their consumers opinion in order to analyse what the consumers think about their products and/or the services, which permit the company to upgrade them with a high certainty value that the changes will be accept by a large majority of their target audience. Another possibility is the analysis of opinion related to new products and services that the consumers would like to see in the future, leading to a challenging task called prediction [4].

3   Ambient Intelligence

The vision concept of Ambient Intelligence (AmI), introduced by the European Commission's Information Society Technologies Advisory Group (ISTAG) in 2001 [6], was based on the principle that at some point in time, humans will be surrounded by intelligent interfaces supported by computing and networking technology. This network of technologies will be present on every aspect of our life, embedded in everyday objects, from furniture, to vehicles, or materials. In the '40s and '50s the attention was centred on the hardware, in the '60s that attention shifted to computers, '70s and 80s were focused on networks, and from the '90s till the present day the attention is centred on the web [11].

The potential business opportunities for AmI were identified as a job for the future generations of industrialists and entrepreneurs and, in order to achieve this vision, intelligence must be provided to the environment (either in the context of intelligent homes, intelligent cities, or even intelligent factories), and for this reason this concept is not possible without Artificial Intelligence (AI). AI researchers must be aware of the need to integrate their techniques with other scientific communities techniques (automation, communication, machine learning, computational intelligence, natural language, knowledge representation, computer vision, intelligent robotics) [11].

3.1   Industry as a Smart Environment

With the advances and implementations of industry 4.0 is interesting to look to the industry sector as a SmE. By definition, a SmE is an ecosystem of interacting objects that has been enriched with technology (sensors, actuators, information terminals, and other devices interconnected through a network) that have the capability to self-organise, provide services and manipulate complex data. This physical space is smart in nature, and that smartness results from a interaction of different devices and computing systems, and aims to enhance the services that can provide to humans. This vision is still applicable when we start to look at industry as a SmE.

Similarly to AmI, three computational areas must converge in order to develop a truly SmE: ubiquitous computing, intelligent systems, context awareness. Ubiquitous computing is responsible for providing a seamless interface between the environment and its users, and the integration of the system into the everyday objects must be simple, natural, and a non intrusive experience for the users. Intelligent systems are responsible for inferring the context of the environment and understand patterns based on the users behaviour (techniques such as data mining, statistical analysis, machine learning, or optimisation methods, take a huge role on this scenario). Context awareness is the adaptation of the environment, and is vital to have a system that shifts around needs by perceiving context (using sensors) and being capable to act around it (using actuators) [2].

3.2   Sensors

In order to fulfil the necessities of SmE, in particularly the context awareness for perceiving the environment, sensors are fundamental. AmI or SmE are often associated with intelligent sensors that are embedded in our environment, that are usually conceived for detecting or measuring motion, light, temperature, humidity, and other conditions that are descriptive of the environment [5].

The role of sensor networks in a SmE is to furnish the higher levels of the system with answers to the following necessities [10]: entity identification; provide a time frame for location and object associations that help to determine context; recognise activities, interactions and relations; recognising behavioural patterns; and tracing information flow trough multiple modalities.

4   Intelligent Virtual Sensor for Industry

In order to fulfil the capabilities of a SmE, there is a constant need for knowledge creation. This process is better explained by the Data-Information-Knowledge-Wisdom pyramid, also known as pyramid of knowledge. The concept introduced by Ackoff [1], divides the knowledge creation process into four categories: data, information, knowledge, and wisdom. The process of knowledge creation begins with data, and despise the internal volume of data that any organisation might possess regarding their consumers, in order to enhance their approximation with consumers there is a need to look to external sources of data. The external data can be used to improve their processes and products, and provide the knowledge that organisations seek about their consumers or other external factors that can help any decision making process [14].

There is little doubt about the importance of sensory technology to the SmE and, in particular, to the implementation and the success of industry 4.0. These physical devices are often a tool for collecting internal data of the environment that they are inserted. However, sensors can be more than that, and in our vision they can act as a window for the exterior world that organisations cannot ignore. Inspired by previous work [3], which contains the model definition of a virtual social sensor that uses data gathered from online social network platforms to create knowledge for SmE and organisations about behavioural and preferential characteristics, the proposed model present on this work shares the same virtual characteristic and overview regarding the external factors that organisations can be currently neglecting.

The proposed model can act as an 'window to the exterior', by linking external sources with industry processes, allowing organisations to understand the preferences of their consumers, and creating an approximation to them trough smart production of their products and/or services.

4.1 Description of the Intelligent Virtual Sensor Model

Our model proposal, represented by Fig. 1, is an Intelligent Virtual Sensor that can act as a knowledge bridge between external data sources and the industry. By gathering external data, that can be present on online, or other types of, platforms containing a continuous flux of updated data, is possible to create a knowledge layer directed to the industry, which can cause an increase of 'intelligence' associated with the smart manufacturing processes.

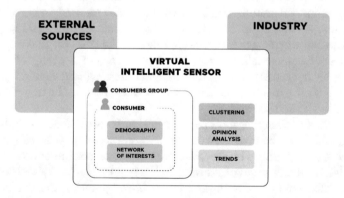

Fig. 1. Intelligent virtual sensor model connecting external data sources with industrial processes

The virtual aspect of the sensor means that can perform actions other than only collecting data. For example, the different types of data that the sensors will gather could be from multiple data sources, and is important to understand and create relations and connections between them. From demographic characteristics, to clustering of consumers based on custom metrics, understanding and evaluate interests, or even analyse opinion and other forms of feedback, the intelligent aspect of the sensor can answer some questions for the organisation that can influence and aid their manufacturing process: 'What type of products our consumers like?'; 'Which demographic groups are interested for certain products'; 'What age group buy more products inside a demographic group?', 'What type of products are bought by customers?'; 'Why are sales increasing or decreasing?'; 'What are ou consumers saying about our products', 'What type of products are our consumers asking for?'.

The clustering capability is one of the modules that the intelligent virtual sensor contains. By being focused on a consumer, or a group of consumers, is possible to perform a set of actions (represented by modules) that can be adjusted to the needs of the industrial processes and the objectives of an organisation. However, some 'default' modules were established in order to construct a knowledge layer that can potentiate the industry 4.0 to a higher level. These modules are:

- Clustering: Responsible to cluster groups of consumers with specific characteristics defined by the organisation, in order to associate different groups of consumers with different products.
- Opinion Analysis: The external feedback factor that can help the industrial processes, as well as the decision making. This model is responsible for generating knowledge based on the captured insights about a consumer, or group of consumers, in order to understand their opinion and interest towards certain products or services. This knowledge is very valuable for the organisations since it can provide them with a closer relationship and understanding of their consumers demands. Such knowledge could be used to better know the consumers, as well as include their opinions the decision making processes. The inclusion of this generated knowledge in the decision making processes, will allow the personalised creation of products, to a group of consumers, fulfilling their necessities and, especially, addressing and delivering products that meet their expectations. This process can reinforce the organisation relationship with their consumers.
- Trends: This challenging module is dedicated to predict new trends that can help organisations be proactive on the market, attending their consumer needs on a short time-frame. The prediction of trends and consumers necessities contains the uncertainty factor whereas it is not possible to evaluate the results, however that could be diminished through the quantity and quality the previous obtained knowledge. To achieve complete prediction, it is necessary to generate and considerate multiple sources of knowledge, such knowledge about consumers, as well as knowledge obtained trough market analysis. The knowledge about the consumers will be acquired through the analysis made in the 'Opinion Analysis' module, and the knowledge of the market will be obtained through the constant analysis of the general market and the knowledge created for acquired data from being a stakeholder of the market. The process of prediction is a very important part and with a huge impact in organisations, possibly being related to their success or failure since influence

the success of the products that could me measure by, for example, the consumers affluence and number of sales, and allowing organisations to be active on the market instead of just reacting to its changes.

One of the motivations for this model, is that the organisations will be able to fulfil their products in an individualized way accordingly to the interests of their consumers. This is only possible because of the knowledge process associated with the data acquired about the consumers. The knowledge layer intended to support the decision making process, which will allow organisations to be more decisive and active when it comes to deliver their products accordingly to their consumers interests (or the market need), and will be able to respond faster to any possible alteration, increasing the relationship with their consumers.

4.2   Vertical Integration of Industry 4.0

The vertical integration of Industry 4.0 means implementing a smart factory that is highly flexible and reconfigurable, and the Intelligent Virtual Sensor can be an active part of this integration. The virtual sensor can be an active part in the IT integration which contains the necessary integrations to the new IT solutions; Can be present in Analytics and Data Management by fulfilling the necessity of process the data stored in a way that could be part of the decision making of the organization; Relatively to the Could-Based Application, the sensor could be in cloud so that he can use their advantages attached to the Industry 4.0; Finally, the sensor could improve the operational efficiency of the company, based on the knowledge layer provided by the intelligent virtual sensor.

5   Conclusion and Future Work

The industry 4.0 came to stay and change the way that organisations works. With the implementation of this new vision, the organisations are focused on better understanding their consumers, knowing their preferences, interests, while determining their index of satisfaction and opinions. The focus on the consumer demands that new types of data need to be gathered, and such data can be obtained from external sources.

By looking to organisations as a smart environment, is possible to enhance the global behaviour of a system in order to provide a high level functionality. To fulfil the capabilities of SmE, the proposed intelligent virtual sensor can create consumer knowledge trough the gathering of external data, in order to provide a closer relationship between organisations and their consumers.

The implementation of this model is not concluded only by the definition of it. There is the intention to continue its definition description, and to perform the implementation process on an industrial context. There is also a need to represent knowledge, and in order to truly fulfil the necessities of the organisations, the intelligent virtual sensor must be able to provide a wisdom layer (created by the knowledge layer) that will be responsible to predict trends or even market changes, allowing organisations to stay on the lead of the market.

References

1. Ackoff, R.: From data to wisdom. J. Appl. Syst. Anal. **16**, 3–9 (1989)
2. Antunes, M., Gomes, D., Aguiar, R.: Towards behaviour inference in smart environments. In: 2013 Conference on Future Internet Communications (CFIC), pp. 1–7. IEEE. https://doi.org/10.1109/cfic.2013.6566324
3. Barbosa, R., Santos, R.: Online social networks as sensors in smart environments. In: 2016 Global Information Infrastructure and Networking Symposium (GIIS), pp. 1–6. IEEE (2016). https://doi.org/10.1109/giis.2016.7814950
4. Batra, M.M.: Customer experience-an emerging frontier in customer service excellence. Compet. Forum **15**(1), 198–207 (2017)
5. Cook, D.J., Song, W.: Ambient intelligence and wearable computing: sensors on the body, in the home, and beyond. J Ambient Intell Smart Environ. **1**(2), 83–86 (2009). http://dl.acm.org/citation.cfm?id=1735835.1735836
6. Ducatel, K., Bogdanowicz, M., Scapolo, F., Leijten, J., Burgelman, J.C.: Istag scenarios for ambient intelligence in 2010 (2001). ftp://ftp.cordis.europa.eu/pub/ist/docs/istagscenarios2010.pdf
7. Gronholdt, L., Martensen, A., Kristensen, K.: The relationship between customer satisfaction and loyalty: cross-industry differences. Total. Qual. Manag. **11**(4–6), 509–514 (2000). https://doi.org/10.1080/09544120050007823
8. Huxtable, J., Schaefer, D.: On servitization of the manufacturing industry in the UK. Procedia CIRP **52**, 46–51 (2016)
9. Huxtable, J., Schaefer, D.: On servitization of the manufacturing industry in the UK. Procedia CIRP **52**, 46–51 (2016). https://doi.org/10.1016/j.procir.2016.07.042. The Sixth International Conference on Changeable, Agile, Reconfigurable and Virtual Production (CARV2016). http://www.sciencedirect.com/science/article/pii/S2212827116307922
10. Pauwels, E.J., Salah, A.A., Tavenard, R.: Sensor networks for ambient intelligence. In: 2007 IEEE 9th Workshop on Multimedia Signal Processing, pp. 13–16. IEEE. https://doi.org/10.1109/mmsp.2007.4412806. http://ieeexplore.ieee.org/lpdocs/epic03/wrapper.htm?arnumber=4412806
11. Ramos, C., Augusto, J.C., Shapiro, D.: Ambient intelligence - the next step for artificial intelligence. IEEE Intell. Syst. **23**(2), 15–18 (2008). https://doi.org/10.1109/mis.2008.19. http://dl.acm.org/citation.cfm?id=1782254.1782282
12. Schlaepfer, R.C., Koch, M.: Industry 4.0. challenges and solutions for the digital transformation and use of exponential technologies (2015). http://www2.deloitte.com/content/dam/Deloitte/ch/Documents/manufacturing/ch-en-manufacturing-industry-4-0-24102014.pdf
13. Wan, J., Cai, H., Zhou, K.: Industrie 4.0: Enabling technologies. In: Proceedings of 2015 International Conference on Intelligent Computing and Internet of Things, pp. 135–140. IEEE (2015). https://doi.org/10.1109/icaiot.2015.7111555
14. Xun, X., YiBai, L.: The antecedents of customer satisfaction and dissatisfaction toward various types of hotels: a text mining approach. Int. J. Hosp. Manag. **55**, 57–69 (2016)

A Framework to Improve Data Collection and Promote Usability

Davide Carneiro[1,2(✉)] and Albertino Vieira[1]

[1] CIICESI, ESTG, Polytechnic Institute of Porto, Felgueiras, Portugal
{dcarneiro,8140346}@estg.ipp.pt
[2] Algoritmi Centre/Department of Informatics, Universidade do Minho, Braga, Portugal

Abstract. Many of nowadays organizations can be said to be knowledge-based. That is, they have relevant decision-making processes that are supported by data and data mining processes. These data may be created/collected by the organization or acquired from external sources (e.g. open data portals). In any case, the quality of the data will, ultimately, be one of the main drivers of decision quality. In this context, it is important that data-producing organizations also produce relevant meta-information characterizing the provenance of the data, its context or the representation standards used. This paper presents a framework to facilitate this process, promoting the inclusion of information concerning representation standards, provenance, trust and permissions at the data level. The main goal is to promote data usability and, consequently, its value for the organizations.

Keywords: Data acquisition · Provenance · Data representation

1 Introduction

In the last years, the collection and storage of data of the most different sources, has become a routine task in many organizations [1]. In technical terms, this is made possible by different technological advances, namely in telecommunications and miniaturization, which resulted in a wide range of different data generating devices. It is now fairly easy to monitor many different variables from an environment and its users [2]. Moreover, there are also nowadays many applications that act as sensors, not in the traditional sense, but as so-called virtual or soft sensors [3], as opposed to the traditional hard sensors. These applications monitor users or other applications, and generate data about its usage.

On the other hand, the are also nowadays many frameworks that facilitate the management of sensors/applications, data acquisition and data storage (e.g. Kaa, AWS IoT, ThingWorx). Many of these frameworks run as a service, making it very easy to start collecting data.

Finally, the last drivers of this trend are modern databases (notably distributed large-scale databases) and data processing frameworks. The former

© Springer Nature Switzerland AG 2019
P. Novais et al. (Eds.): ISAmI 2018, AISC 806, pp. 357–364, 2019.
https://doi.org/10.1007/978-3-030-01746-0_42

allow for (often unstructured) large amounts of data to be stored and accessed [4]. The latter allow for these data to be processed, often in real-time [5], to efficiently provide valuable insights into the organization or the domain being studied, namely through the creation of models that systematize implicit rules or behaviors [6]. Moreover, a plethora of open source data processing frameworks exists nowadays, which allow for anyone to be able to quickly prototype data processing pipelines, without the need for powerful computers [7]. Finally, there are also nowadays many open data initiatives, which provide online datasets that any organization can use.

All this resulted in the democratization of data acquisition and processing initiatives, making them available to every organization or individual. The main positive consequence of this trend is the unprecedented increase in the amount of data collected and stored, which supports the finding of insights that were previously unattainable. However, there are also some significant drawbacks.

Namely, many of these data are collected and stored without a proper methodology to ensure its quality [8]. Indeed, aforementioned platforms such as Kaa allow for data to be stored but do not necessarily enforce important aspects such as provenance or trust measures, nor do they allow for the quality of the data to be queried. Moreover, they are cumbersome, requiring a significant configuration effort for setting up a data collection environment. Finally, existing frameworks are mostly turned towards the IoT paradigms, i.e., the offer functionality to acquire data from devices (e.g. they provide SDKs for mobile devices).

Since, in data analytics, the quality of the analysis is dependent on the quality of the data analyzed, data quality can have a significant impact on the organization's decision-making processes. The lack of these methodologies and other standards during the process of data acquisition and storage also make it difficult to integrate data from different sources and to efficiently use it.

The main aim of this work is to propose a framework that enforces good practices from the first step of the data analytics pipeline, i.e. data collection. Namely, this framework enforces the notions of trust, provenance, vocabularies and permissions, all in a fine-grained manner and at the data-level. An organization can easily create a new public or private instance of the framework and start collecting and storing data. Data collected with this framework can be more easily analyzed, interpreted, reused and integrated, resulting in the development of better data-driven models for knowledge organizations. In this paper we explore the potential of this framework for the collection and reuse of data in the context of intelligent environments, by researchers or other practitioners.

2   Architecture

One of the requirements of this project is that new instances of the framework can be easily started so that the software environment to support knowledge organizations can be started off-the-shelf, in any environment. To this end, the architecture of the framework is based on Docker [9]. Docker allows for an application to be packaged with all of its dependencies and then run in disparate environments. Applications are thus seen as portable and self-sufficient containers, that can run on the cloud or on-premises.

One of the core concepts is that of a container. A docker *container* represents the execution of an application, process or service. It is defined by an *image*, that is a package with all the dependencies (e.g. frameworks) and information needed (e.g. deployment and execution configuration) to create the container. Multiple instances of a container (created from the same image) can exist simultaneously. Containers can also be linked to communicate with one another. A container can be built from a *Dockerfile*: a text document that contains all the commands one would normally execute manually to build the image. This file allows for images to be automatically built.

Multi-container applications (as the one detailed in this paper) can be configured using docker's Compose tool. This tool allows to define a single application based on multiple images and then build and deploy the resulting application to a docker host. Finally, a docker application can be scaled using multiple docker hosts, coordinated in a cluster and acting as a single virtual host.

The proposed framework is supported by four containers (Fig. 1). The first container runs a mongo database, configured with a set of collections to enforce data representation standards, provenance and permissions, as described in Sects. 2.1 to 2.3. Data is stored through this container, in a *volume*: a specially-designated directory on the host's filesystem, to persist data, independent of the container's life cycle.

The second container runs a node.js server, with some additional modules pre-installed, namely to enforce the adherence to data representation standards and permissions. The aim of this container, which is linked to mongo, is to expose a REST API that allows for data to be stored, accessed and managed from any client application. The REST API is pre-configured and already created when the image is started. The API also validates, at an application level, aspects such as authentication, data validity (prior to storing it in the database, by checking the associated validator) and access permissions (prior to returning data to the client) (Sect. 2.1).

The third container runs a Spark image. The main goal of this container is to receive streaming data, to process it (if necessary) and to store it in the database. It uses Spark's Streaming API to consume continuous data streams in an efficient manner. The container is also linked to the mongo container.

Finally, the fourth container runs a swagger-ui image. The purpose of this container is to allow the users of the framework to visualize and interact with the API's resources, to facilitate the development of applications.

Sections 2.1 to 2.3 detail how this architecture implements some of the most important requirements for data usability in an organization [10], namely data representation standards, provenance, trust and permissions management.

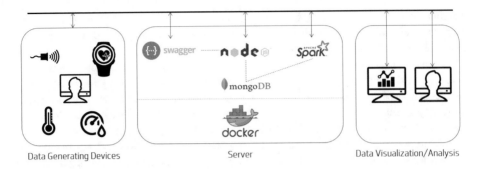

Fig. 1. Architecture of the proposed framework.

2.1   Data Representation Standards

A standards-based data representation reduces syntactic and semantic hetero-geneity in the data and, consequently, improves interoperability and data usabil-ity. Depending on the domain, different approaches can be followed to implement such standards (e.g. vocabularies, ontologies).

In this work, data usability from a representation standards point of view is achieved by means of two different approaches: (1) creation of pre-determined collections on the database, with specific validators; and (2) mandatory use of dictionaries associated to the data.

The image of the mongoBD container is configured with specific commands to create a group of collections that define the structure of the database and support the definition of data dictionaries as well as the control of provenance and permissions. The following collections are defined when the mongoDB container starts:

- Users - this collection uniquely identifies each user of the framework. Depend-ing on the permissions, a user can create/manage data as well as other con-cepts such as sensors, subjects, sessions or variables;
- Subjects - data is often collected concerning specific subjects, which may include people, locations, environments, animals, among others. Subjects are created by Users. Any subject from which data is collected is described in this collection;
- Sessions - a Session describes data that can be aggregated, generally by a prox-imity in time, space or variables used (e.g. a period of 1 h in which biofeedback data was collected from an individual). A session is created by a User, may refer to a Subject and must describe the context (e.g. physical, temporal, goal) in which data was collected;
- Sensors - a Sensor describes a source of data. This definition admits both hard and soft sensors, as well as any other sources (e.g. manual annotations of data by the researcher). Sensors are created by Users and should thoroughly describe the source of the data;

- Variables - this collection describes the Variables that can be extracted from Sensors. For instance, different variables can be extracted from a blood volume pulse Sensor (e.g. inter-beat interval, heart rate, pulse amplitude). Each variable must also contain a document in MongoDB's implementation of the JSON Schema, with the meta-data of the variable. Variables are created by Users and refer to a Sensor;
- Data - this collection stores, in each document, an instance of one Variable, collected from a Sensor, in the context of a Session, by a User, from a Subject. Documents in this collection are validated during insertion against the schema defined in the corresponding document on the Variables collection, to ensure its validity;
- Permissions - this collection models the permissions of each user concerning each of the remaining collections and their documents, and the respective roles. For instance, a User can have the role *Owner* regarding a Session, although he may not own the Variables from which data was collected in that Session. Read permissions can be attributed at several levels: sensor, variable, session and data-level. This makes it possible for a User to control the level of access in a fine grained manner.

Each collection is created with the corresponding validator, that ensures that only documents with the correct structure are stored in each collection. Validators are defined using MongoDB's implementation of JSON Schema.

2.2   Provenance and Trust

The quality of the data is one of the most important aspects influencing the quality of the decision-making processes that are based on that same data. That is, data with low quality will most likely result in poor decision-making. The quality of the data can, sometimes, be estimated from the data itself (e.g. missing values, invalid values). However, it can also be quantified from their provenance and the trust one puts on this provenance [11]. If the provenance of data is known by the users of the data, these can better interpret and judge the quality of the data [12]. This is especially important in a time in which so many data sources exist on the web, for which the provenance is not always clear.

In this platform, provenance can be internal (when data was originally produced by a user of the platform) or external (when data results of a collection or transformation process whose source is external to the platform.

Internal provenance is defined at a data-level. That is, the provenance of each individual data unit is known and traceable: each document in the Data collection identifies the Session (which identifies the responsible User and describes the context and goal of the session), the Variable (which also identifies the Sensor) and the Subject from which the data was acquired.

A data user can thus query the provenance of each individual data unit: who was responsible for its collection, what sensor was it collected from, which variable (from this sensor) does it represent, which subject does it relate to and in which context and with what goal was it collected.

Trust emerges from this provenance repository: if a user, or a sensor or even a variable is identified as producing bad data, all related data may eventually be regarded as not trustworthy. This is very important to identify potential bad data. Conversely, a user or sensor that consistently produces that that is regarded as sound has a higher level of trust.

2.3 Permissions

Permissions and their management become necessary whenever multiple users collaborate on a system and when different groups of users may have different access policies (e.g. not all users can access all data).

Permissions are often defined at an application level, in parallel with authentication mechanisms. For instance, a user that has read permission to a database or collection can read all the data on that database or collection. These are coarse-grained permission systems and are rather rigid. This platform considers a fine-grained permission system, in which permissions can be defined at different levels. For instance, a user may mark all data collected from a sensor as being public (which includes all data collected from all its variables). Alternatively, a user may attribute specific permissions to variables, sessions or even individual data items. The latter, for example, allows the user to publicly share data from one variable of part of a session while withholding the rest.

3 Use Case

The proposed framework can be used in virtually any knowledge-based organization or domain. In this section we illustrate two use cases in which data is being collected. We highlight the variables used and the advantages, for each domain, that the fusion of the information from these different sources and its joint analysis may encompass for domain experts.

The first use case concerns the collection of data from the management of a public lighting network and from a weather station. In this use case, researchers are collecting and combining data from different variables including the temperature, voltage, power and dimming on each of the 305 luminaries of the network, as well as ambient data such as temperature, humidity and wind, collected from weather stations or weather data services. Some of these variables are depicted in Fig. 2, which shows data for a specific day.

These data are used by researchers to better understand the relationship between these different variables, namely the relationship between ambient temperature, luminary temperature, dimming and power consumption. The goal is to develop an intelligent public lighting management system that can take into account these different sources of information to decrease energy consumption while maintaining service quality.

Figure 3, in which data was normalized to be represented together, shows the different dimming settings of the luminaries throughout the day and how this affected power consumption. In this domain, information is kept about the

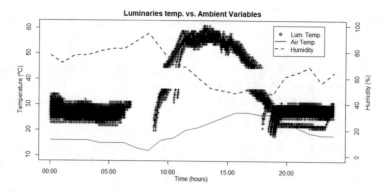

Fig. 2. Plot of data from 307 sources in one session of 24h: 305 temperature sensors in luminaries (black crosses), air temperature (solid red line) and humidity (dashed blue line).

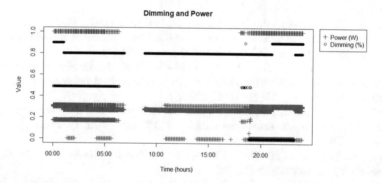

Fig. 3. Plot of data from 2 different sensors of 305 devices (luminaries): dimming and power of the luminaries throughout the day (normalized).

provenance of each individual piece of data, as proposed in this work. Namely, for every value of every variable, researchers now when it was collected, which luminary, weather station or online data service it was collected from, who is the responsible researcher and in which context it was collected.

4   Conclusions

This paper details a framework for facilitating and improving the quality of data collection processes, incorporating concepts such as data representation standards, provenance, trust and permissions management. This is done at a data-level, with the goal to facilitate data reuse and integration, and consequently its value. The framework promotes the reuse of concepts, facilitates the integration of data and the validation of the data through rules defined in json schemas. This increases data quality and decreases the necessity for data cleaning since it allows for data to be selected according to its quality, that is, data quality can be queried. Read permissions can be attributed at several levels: sensor, variable, session and data-level. This makes it possible for a user to control the level of access in a fine grained manner. In summary, validity and provenance is ensured at the data-level, for each individual instance of data. This may be relevant for any knowledge-based organization or for research initiatives that rely significantly on data from multiple and sometimes external sources, as in the field of Intelligent Environments.

Acknowledgement. This work is co-funded by Fundos Europeus Estruturais e de Investimento (FEEI) through Programa Operacional Regional Norte, in the scopre of project NORTE-01-0145-FEDER-023577.

References

1. Gudivada, V.N., Baeza-Yates, R.A., Raghavan, V.V.: Big data: promises and problems. IEEE Comput. **48**(3), 20–23 (2015)
2. De Paz, J.F., Julián, V., Villarrubia, G., Marreiros, G., Novais, P.: Ambient intelligence–software and applications. In: 8th International Symposium on Ambient Intelligence (ISAmI 2017), vol. 615. Springer (2017)
3. Gonzaga, J., Meleiro, L.A.C., Kiang, C., Maciel Filho, R.: Ann-based soft-sensor for real-time process monitoring and control of an industrial polymerization process. Comput. Chem. Eng. **33**(1), 43–49 (2009)
4. Diallo, O., Rodrigues, J.J., Sene, M., Lloret, J.: Distributed database management techniques for wireless sensor networks. IEEE Trans. Parallel Distrib. Syst. **26**(2), 604–620 (2015)
5. Marz, N., Warren, J.: Big Data: Principles and Best Practices of Scalable Realtime Data Systems. Manning Publications Co. (2015)
6. Tassa, T.: Secure mining of association rules in horizontally distributed databases. IEEE Trans. Knowl. Data Eng. **26**(4), 970–983 (2014)
7. Zaharia, M., et al.: Apache spark: a unified engine for big data processing. Commun. ACM **59**(11), 56–65 (2016)
8. Fan, W.: Data quality: from theory to practice. ACM SIGMOD Record **44**(3), 7–18 (2015)
9. Merkel, D.: Docker: lightweight linux containers for consistent development and deployment. Linux J. **2014**(239), 2 (2014)
10. Freitas, A., Curry, E.: Big data curation. In: New Horizons for a Data-Driven Economy, pp. 87–118. Springer (2016)
11. Sänger, J., Richthammer, C., Hassan, S., Pernul, G.: Trust and big data: a roadmap for research. In: 2014 25th International Workshop on Database and Expert Systems Applications (DEXA), pp. 278–282. IEEE (2014)
12. Moreau, L., et al.: The provenance of electronic data. Commun. ACM **51**(4), 52–58 (2008)

MyDentist: Making Children with Autism Familiar with Dental Care

Mariasole Bondioli[1], Maria Claudia Buzzi[2], Marina Buzzi[2], Maria Rita Giuca[3],
Francesca Pardossi[3], Susanna Pelagatti[1(✉)], Valentina Semucci[4],
Caterina Senette[2], Fabio Uscidda[1], and Benedetta Vagelli[4]

[1] University of Pisa, Lungarno Pacinotti 43, 56124 Pisa, Italy
{mariasole.bondioli, susanna.pelagatti}@di.unipi.it, fabio.usc@gmail.com
[2] IIT-CNR, via Moruzzi 1, 56124 Pisa, Italy
{claudia.buzzi,marina.buzzi,caterina.senette}@iit.cnr.it
[3] Dipartimento di Patologia Chirurgica, Medica, Molecolare e dell'Area Critica,
Ambulatorio di Odontoiatria Pediatrica, Ospedale Santa Chiara, Pisa, Italy
mariarita.giuca@med.unipi.it, f.pardossi@yahoo.it
[4] UFSMIA Pisa, ASL Nordovest, Via Garibaldi, Pisa, Italy
{v.semucci,benedetta.vagelli}@uslnordovest.toscana.it

Abstract. Children with autism perceive sensory experiences differently and have problems accepting unknown social contexts. In a dental care setting, there are many strong sound-visual stimuli, which are usually not found in known environments such as home or school. This usually upsets a child with autism, often forcing dentists to administer an anesthetic even to perform normal dental hygiene. In recent years, technology-enhanced systems and apps have been developed to help people with autism adapt to new contexts and cope with distressing social situations. In this paper, we present a web application (MyDentist) which provides dentists, parents and children with a set of resources to teach children with autism the proper oral procedures at home and correct behavior during dental visits. MyDentist includes multimodal games and activities to lighten children's tension and to make them familiar with dental procedures. Games and activities can be personalized according to child's needs and preferences. Professionals may also take and collect photos and videos during dental visits, making them accessible to parents and caregivers in the private profile of each child.

Keywords: Autism · Web application · Dental care · Dental anxiety

1 Introduction

Autism is a spectrum of neuro-developmental conditions that seriously compromise the way people communicate with and relate to other people and their ability to make sense of the world around them. In recent years, attention has been increasingly focused on the use of ICT to help people with autism in different contexts [13]. This interest is motivated by the rapidly growing prevalence

© Springer Nature Switzerland AG 2019
P. Novais et al. (Eds.): ISAmI 2018, AISC 806, pp. 365–372, 2019.
https://doi.org/10.1007/978-3-030-01746-0_43

of autism and by the positive effects of the interaction of people with autism with computers and predictable environments [18]. In 2010, the overall prevalence of ASD (Autistic Spectrum Disorder) measured by the network of Centers for Disease Control and Prevention in the US was 14.7 per 1000 (one in 68) children aged 8 years [8]. As a matter of fact, prevalence has increased by more than an order of magnitude in the past 15 years. At present the only possible intervention is based on early diagnosis based on clinical observation (usually, in Italy at around age 1.5/2 years), followed by intensive treatment. Research suggests that behavioral programs implemented early and in an intensive manner can be efficacious in improving cognitive, adaptive, and social–communicative outcomes in young children with ASD [20]. Moreover, autism is characterized by a remarkable heterogeneity on the behavioral level, with substantial individual differences [11]. Dental health is a very serious issue for people with autism. In a dental care setting, there are many strong sound-visual stimulations that are different from those in any other setting. This usually upsets a patient with autism, often forcing dentists to administer an anesthetic in order to carry out dental procedures. However, the positive attitude of people with autism regarding technology has been used to simplify oral care with positive results [2,12].

In this paper, we describe MyDentist, a web application designed to support dental professionals in teaching ASD children to perform proper dental care at home and in dental clinics. The main goal is to train children with autism to take care of their own oral hygiene and to accept the dental clinic environment and procedures for prevention procedures, thus drastically reducing dental problems. MyDentist design is based on a three-month study involving ten children with autism, age 6–12 years (age range covering the permanent dentition process) at a public hospital in Pisa [3].

The paper is organized as follows. Section 2 discusses the related work on experiences using ICT to simplify dental care in autism. The requirements emerging from the preliminary study are summarized in Sect. 3. Section 4 provides an overview of the application and Sect. 5 concludes.

2   Related Work

Several studies in literature underline the positive effects of the daily use of computer-based intervention on different life aspects of ASD people [6,14,17,18]. Indeed, to this day, the ICT field has already been widely investigated and exploited in connection with many different educational approaches for ASD people. A wide set of digital tools has led to positive reactions in areas such as improving language, encouraging vocalization and learning appropriate forms of communication [4,7,10]. Moreover, many software programs have been developed and tested to facilitate the intervention of therapists and teachers involved in the educational path of ASD children.

The literature underlines the good results obtained in learning new skills using the video modeling technique [22], augmented reality [5], software to facilitate communication [15] and other different applications favoring interaction and communication between therapists, children and parents in normal or clinical context [1,9] . The important results by the literature related to computer intervention used with ASD people focus on improving autonomy and properly manage socially stressful situations [16,19,21]. Among all the stressful situations in ASD people's daily life, dental care contexts have rarely been the target of research intervention mediated by computer. In fact, few studies have attempted to explore the full potential of ICT in the domain of dental care. Isong et al. [12] conducted a randomized study with ASD children to test different anxiety reactions to the dental intervention with or without the introduction of electronic screen media - a DVD reader and Google glass - during the visit. Isong et al. have obtained some results similar to those presented herein, but in their research they do not take into account the reproducibility of the tested approach and moreover, they did not provide any form of personalization of the materials. A lack of personalized tools can also be found in the study conducted by Berry [2] in which the researcher proposes a familiarization method based on some digital tools, such as the reproduction of photos of dental clinics and staff, and social stories to memorize the important elements of this activity, to distract the children during the dental visits. Unfortunately, even in this case, the possibility of customizing the resources offered to the child is not provided.

To sum up, a wide knowledge of the positive effects of technological interventions in the life of people with ASD emerges from the literature, but we identified little research in the specific field of personalized digital tools to facilitate ASD dental care. Moreover, in most cases, the interest in this domain still focuses only on the positive results offered by a limited set of digital tools, e.g., video modeling, without investigating other digital resources that may better involve patients, such as cognitive games.

3   Participative Requirement Analysis

We analyzed the problem of using technological support to reduce anxiety and teach various dental care skills during a 3-month study involving ten children 6–12 years old, their parents/caregivers and a team of dentists at the public Santa Chiara hospital in Pisa [3]. During this process, we worked with ASD therapists in a public ASD therapy center (UFSMIA ASL NO Pisa) to select the proper activities to include in MyDentist and determine how they should appear.

In the study, the goal for each child was to learn how to perform proper oral care at home and how to behave properly during visits at the clinic in order to undergo professional oral hygiene and dental care. The approach followed by the medical team took several visits (the number could vary depending on the child's response) in which the child could explore and get used to the new environment and to the medical staff. The visits were preceded by an informal meeting with parents/caregivers to explain the goal of the project and to understand the special needs of each child. A questionnaire was administered to the caregivers to profile children' oral health, dental hygiene habits, sensory disturbances, autism condition, etc. [3]. During the visits, the dentist (always the same in all the visits of a given child) worked first on creating a relationship with the child, taking into account his/her special needs and then gradually moving on to dental care procedures.

In this process, a very important aid to reducing anxiety was the possibility of taking pictures and videos during the visits, to be seen there or later at home. Multimedia material recorded during each visit was elaborated in interactive PDF social stories recalling the highlights of each visits. Pictures, videos and the PDFs were made available to parents/caregivers to be used at home in order to prepare the child for the next visit. Furthermore, the basic abilities for dental care at home were taught using video models and digital activities. In each visit, a set of suitable resources for activities reinforcing previous concepts and preparing the child for the next visit were given to parents and caregivers to be used at home. Some of them were general, others were personalized to meet individual child's needs.

The study with children with autism allowed us to better identify the most accepted and suitable digital tools to provide in MyDentist, and how to simplify the children access to the digital activities, e.g. through a calendar highlighting past and future dental visits. Furthermore, dentists helped us to improve MyDentist usability, e.g. the creation/management of pictures and videos related to each patient and the production of PDF stories that is automatically performed through simple guided steps. Additional information on the three-month study carried out to evaluate if and how using technological support to reduce anxiety and teach dental care skills in ASD children can be found in [3].

4   MyDentist: A Web Application

MyDentist app is a dynamic Rich Internet Application based on distributed Web architecture. Specifically, it is coded in PHP, AJAX and HTML5 and relies on a MySQL database. All the software's functions are implemented using jQuery and JSON libraries, which enable the creation of abstractions for low-level interaction, advanced effects and high-level widgets. The AJAX technology allows the efficient and easy exchange of data, so the HTML5 client user interfaces are constantly updated. MyDentist works as a big repository, a "toolkit" where materials are arranged in two main sets: one set includes pre-packaged general materials available for all the children as needed; the other set includes each patient's materials (dynamically created) containing multimedia collected during each visit and classified by visit date.

MyDentist provides pre-defined games but also a game-engine that allows to easily create three types of new games using personal resources (photos and images): sequences, puzzles, memories. The game-engine may be used autonomously by the caregiver with minimal effort through simple wizards.

Two access profiles are available: one for the dentist's use and the other for children and parents. The main tasks of the dentist profile are:

Create and delete patient profiles. The dentist can create accounts for patients and delete profiles no longer in use (see Fig. 1).

Fig. 1. Dentist's profile: Home: patient management

Fig. 2. Dentist's profile: patient resource management

Fig. 3. Dentist's profile: Instant capture of videos and photos (left) Visits calendar (right)

Record information about each patient. In each patient profile, the dentist can record biographical data, diagnostic data, information on previous dental treatment and occurrence of specific sensory disturbances.

Add and manage material related to patients. Digital resources can be uploaded and associated with one or more patients using a simple interface (see Fig. 2). Photos and videos of a visit are automatically created using the application (managing the tablet's cam) and unequivocally associated with the child (see Fig. 3. left). The dentist can select resources (photos, videos, audio files and other documents), connect them to a specific child or use them to personalize games and activities.

Create and assign customized activities. MyDentist includes a repertoire of pre-packaged games and activities related to different dental procedures that can be personalized according to the patient's needs. Examples of activities are memory, puzzle, sequencing and matching. Typical customization includes selecting the difficulty level, the images used, the sounds and colors. Using the repository, the dentist can select a different set of activities for each patient, taking into account the cognitive level and the skills to be taught.

Schedule the visits and manage materials. For each patient, the dentist can access an interactive calendar, creating an entry record for each appointment containing all the material related to the visit: the homework activity to prepare for the visit and the multimedia resources recording the visit. Each appointment can contain private notes not accessible to the patient (see Fig. 3. right).

The patient profile allows access to all the interactive activities that the dentist has set up for him/her, proposed through an accessible and usable interface. In particular, using their interface, a patient can:

Play personalized games. Games selected and personalized for a child are accessible in a separate section of a simple interface (without distracting elements) in which the child can play.

Explore multimedia resources. The application gives a child access to all the materials collected and selected for him/her by the dentist. He/she can explore photos, play videos and audios through simple use galleries (see Fig. 4). This activity is usually performed at home before the visit in the dental clinic to remember and familiarize oneself with dental care procedures.

Remember previous visits and prepare for future ones. In that specific section, the child can remember what was done in the last visit and prepare for procedures in the next visit. This is a sort of digital Personal Visual Diary in which the child can navigate among past and future visits, finding all of the material inserted by the dentist.

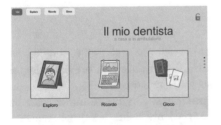 

Fig. 4. Child's profile: home (left) and explore child's resources (right)

5 Conclusions

Routine hygiene practices may be difficult to learn for children with low functioning autism. For this reason, they often require dental care in childhood and adolescence. However, people with autism experience great difficulty in receiving dental interventions due to their different perception of environmental stimuli. Familiarization with place, people, tools as well as care protocol sequences might facilitate the delivery of dental care. This study investigates protocols for teaching these concepts to children with autism mediated by technology. Specifically, a web Application, MyDentist, was created by both collecting user requirements with dentists and psychologists, and observing several children with autism receiving dental care with the support of technology. MyDentist facilitates the work of the dentist, visit scheduling and selection of personalized multimedia content produced in a visit. The effectiveness of this tool for reducing anxiety and stress in young patients, facilitating the dental intervention, remains to be confirmed by objective data coming from future tests involving two sets of ASD children (one used as a control group). Technology can have an important role in accessible teaching and avoiding a distressing intervention. Ambient intelligence and Internet of Things (IoT) technologies can leverage the burden of work that dentists face when taking care of children with autism. Future work will define intelligent dental rooms equipped for facilitating the delivery of dental care to special needs patients.

Acknowledgements. We would like to thank the *Associazione Autismo Pisa*, the *Azienda Ospedaliera Universitaria Pisana* and the *Azienda Sanitaria Locale Toscana Nordovest* for the support to the project.

References

1. Ayres, K.M., Maguire, A., McClimon, D.: Acquisition and generalization of chained tasks taught with computer based video instruction to children with autism. In: Education and Training in Developmental Disabilities, pp. 493–508 (2009)
2. Barry, S.M.: Improving access and reducing barriers to dental care for children with autism spectrum disorder. PhD thesis, University of Leeds (2012)
3. Bondioli, M., Buzzi, M.C., Buzzi, M., Pelagatti, S., Senette, C.: ICT to aid dental care of children with autism. In: Proceedings of the ASSETS 2017 (2017)
4. Bosseler, A., Massaro, D.W.: Development and evaluation of a computer-animated tutor for vocabulary and language learning in children with autism. J. Autism Dev. Disord. **33**(6), 653–666 (2003)
5. Casas, X., Herrera, G., Coma, I., Fernández, M.: A kinect-based augmented reality system for individuals with autism spectrum disorders. In: Proceedings of the GRAPP/IVAPP, pp. 440–446 (2012)
6. Chen, S., Bernard-Opitz, V.: Comparison of personal and computer-assisted instruction for children with autism. Ment. Retard. **31**(6), 368–376 (1993)
7. Coleman-Martin, M.B., Wolff-Heller, K., Irvine, K.L., Cihak, D.F.: Using computer-assisted instruction and the nonverbal reading approach to teach word identification. Focus Autism Other Dev. Disabl. **20**(2), 80–90 (2005)

8. Centers For Disease Control and Prevention (CDC): Prevalence of autism spectrum disorder among children aged 8 years – autism and developmental disabilities monitoring network, 11 sites, unites states, 2010. MMWR Surveill. Summ. 63(2), 1–21, March 2014

9. Costa, M., Costa, A., Juliãn, V., Novais, P.: A task recommendation system for children and youth with autism spectrum disorder, pp. 87–94 (2008)

10. Davis, M., Dautenhahn, K., Nehaniv, C., Powell, S.: Touch story: towards an interactive learning environment for helping children with autism to understand narrative. In: Designing accessible technology, 785–792 (2005)

11. Vivanti, G.: Predictors of outcomes in autism early intervention: why don't we know more? Front. Pediatr. **2**, 58 (2015)

12. Isong, I.A.: Addressing dental fear in children with autism spectrum disorders: a randomized controlled pilot study using electronic screen media. Clin. pediatr. **53**(3), 230–237 (2014)

13. Goldsmith, T.R., LeBlanc, L.A.: Use of technology in interventions for children with autism. J. Early Intensiv. Behav. Interv. **1**(2), 166 (2004)

14. Jordan, R.: Computer assisted education for individuals with autism. In: Autisme France 3rd International Conference (1995)

15. De Leo, G., Leroy, G.: Smartphones to facilitate communication and improve social skills of children with severe autism spectrum disorder: special education teachers as proxies. In: Proceedings of the 7th International Conference on Interaction Design and Children, pp. 45–48. ACM (2008)

16. Moore, D., Cheng, Y., McGrath, P., Powell, N.J.: Collaborative virtual environment technology for people with autism. Focus Autism Other Dev. Disabil. **20**(4), 231–243 (2005)

17. Moore, M., Calvert, S.: Brief report: vocabulary acquisition for children with autism: teacher or computer instruction. J. Autism Dev. Disord. **30**(4), 359–362 (2000)

18. Murray, D.: Autism and information technology: therapy with computers. In: Powell, S., Jordan, R. (eds.) Autism and Learning: A Guide to Good Practice, pp. 100–117. David Fulton, London (1997)

19. Passerino, L.M., Santarosa, L.M.C.: Autism and digital learning environments: processes of interaction and mediation. Comput. Educ. **51**(1), 385–402 (2008)

20. Rogers, S., Wallace, K.: Intervention for infants and toddlers with autism spectrum disorders. In: Amaral, D.G., Dawson, G., Geschwind, D.H. (eds.) Autism Spectrum Disorders, pp. 1081–1094. Oxford University Press, New York (2011)

21. Da Silva, M.L., Goncalves, D., Guerreiro, T., Silva, H.: A web-based application to address individual interests of children with autism spectrum disorders. Procedia Comput. Sci. **14**, 20–27 (2012)

22. Tereshko, L., MacDonald, R., Ahearn, W.H.: Strategies for teaching children with autism to imitate response chains using video modeling. Res. Autism Spectr. Disord. **4**(3), 479–489 (2010)

Analysis of the Use of ICT in Compulsory Bilingual Secondary Education in Andalusia Public Schools. Case Study

Pilar Maldonado-Manso[1] , Julio Ruiz-Palmero[2]([✉]) , Melchor Gómez-García[3] , and Roberto Soto-Varela[4]

[1] Education Counseling Junta de Andalucía, 29071 Málaga, Spain
[2] Department of Didactics and School Organization, Education Faculty, Málaga University, 29071 Málaga, Spain
`julio@uma.es`
[3] Department of Didactics and School Organization, Education Faculty, Autonomous University of Madrid, 28071 Madrid, Spain
[4] Department of Mathematics and School Organization, Education Faculty, Autonomous University of Madrid, 28071 Madrid, Spain

Abstract. This research focuses on an Andalusia public secondary education centre that studies the use of information and communication technologies (ICT) in the teaching-learning process and in bilingual education in particular. Its importance is due to the special impact that the Organic Law 8/2013 on the Improvement of the Quality of Education (LOMCE 2013) makes on ICT and the promotion of multilingualism with a view to transforming the education system. The purpose of this study is to verify the actual use of ICT by teachers who teach the non-linguistic areas included in the bilingual project of the centre and relate it to the characteristics of the teachers themselves and the benefits that the use of ICT can provide in this mode of teaching for the students. The knowledge of the factors involved in the use of ICT in the centre, both for and against them, would allow the design of improvement strategies that encourage the necessary changes to optimize and make the educational use of these new technologies more effective, converting them from mere ICT to real LKT (learning and knowledge technologies).

Keywords: ICT · Bilingual education · Competences

1 Introduction

1.1 ICT and Key Competences

The origin of the concept of competence is often stated in an article by North American cognitive psychologist, business management specialist McClelland (1973), who (using "competency" rather than "competence") considered it as a way of assessing what actually causes superior performance at work, including not only practical but also social skills. In the field of education, competences were subsequently introduced through

© Springer Nature Switzerland AG 2019
P. Novais et al. (Eds.): ISAmI 2018, AISC 806, pp. 373–380, 2019.
https://doi.org/10.1007/978-3-030-01746-0_44

vocational training with the national vocational qualifications system in the United Kingdom and similar movements in several Anglo-Saxon countries. In Spain, and with regard to basic education (primary and compulsory secondary education), the introduction of the concept of basic or key competence was carried out following European recommendations (Bolívar 2008).

The European Education Information Network (Eurydice 2002) considers that a common European approach is needed to identify the competences that all citizens need to acquire, by carrying out a study on key competences, in order to establish their development, characteristics and evaluation. At the same time, the Organisation for Economic Co-operation and Development (OECD 2005) is developing the Project for the Definition and Selection of Competencies (DeSeCo), which defines competence as the ability to respond to demands and carry out tasks appropriately. It arises from the combination of practical skills, knowledge, motivation, ethical values, attitudes, emotions and other social and behavioural components that are mobilised together to achieve effective action. In this project, competence is considered to be key when it fulfils three conditions: contributing to the achievement of results of high personal or social value, being applicable to a wide range of relevant contexts and environments, and enabling people who acquire it to successfully overcome complex demands (Rychen and Salganik 2006). Likewise, the fundamental competences are those that all human beings need to meet the demands of the different contexts of their lives as citizens.

1.2   ICT and Bilingual Education

The European Union defines the promotion of multilingualism as an essential objective for the creation of a European project. The LOMCE (2013) in its preamble is committed to the curricular incorporation of a second foreign language (L2). In Andalusia, the bilingual programme started in 1998 with first French-language sections like L2. Later, The Plan for the Promotion of Plurilingualism in Andalusia (2005–2008), approved by the Governing Council's Agreement of 22 March 2005 (2005), was framed within the European objectives and included the development and extension of bilingual education programmes with the creation of bilingual centres.

The Strategic Plan for the Development of Languages in Andalusia called Horizon 202020209 has been recently approved by the Governing Council's Agreement of 24 January 2017 (2017), which seeks to consolidate the achievements of the bilingual programme, improve teacher training and student achievement, enhance language exposure beyond the classroom and foster interculturality. In this context, the Department of Education of the Regional Government of Andalusia offers a Plurilinguishing Portal10 that offers the whole educational community all kinds of materials, resources, regulations, etc. in relation to plurilingualism in Andalusia, including didactics units developed by teachers from CLIL approach. In addition, all this requires certified teachers with B2 level or higher in L2, according to the Common European Framework of Reference for Languages (Council of Europe 2002).

For this reason, the Order of 31 July 2014 approving the Third Andalusian Lifelong Learning Plan for Teachers (2014) provides language updating courses for teachers, in coordination with official language schools, and training in the pedagogy and didactics of CLIL approach. The integration of ICT in the bilingual projects of Andalusia has played a decisive role in the development of multilingualism in our community. Torres Caño (2014) points out that "the teachers involved in the bilingual Andalusian projects have integrated ICT into their teaching practice in two specific and interrelated fields: on the one hand, through the development of teaching materials, and on the other hand, through CLIL methodology". He highlights the common characteristics of ICT and CLIL: "a student-centred teaching methodology that addresses the diversity of learning styles, is based on the realization of projects and tasks, promotes student autonomy and supports cooperative teaching-learning processes". And they also share the 4 C's of which Coyle, Hood and Marsh (Coyle et al. 2010) speak: Content, Communication, Cognition and Culture. ICTs encourage the introduction of attractive methodologies for students such as gamification (Marín Díaz 2015) and flipped classroom (Tourón and Santiago 2015), which furthermore contribute to diversity.

2 Aim of the Research

2.1 Main Objective

To analyse the influence of ICT on the teaching-learning process in bilingual teaching in Compulsory Secondary Education in an Andalusia public school.

2.2 Secondary Objectives

- To verify the real use of ICT in non-linguistic areas for bilingual teaching.
- To relate the characteristics of the teachers involved to the actual use of ICT.
- To associate the use of ICT with the use of innovative methodologies.
- To link the use of ICT with improvements in students' key competences.

3 Methodology

The design of this paper corresponds to a descriptive and applied study, field research about a group of students and teachers of a bilingual compulsory secondary education school (CSE). No single variable is manipulated or controlled; only the current use of ICT in a single high school is described at the present time, taking a transversal approach, due to the study taking place at a single moment in time. It is an empirical research, based on relationships and associations between the observed variables.

The method used is survey research, since it is "capable of answering problems both in descriptive terms and in relation to variables, after the collection of systematic information. With this information the researcher tries to: (a) describe the conditions of the existing nature, (b) identify patterns with whom these conditions can be compared and (c) determine the relationships that exist between concrete events". (Buendía Eisman et al. 1998, p. 120).

3.1 Research Context and Population Studied

The research has been carried out in an Andalusia public high school in the bilingual English-Spanish teaching system with different NLAs depending on the level of education.

3.2 Data Collection and Data Analysis Techniques

Data Collection Techniques Surveys were designed for the two groups analysed, students and teachers. Although Martínez Olmo (2002) recommends that there should be no more than thirty questions. The initial questions correspond to socio-demographic data, which is necessary to "describe globally the group of people who have answered the survey, and later, to make a differentiated analysis of the answers" (Fernández Núñez 2007). In this way, both have included short open-ended, multiple choice and Likert type questions grouped in different categories.

Data Analysis Techniques The data collected have been analysed using descriptive statistics, using both graphical (histograms, cyclograms) and numerical descriptors (tables) (Buendía Eisman et al. 1998).

4 Results

4.1 Teacher Survey Results

The data obtained from the Teacher Survey corresponds to nine teachers, six women and three men, five of whom are between 35 and 45 years old, three of whom are under 35 and one over 45 years old. The non-linguistic areas imparted partially in English and at different CSE levels are Biology and Geology (two teachers, 1st and 3rd year), Mathematics (1st and 2nd, year, three teachers), Geography and History (two teachers, 3rd and 4th year), Physics and Chemistry (2nd year, one teacher) and Physical Education (4th year, one teacher).

Chart 1 shows their careers as teachers and their careers in the bilingual education system. It can be observed that, although most of them have been teaching for more than seven years, only three of them have accumulated more than three years as bilingual teachers. Although most of them have carried out more than five educational ICT training activities, only two of the teachers have attended any of them related to CLIL, a bilingual teaching approach. These activities are mostly live, since only three teachers have carried out e-learning, and two of them are those teachers who have been trained in the CLIL approach and also coincide with those who have taught in the bilingual education system in the past (Fig. 1) .

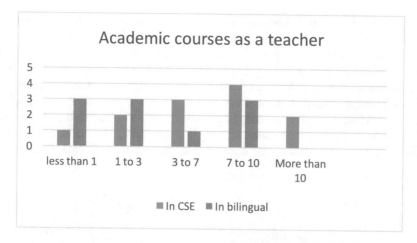

Fig. 1. Comparison between career as a teacher and as a bilingual teacher

With regard to the use of ICT, everyone recognizes using them but divides them into three equal groups in terms of frequency of use: daily, once or twice a week and two or three times a month. Most of them use resources (videos, images, presentations, etc.) and educational websites but none of them use social networks in their teaching.

Those who produce their own ICT resources use mostly software for making slideshows or video presentations. Most of the groupings used are couples and small groups. Everyone agrees that they facilitate the active learning of students and most believe that they motivate them, facilitate cooperative and collaborative work and foster attention to diversity. In addition, everyone agrees that a single use of ICT does not imply a significant change in student learning and they do not contribute more than speed and ease of access to content if they are not accompanied by changes in traditional methodology. However, these changes require training and innovative spirit on behalf of the teacher.

4.2 Student Survey Results

The data obtained from the student survey corresponds to one hundred and thirteen students, divided into five different groups of second year of CSE from the same school: three groups of twenty-four students, one of twenty-one, and one of twenty. This number of students, low for the second year of CSE, shows the number of students present on the day the questionnaire was taken. As mentioned above, all students belong to the bilingual modality of English, in the subjects of Physics and Chemistry and Mathematics, including repeaters and SEN students.

55.8% are girls and the rest are boys. 91% are 13 or 14 years old. 8.8% (10 students) are repeaters of 2nd year of CSE and 7.9% (9 students) have repeated a different second year course, these two groups having no element in common. The different second course that has been repeated the most is 1st year of CSE (6 students), followed by 2nd and 4th years of Primary Education, with one student each, and another student who has repeated 6th year of Primary and 1st year of CSE.

Although all of them, except for the ten CSE second year repeaters, have studied the bilingual education system in 1st year of CSE at their current school, the vast majority of them have not been in a bilingual education system in primary education, which means that the levels of linguistic competence in English are as diverse as possible, from students with native English-speaking families to students who have difficulty even with the linguistic competence in Spanish. Taking into account all of this, chart 2 shows the student responses to: Do you feel comfortable in your bilingual school? (Fig. 2)

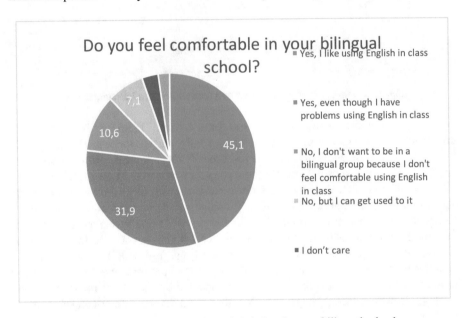

Fig. 2. Students' opinions about their belonging to a bilingual school.

Only 17.7% are not comfortable in a bilingual high school and 40% of these students could get used to it, so it is not a serious problem to be in the bilingual mode. Only four of them do not care and the rest, 78.8% are good at using English in class, although 40% of these students admit that they find it difficult to use L2 and 2% say that it depends on the subject used. Surprisingly, most of the students surveyed pointed out that it is not a problem to be in a bilingual high school even though their experience in this type of teaching is reduced for the majority of them to the previous academic year (1st year of CSE).

5   Discussion and Conclusions

The main objective of this research was to analyse the influence of ICT on the teaching-learning process in bilingual teaching in Compulsory Secondary Education in an Andalusia public school. To achieve this, a descriptive study of the current situation of the issue has been carried out to link the different variables involved. The method used is a survey research. Qualitative data have been collected through questionnaires for teachers and students. The teachers involved have been the nine who teach bilingually in the institution, i.e. the non-linguistic areas (NLAs) that use English as a second language in their educational practice.

The first secondary objective has been achieved due to the fact that the real use of ICTs in bilingual teaching in non-linguistic areas has been proven. New technologies are effectively used in the teaching of bilingual NLAs in CSE at the school. One of the most likely solutions to these problems would be the use of mobile phones or laptops belonging to the student body, since it has been proven during the investigation that the vast majority of students have access to them. Thankfully, in recent years, the current context in students' homes has come closer to the so-called myth of democratizing education with ICT (Cabero 2008), according to which, everyone is connected.

Firstly, the possibility that gender or age of teachers may influence the use of ICTs in NLAs has been ruled out. While it is true that older teachers use them to a greater extent and more frequently, this is not because of their age, but because they have had more time to accumulate more teaching experience. This shows that after the increase in the physical presence of ICT, the need for its real and effective implementation is teacher training (Cabero 2010).

The third secondary objective was to associate the use of ICT with the use of innovative methodologies, and this has been achieved by analysing teacher training in CLIL approach as well as their views on the introduction of new methodologies in a parallel way to ICT use. Thus, the data shows that teachers who have carried out teaching training in the CLIL approach often introduce ICT into teaching practice.

To conclude, it has also been possible to establish a link between the use of ICT and improvements in key competences of students, the fourth and final secondary objective set at the beginning of the research. Obviously, digital competence is developed by the simple use of electronic devices and digital resources. The importance of the need to work on this competence should not be underestimated. Despite the fact that students are often called "digital natives", due to their youth, Cabero (2010) points out that "the distance between natives and digital immigrants is not only based on time-based variables, but also on the cognitive acquisition of technology, in this sense the technological competences of the students are less than it could be expected". Indeed, as the data from this study shows that ICT plays an important role in the lives of students.

References

Acuerdo de 22 de marzo de 2005, del Consejo de Gobierno, por el que se aprueba el Plan de Fomento del Plurilingüismo en Andalucía, 65, BOJA, n° 1 (2005)

Acuerdo de 24 de enero de 2017, del Consejo de Gobierno, por el que se aprueba el Plan Estratégico de Desarrollo de las Lenguas en Andalucía Horizonte 2020, 24, BOJA, n° 1 (2017)

Bolívar, A.: El discurso de las competencias en España: educación básica y educación superior. Revista de Docencia Universitaria, Número monográfico II, pp. 1–23 (2008)

Buendía Eisman, L., Colás Bravo, M.P., Hernández Pina, F.: Métodos de investigación en psicopedagogía. McGraw Hill, Madrid (1998)

Cabero, J.: La formación en la sociedad del conocimiento. Indivisa: Boletín de Estudios e Investigación, 10, 13–47 (2008)

Cabero, J.: Los retos de la integración de las TICs en los procesos educativos: límites y posibilidades. Perspect. Educ. **49**(1), 32–61 (2010)

Consejo de Europa, & Subdirección General de Cooperación Internacional del Ministerio de Educación, Cultura y Deporte. Marco Común Europeo de Referencia para las Lenguas: aprendizaje, enseñanza, evaluación. Madrid: Artes gráficas Fernández Ciudad, S.L (2002)

Coyle, D., Hood, P., Marsh, D.: Content and Language Integrated Learning. Cambridge University Press, Cambridge (2010)

Eurydice. Key competencies. A developing concept in general compulsory education (Eurydice European Unit, European Commission, Directorate-General for Education and Culture). Brussels, Belgium (2002). http://bookshop.europa.eu/en/key-competencies-pbEC3212295/

Fernández Núñez, L.: ¿Cómo se elabora un cuestionario? Butlletí La Recerca (2007)

Instrucciones de 22 de julio de 2016 conjuntas de la Dirección General de Innovación y de la Dirección General de Formación Profesional Inicial y Educación Permanente, sobre la Organización y Funcionamiento de la enseñanza bilingüe para el curso 2016/2017 (2016). https://www.adideandalucia.es/normas/instruc/Instrucciones22julio2016EnsenanzaBilingue.pdf

Ley Orgánica 8/2013 para la Mejora de la Calidad Educativa (LOMCE), 295, BOE, n° I (2013)

Marín Díaz, V.: La gamificación educativa. Una alternativa para la enseñanza creativa. Digit. Educ. Rev. **27**, 1–4 (2015)

Martínez Olmo, F.: El cuestionario: un instrumento para la investigación en las ciencias sociales. Laertes, Barcelona (2002)

McClelland, D.: Testing for competence rather than for intelligence. Am. Psychol. **28**(1), 1–14 (1973)

OCDE: La definición y selección de competencias clave. Resumen ejecutivo (DeSeCo) (2005). http://deseco.ch/bfs/deseco/en/index/03/02.html

Orden de 28 de junio de 2011, por la que se regula la enseñanza bilingüe en los centros docentes de la Comunidad Autónoma de Andalucía, 135, BOJA, n° 1 (2011)

Orden de 31 de julio de 2014, por la que se aprueba el III Plan Andaluz de Formación Permanente del Profesorado, 170, BOJA, § 1 (2014)

Rychen, D.S., Salganik, L.H.: Las competencias clave para el bienestar personal, social y económico. Ediciones Aljibe, Archidona (2006)

Torres Caño, P.: Integración de las TIC en los proyectos bilingües en Andalucía. Andalucía educativa. Revista Digital de la Consejería de Educación (2014). http://www.juntadeandalucia.es/educacion/webportal/web/revista-andaluciaeducativa/contenidos/-/contenidos/detalle/nuevas-miradas-sobre-las-ticaplicadas-en-la-educacion-julio-cabero-almenara-1

Tourón, J., Santiago, R.: El modelo Flipped Learning y el desarrollo del talento en la escuela. Rev. Educ. **368**, 196–231 (2015)

Personal Learning Environments (PLE) on the Bachelor's Degree in Early Education at the University of Granada

Eduardo Chaves-Barboza[1], Juan Manuel Trujillo-Torres[2(✉)],
Francisco Javier Hinojo-Lucena[2], and Pilar Cáceres-Reche[2]

[1] National University, Heredia, Costa Rica
echav@una.cr
[2] University of Granada, Granada, Spain
{jttorres,hinojo,caceres}@ugr.es

Abstract. It is studied the devices which are incorporated in their personal learning environments (PLE) a university student population in teacher training. Additionally, the time spent on activities related to ICTs and the desired characteristics by this population in a PLE. For that, it was applied a questionnaire with Likert scales in a sample of 520 students in 12 groups in the Bachelor's Degree in Early Education of the University of Granada, Spain. The data have been examined through descriptive and inferential statistic (confidence intervals at 95%), correlation tests (τ Kendall's coefficient) and analysis of variance (Kruskal-Wallis test), with levels of bilateral significance at 95%.

The results show that the devices largely incorporated in PLE of the students are laptops and smartphones, university platforms have not been incorporated to the students' PLE, students prefer PLE foster the development and the professional career, which is customizable, interactive and useful to produce, edit and publish information.

Keywords: Personal learning environment · Higher education
ICTs

1 Introduction

The personal learning environment (PLE) is a theoretical-pedagogical concept built by a community of technologists and educators to study the impact of the Web 2.0 technology in people's learning, given their inbuilt learning capacity and their everyday relationship with technology tools. As PLE it is denoted the biotechnology system accessed, organised, adequate and used by a person to self-regulate their learning, this system is comprised of personal contacts, supports groups, social and corporate networks, ICTs tools, educational resources, information sources, inter alia.

In accordance, the study of the students PLE gains traction on the professional training and development processes, because of the implicit influence that its development and management has in these processes [1–3]. To give a contribution in this respect, this work present as its objective characterize the PLE of a trainee teacher

P. Novais et al. (Eds.): ISAmI 2018, AISC 806, pp. 381–388, 2019.
https://doi.org/10.1007/978-3-030-01746-0_45

population, this intended to increase the efficiency of the PLE's role in the teachers' initial training and giving an empirical contribution to the issue.

2   State of the Art

Nowadays, there is an agreement about the importance of the personal learning environment in the formal educational sphere, as much for initial higher education as for professional progress, as much for primary level education; as for secondary level. Some advantages highlight for the great flexibility and ease of use, the possibilities of customization and adequacy, the access given to unlimited web 2.0 tools, the prospective that they have for the collaborative work and social relationships, the active role conferred to the subject in formation, as well as the emphasis given to the development of their own responsibility and control and the possibilities offered for the formal, non-formal and informal education [4–6].

3   Methodology

The study population is composed of active students on the Bachelor's Degree in Early Education at the University of Granada. This population has a total of 877 members, between these 829 (94.53%) are women and 48 (5.47%) men, with a sampling error at 5%. The average age is 23.45 ± 0.14 years.

The populations are divided into 14 groups of students of homogeneous size, 5 groups are from the first grade, 5 from the second and 4 from the third. As a way of saving costs and time in the application of the questionnaire, the advantage of this existing concentration of students is taken to generate a framework of conglomerates. The questionnaire has been applied to 12 groups of students (primary sampling unit) randomly selected and it has reached a total of 250 students (elemental sampling units), and this allows the calculation of the confidence intervals (CI) with a confidence level of 95%.

The questionnaire applied has as initial questions gender and age. It continues with questions that refer to the possession of devices intended for personal use (dichotomous variable) and the number of hours per week dedicated to different activities (scalar variable). It concludes with questions about the desirable characteristics of a PLE (ordinal reagents, Likert type scale, with four options).

Over the data are applied nonparametric tests for the analysis of correlations, in particular Kendall's τ coefficient. Also, the analyses of variances are performed, through the Kruskal-Wallis test, in order to stablish whether there are significant differences between different levels of Likert scales. In any event, with a level of bilateral significance of 95%.

The questionnaire has been validated by ten experts in educational technology. The information offered by the experts has been analysed qualitatively through a process of codification and categorization which allowed the suggestions of the experts to be taken into account, and it has also been analysed quantitatively through the Osterlind congruence index. It is noted that each item of the questionnaire has a congruence index greater than .5, which test the validity of the instrument. Moreover, it has been determined that the questionnaire has a Cronbach alpha of .927, which tests its excellent reliability.

4   Analysis of the Results

Figure 1 shows that the device most incorporated into the student PLE is the laptop, followed by smartphones; more in-depth research about the way students use these devices is recommended. The next on the list, the desktop computer is incorporated into the PLE by almost half of the students. with this it is concluded that portability and ubiquity are important elements in the PLE of students. Greater power, reliability, accessibility and better desktop computer features recede into the background.

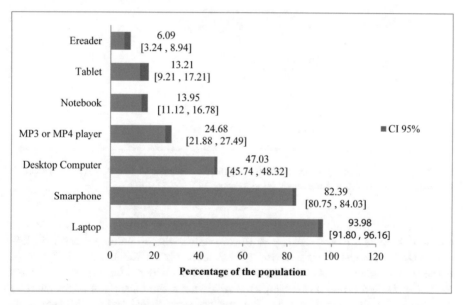

Fig. 1. Percentage of students per incorporated device in their PLE, with confidence intervals. **Note**: own development.

The incorporation of different devices to the students' PLE have slights correlations between them, $p < .001$, $-.18 < \tau < .20$, IC = $[-.24, .26]$. This fact is interesting, because it means that neither one nor the other of the devices are complementary or substitute each other, for the study population. This shows us that it is possible to foster the simultaneous use of several devices and that it is necessary to promote complementarity as a way of promoting more effective use as well as a greater scope for the resources being accessible to students.

On the other hand, devices such as MP3 players, Notebooks, Tablets and Ereaders have relatively little presence in students' PLE, investigating the causes of this fact is suggested and as much as possible fostering, in the surveyed population, the use of these important and versatile devices.

In the Fig. 2 it can be observed that students spend similar quantities of time using the PLE tools, using the computer and surfing the Internet. The time spent in visiting University platforms is considerably smaller.

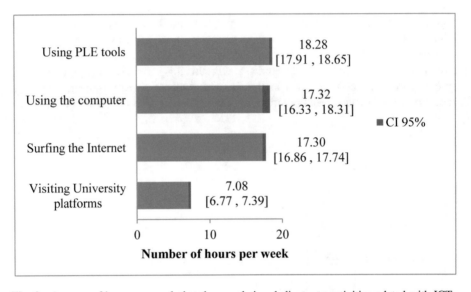

Fig. 2. Average of hours per week that the population dedicates to activities related with ICTs, with confidence intervals. **Note**: own development.

There is a significant and positive correlation between surfing the Internet and the use of PLE tools, $.57 < \tau < .67$, $p < .01$ two-tailed with a confidence level of 95%. A substantial amount of the student's PLE tools are on the Internet is indicated. However, the relation between PLE tools and the University platforms is relatively low, $.24 < \tau < .34$, indicating that these platforms are not a significant part of the students' PLE tools. The fact that university platforms are little visited and are not part of the PLE means that reinforcement is needed, making an accurate diagnosis of these platforms in order to determinate how they could be more appealing for students is recommended.

Chart 1 shows a strong positive correlation between surfing the Internet and using the computer, there is a significant relationship between the use of this device and the use of PLE tools together with the visit to university platforms. This allows stablishing that this device is crucial for the web surfing, and important for the use of PLE tools and the use of the services offered by the University in its platforms.

There is a significant relationship between surfing the Internet and the use of PLE tools, indicating that a substantial amount of student's PLE tools are on the Internet. Nevertheless, the relationship between PLE tools and university platforms is relatively low, indicating that these platforms are not fully implemented in students' PLE.

Chart 1: Relation between weekly hours that population dedicate to activities related with ICT

Desirable characteristics in a PLE	Kendall's τ correlation coefficient* and confidence intervals**		
	Using the computer	Surfing the Internet	Using PLE tools
Surfing the Internet	.62 [.57, .67]		
Using PLE tools	.42 [.37, .47]	43 [.38, .48]	
Visiting a University platform	.32 [.27, .37]	.31 [.26, .36]	.29 [.24, .34]

*In all cases $p < 0,01$, two-tailed. **With a confidence level of 95%. Note: own development.

In line with the Chart 2, a PLE fostering the professional development and career, and being customizable, interactive and useful in order to produce, edit, show and publish information, is preferred. Likewise, a PLE having varied, miscible, customizable and easy to use tools, is desirable. Considering these desirable characteristics in PLE in order to analyse university platforms, and achieving making them more related to students' interest, needs and expectations, is recommended.

In all the studied characteristics, significant differences are shown between the frequency levels, in favour of "agree" and "strongly agree" options, according to the Kruskal-Wallis test, $p < .05$, two-tailed and a level of significance of 95%. In addition, as it can be observed in Chart 3, all the desirable characteristics in a PLE have significant and positive correlations between each other, which are an indicative of the relevance and the transcendence of these characteristics.

It should be emphasized, that the four characteristics with higher frequency levels (see Chart 2) have strong bivariate correlations ($\tau > .58$) between each other. In other words, having tools easy to use, having a range of tools, being interactive and fostering personal development, are very desirable complementary characteristics in a PLE.

To a lesser extent, allowing producing and publishing information, showing the information in a variety of ways and mixing different tools are very desirable characteristics by students in the population, and they are also strongly and positively correlated between them ($\tau > .55$).

Another strong correlation ($\tau = .58$) is observed between the characteristic of allowing mixing different tools and having a place to show and to update the curriculum.

Chart 2: Desirable characteristics in a PLE, according the students

Characteristics	Relative frequencies and confidence intervals*				
	Strongly disagree	Disagree	Agree	Strongly agree	NA**
Having tools easy to use	0.97 [0.12, 1.81]	0.97 [0.12, 1.81]	35.91 [31.78, 40.03]	61.97 [57.80, 66.14]	0.19
Having range of tools to choose	0.39 [−0.15, 0.92]	0.77 [0.02, 1.53]	37.45 [33.29, 41.61]	61.20 [57.01, 65.38]	0.19
Being interactive	0.39 [−0.15, 0.92]	2.12 [0.88, 3.36]	39.00 [34.80, 43.19]	58.30 [54.06, 62.54]	0.19
Fostering personal development	0.19 [−0.18, 0.57]	2.32 [1.02, 3.62]	37.91 [33.74, 42.09]	59.38 [55.16, 63.60]	0.19
Allowing to edit, to produce and to publish information	0.58 [-0.07, 1.23]	6.95 [4.76, 9.14]	46.91 [42.62, 51.20]	45.37 [41.09, 49.65]	0.19
Showing the information in a variety of ways	0.77 [0.02, 1.53]	7.34 [5.09, 9.58]	51.93 [47.64, 56.22]	39.77 [35.56, 43.98]	0.19
Allowing to mix different tools	0.39 [−0.15, 0.92]	7.72 [5.43, 10.02]	53.47 [49.19, 57.76]	38.22 [34.05, 42.40]	0.19
Having a place to show and to update the curriculum	0.39 [−0.15, 0.92]	7.56 [5.28, 9.84]	49.22 [44.92, 53.53]	42.64 [38.38, 46.90]	0.19
Allowing changing the aspect and customizing it	0.77 [0.02, 1.53]	9.07 [6.60, 11.54]	45.37 [41.09, 49.65]	44.59 [40.32, 48.87]	0.19

*With a confidence level of 95%. **Relative frequencies of no answer. **Note**: own development.

Chart 3: Relationship between the desirable characteristics in a PLE

Desirable characteristics in a PLE	Kendall's τ correlation coefficient* and confidence intervals**							
	1	2	3	4	5	6	7	8
1 Having tools easy to use								
2 Having a range of tools to use	.75 [.70, .80]							
3 Being interactive	.66 [.61, 71]	.75 [.70, .80]						
4 Fostering personal development	.58 [.53, .63]	.63 [.58, .68]	.65 [.60, .70]					
5 Allowing changing the aspect and customizing it	.34 [.29, .39]	.39 [.34, .54]	.39 [.34, .54]	.43 [.38, .48]				
6 Having a place to show and to update the curriculum	.40 [.35, .45]	.39 [.34, .44]	.42 [.37, .47]	.41 [.36, .46]	.49 [.44, .54]			

(*continued*)

Chart 3: (*continued*)

Desirable characteristics in a PLE	Kendall's τ correlation coefficient* and confidence intervals**							
	1	2	3	4	5	6	7	8
7 Allowing to mix different tools	.33 [.28, .38]	.40 [.35, .45]	.42 [.37, .47]	.49 [.44, .54]	.50 [.45, .55]	.58 [.53, .63]		
8 Showing the information in a variety of ways	.39 [.34, .44]	. .42 [.37, .47]	41 [.36, .46]	.38 [.33, .43]	.50 [.45, .55]	.49 [.44, .54]	.61 [.56, .66]	
9 Allowing to edit, to produce and to publish information	.40 [.35, .45]	.45 [.40, .50]	.48 [.43, .53]	. 45 [.40, .50]	.47 [.42, .52]	.47 [.42, .52]	.55 [.50, .60]	.57 [.52, .62]

*In all the cases $p < 0,01$, two-tailed. **With a confidence level of 95%. Note: own development.

5   Conclusions and Recommendations

The following conclusions and recommendations result from the survey conducted:

1. Laptops and smartphones are the most accessible devices for the studied population. Investigating the way of how students use these devices is recommended.
2. For the study population, devices such as laptops, smartphones, desktop computers and others, are neither complementary nor substitutes. Promoting complementarity as a way of fostering a more effective and greater extent use of these devices is recommended.
3. For the study population, computers are devices of great importance in order to surfing the Internet, the use of PLE tools and to visit university platforms. A more in-depth research about the use of devices is recommended as well as encouraging the population to use a greater variety of them.
4. Students dedicate little time to the visit to the University platforms and these ones are poorly integrated in student's PLE. Making a diagnosis on these platforms, in order to make them more appealing is recommended. For that purpose, considering the desirable characteristics in a PLE is needed.
5. Students prefer that their PLE are interactives, foster the development and professional career, simultaneously they desire them to have a range of miscible, customizable and easy to use tools.
6. Students want their PLE to allow producing and publishing information, empowering show information in a variety of ways and mixing different tools.

References

1. Fiedler, S., Väljataga, T.: Personal learning environments: a conceptual landscape revisited. eLearning Papers **35**, 1–16 (2013)
2. Johnson, M.W., Sherlock, D.: Beyond the personal learning environment: attachment and control in the classroom of the future. Interact. Learn. Environ. **22**(2), 146–164 (2014). https://doi.org/10.1080/10494820.2012.745434
3. Kop, R., Fournier, H.: Developing a framework for research on personal learning environment. eLearning Papers **35**, 1–16 (2013). https://www.openeducationeuropa.eu/sites/default/files/legacy_files/asset/In-depth_35_4.pdf
4. Chaves, E., Trujillo, J.M., López, J.A.: Autorregulación del aprendizaje en entornos personales de aprendizaje en el grado de educación primaria de la Universidad de Granada. España. Formación Universitaria **8**(4), 63–76 (2015). https://doi.org/10.4067/S0718-50062015000400008
5. Marín, V., Salinas, J., de Benito, B.: Research results of two personal learning environments experiments in a higher education institution. Interact. Learn. Environ. **22**(2), 205–220 (2014). https://doi.org/10.1080/10494820.2013.788031
6. Rahimi, E., Van den Berg, J., Veen, W.: Facilitating student-driven constructing of learning environments using Web 2.0 personal learning environments. Comput. Educ. **81**, 235–246 (2015). https://doi.org/10.1016/j.compedu.2014.10.012

Supervising Attention in an E-Learning System

Dalila Durães[1,3]([✉]) [iD], Javier Bajo[1] [iD], and Paulo Novais[2] [iD]

[1] Department of Artificial Intelligence, Technical University of Madrid, Madrid, Spain
d.alves@alumnos.upm.es, jbajo@fi.upm.es
[2] Algoritmi Center, University of Minho, Braga, Portugal
pjon@di.uminho.pt
[3] CIICESI, ESTG, Polytechnic Institute of Porto, Felgueiras, Portugal

Abstract. Until now, the level of attention of a worker has been evaluated through his/her productivity: the more one produces, the better his/her attention at work. First, the worst aspect about this approach is that it only points out a potential decrease of attention after a productivity loss. An approach that could point out, in advance, upcoming breaks in attention could allow active/preventive interventions rather than reactive ones. In this paper we present a distributed system for monitoring attention in teams (of people). It is especially suited for people working with computers and it can be interesting for domains such as the workplace or the classroom. It constantly analyzes the behavior of the user while interacting with the computer and together with knowledge about the task, is able to temporally classify attention.

Keywords: Distributed intelligent system · Attention · Behavioral biometrics

1 Introduction

In the field of computer science, an intelligent environment is a digitally augmented physical world where sensor-enabled and networked devices work continuously and collaboratively to make the lives of the inhabitants more comfortable. With this technological evolution, job offers have changed, bringing along many significant and broad changes. Some of the most notorious ones can be pointed out by the emergence of indicators such as attentiveness which, in extreme cases, can compromise the life and well-being of the workers. In more moderate cases it will impair attention, general cognitive skills and productivity. In addition to these factors, many of these jobs are the so-called desk-jobs, in which people frequently sit for more than 8 h [1].

Until now, the level of attention of a worker has been evaluated through his/her productivity: the more one produces, the better his/her attention at work. While the true nature of this relationship is yet to be thoroughly studied (properly contextualized in each work domain), there are other issues that need to be addressed. First, the worst aspect about this approach is that it only points out a potential decrease of attention after a productivity loss. This means that the "damage" is already done and that it is most likely too late for the worker to cope with whatever caused the attention loss. An approach that could point out, in advance, upcoming breaks in attention (e.g. through

© Springer Nature Switzerland AG 2019
P. Novais et al. (Eds.): ISAmI 2018, AISC 806, pp. 389–396, 2019.
https://doi.org/10.1007/978-3-030-01746-0_46

the observation of behavioral patterns) could allow active/preventive interventions rather than reactive ones [2].

In this paper we present a distributed system for monitoring attention in teams (of people), in line with the vision of intelligent environments [3]. It is especially suited for people working with computers and it can be interesting for domains such as the workplace or the classroom. It constantly analyzes the behavior of the user while interacting with the computer and, together with knowledge about the task, is able to temporally classify attention.

This work may be very interesting for team managers to assess the level of attention of their teams, identifying potentially distracting events, hours or individuals. Moreover, distraction often appears when the individual is fatigued, bored or not motivated. This tool can thus be an important indicator of the team, allowing the manager to act accordingly at an individual or group level. In the overall, this tool will support the implementation of better human resources management strategies.

1.1 Previous Work

Part of this framework was implemented in previous work. The first version focused on the analysis of the individuals' interaction patterns with the computer, including features such as mouse velocity or acceleration, click duration, typing speed or rhythm, among others. For a complete list of features as well as the process of their acquisition and extraction, please see [4]. However, a limitation was also identified in this previous work. In fact, a user that opened a no work-related application and did not interact anymore with the computer until the end of the task had 0% of attention. On the other hand, if the user opens a work-related application and does not interact with the computer after that, the user's attention will be classified as 100% when he is most likely not even at the computer.

The present work adds a new feature to this previously existing framework, by providing a precise measure of attention based not on the tasks work-related patterns but also on the key typing or mouse movement pattern. It thus constitutes a much more precise and reliable mechanism for attention monitoring, while maintaining all the advantages of the existing system: nonintrusive, lightweight and transparent.

2   Architecture

From the architecture of the developed environment described in Fig. 1 it is possible to collect data that describe the interaction with both the mouse and the keyboard in the devices in which students work. Three parts compose the module's architecture.

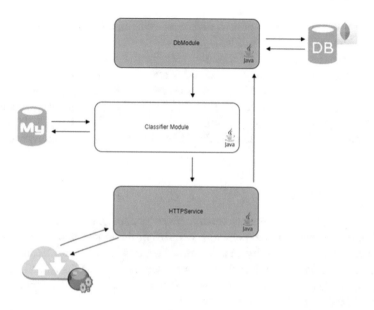

Fig. 1. System architecture.

In the devices operating by the students, it's installed software that generates raw data, which store locally until it is synchronized with the web server in the cloud. In this step it's encodes each event with the corresponding necessary information (e.g. time-stamp, coordinates, type of click, key pressed). These data are further processed, stored and then used to calculate the values of the behavioral biometrics. Mouse movements can also help to predict the state of mind of the user, as well as keyboard usage patterns.

The first part (HTTP Service) is responsible for all database connections, with the needed configuration to access to the MySQL database and the DB Module feature to access to the Mongo DB, and the needed transactions with it.

The second part (DbModule) is the REST connection implementation. Here were built all the necessary connection the access to the system. After the raw data is stored in a data store engine, the analytic layer processes the data received (from the storage layer) in order to evaluate those data according to the metrics presented. Some data preparation tasks are also carried out in this phase, such as removing outliers (e.g. the backspace key being continuously pressed to delete a group of characters is not a regular key press). The system receives this information in real-time and calculates, at regular intervals, an estimation of the general level of performance and attention of each student.

The third part (Classifier Module) is features implementation of attention classifi-cation. This classifier will handle the treatment of the data into useful information, doing for instance the necessary calculations to reach the attention index. All the data analysis and processing were developed in this part. In the classification profile layer the indi-cators are interpreted. Based on data from the attentiveness indicators and building the meta-data that will support decision-making, the system will classify the user profile. When the system has a sufficiently large dataset that allows making classifications with precision, it will classify the inputs received into different attention levels in real-time, creating each student learning profile.

Finally, the user attention information send back to the user managers, where is displayed in the visualization layer in the users managers, and it can be used to personalize instructions according to the specific user, enabling the administrator to act differently with different users as well as to act differently with the same user, according to his/her past and present level of attention. In the visualization layer it is possible to obtain some graphical modules that allow the display of information in an intuitive way to the user.

2.1 Data Acquisition

As previously mentioned, the early version of this system acquired data describing the interaction of the user with the computer in terms of the mouse and the keyboard [5]. Data acquisition is supported by an application installed in the data-generating devices (the computers of the users). It is thus a distributed data collection system, which has now been extended to acquire a new type of data about each user: the application that the user is interacting with, at each time.

Thus, at regular intervals (around 5 min), the server receives a new set of data about each user. This raw data includes all the important interaction events: when keys were pressed down or released, when the mouse moved (and where to), when clicks started or ended and when the user switched to a given application. This data is then transformed by the server as described in [5], to generate 15 features that describe the performance of the interaction with the computer.

In previous work [6] to identify the work-related applications, we analyzed all different applications used by all users and labeled each one as belonging to the task or not. We then quantified the amount of time that each student spent interacting with applications related to the task versus other applications. To do this we measured the amount of time, in each interval, that the user spent interacting with work-related applications. The algorithm thus needs knowledge about the domain in order to classify each application as belonging or not to the set of work-related applications. This knowledge is provided by the teacher and is encoded in the form of regular expressions. The teacher uses a graphical interface to set up rules such as "starts_with Access" or "contains Access", which are then translated to regular expressions that are used by the algorithm to determine which applications are work-related and which are not. Whenever an application that does not match any of the known rules for the specific domain is found, the application name is saved so that the teacher can later decide if a new rule should or should not be created for it. By default, applications that are not considered work-related are marked as "others" and count negatively towards the quantification of attention.

2.2 Features Extraction

Firstly, it is necessary to know the interaction with the mouse and with the keyboard that each task will have. In some tasks the interaction with the mouse will be higher while in others that will happen with the keyboard, and finally in others, it will practically be the same level of interaction. However, it is difficult to know, a priori, what the exact percentage of interaction of the mouse or the keyboard will be. To know these values, we first count the number of times that each key is pressed and the number of times that the mouse is clicked down for each user.

3 Validation

To validate the proposed system, we have been using it for the past months in the Caldas das Taipas High School, located in northern Portugal. In the Portuguese academic context, this system gains increased relevance as current policies move towards the creation of larger classes, which make it increasingly difficult for the teacher to individually address to each student. To validate this system we are following several cohorts of students during their academic activities. This data collection process will allow assessing the influence on attention of aspects such as: breaks, time of day, class contents, and class objectives, among others. For this purpose, a group of 14 (all girls) students were selected to participate in this experience. Their average age is 15.9 years old (SD = 1.5 years). The experiment was applied in a lesson, where they had access to an individual computer and 100 min to complete the task. Students received, at the beginning of the lesson, all necessary data with the goals of the task. For this class, the lesson started in the afternoon and students received, at the beginning of the lesson, a document with the goals of the task, which in this case required the use of Microsoft Access and Adobe Acrobat Reader.

This application runs in the background, which makes the data acquisition process, a completely transparent one from the point of view of the student. It collects data from the students' interaction with the mouse and the keyboard, which act as sensors. The Mouse and Keyboard Sensing layers are responsible for capturing information describing the behavioral patterns of the students while interacting with the peripherals.

Secondly, we calculated the percentage of the interaction of the mouse, which is the average number of times that the mouse was pressed in the class taking into account the total of interactions with the mouse and the keyboard of the class. Similarly, we calculated the percentage of the interaction of the keyboard, which is the average number of times that the keys were pressed in the class taking into account the total of interactions with the mouse and the keyboard of the class.

On the order hand, we obtained the amount of time that each student spent at the computer (Task Duration) as well as the amount (and percentage) of time that each student devoted to work and to other activities in the lesson as explained in [6]. However, the time that the student spent in the task-related doesn't indicate the level of attention in some cases, because in some situations the user might have opened the task-related and during that time didn't interact with the computer. In this case, if the level of attention was measured only by the task-related, the student has 100% of attention and in reality his attention level should be 0%. For these situations, it is necessary to analyze the amount of interaction with the mouse and the keyboard and cross these data.

The next step is to calculate the percentage of usage of the mouse for each user. The formula to calculate this value is:

$$IM = \frac{\sum Md}{Max\,Md} \qquad (1)$$

Where:

- IM: is interaction of the mouse;
- Md: is the number of the times that the mouse was pressed by a user;
- MAX Md: is the higher score of usage of the mouse down in the class.

Similarly, for calculating the percentage of usage of the keyboard for each user, the formula for this value is:

$$IK = \frac{\sum Kd}{Max\,Kd} \qquad (2)$$

Where:

- IK: is Interaction of the keyboard;
- Kd: is the number of the time that one key was pressed by a user;
- Max Kd: is the higher score of usage of the Keyboard in the class.

Finally, in order to obtain the level of attention it is necessary to combine the results of the interaction of the behavior biometrics with the results showed in the time that each student spent in the task-related. This level of attention is the relative attention of each student, as this level of attention is compared with the other user of the class. Formula (3) provides the results for each user, presented in Table 1. The formula (3) that calculated these values is:

$$Rel.At = ((PercM * Work\_rel) + (PercK * Work\_rel)) \qquad (3)$$

Table 1. Relative attention of each user in the lesson

User	Relative MouseDown	Relative KeyDown	Work-related	Relative Attention
T2240001	31,45%	30,22%	0,00%	0,00%
T2240002	31,70%	15,59%	78,70%	37,22%
T2240003	16,95%	11,92%	69,73%	20,13%
T2240004	48,43%	26,55%	56,77%	42,56%
T2240005	21,92%	10,92%	56,95%	18,70%
T2240006	43,71%	21,75%	60,60%	39,67%
T2240007	17,06%	7,11%	61,81%	14,94%
T2240008	44,89%	25,05%	83,73%	58,56%
T2240009	25,64%	24,13%	88,58%	44,09%
T2240010	24,51%	15,61%	61,77%	24,78%
T2240011	38,29%	25,50%	83,72%	53,41%
T2240012	32,93%	23,36%	85,60%	48,19%
T2240013	45,25%	51,57%	75,52%	73,11%
T2240014	33,20%	24,65%	70,21%	40,62%

Where:

- Rel.At: is Relative attention;
- PercM: is the percentage of utilization of the Mouse for a user;
- PercK: is the percentage of usage of the Keyboard for a user;

Figure 2 presents the graphically result between percentage of time interacting with work-related applications and relative attention. It can be observed that the relative attention decreases in general, because the system takes in account the interaction of the user with mouse and keyboard. That is, someone who spent 100% of the time using the application that was supposed to but was interacting only 50% of the time, would have a score of attention of 50%.

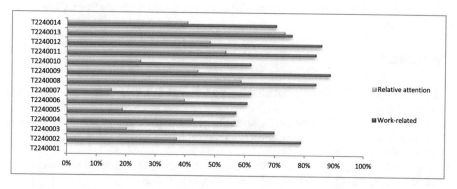

Fig. 2. Comparing work-related interaction with relative attention for each student.

4   Discussion and Conclusions

The task data from the user's interaction with the computer is the most crucial information because it derives most part of the attention level. To obtain the task results, the task rules received in the request are used to get a perception of how much time the user has spent on the applications related with the task rules.

As far as the mouse and keyboard results are concerned, their role is present on how the user interaction is occurring and it helps the manager/teacher to understand if any user, who has the application active, is really working on it or not. To do that, we have used the two features one from the keyboard (key down time) and other from the mouse (time between clicks).

Concerning attention, it is an important theme because it is one of the factors that most influences a person's performance while performing a task. Understanding how attention varies and in which situations the attention varies for each person, it is possible to act in the right moment and right time, to bring the user's attention level to the ideal value.

A framework was proposed to address these issues, especially to monitoring user. Narrowing the scope of the study, a process to detect attentiveness was proposed, through the use of a developed log tool. With this smart environment it is possible to detect potentially negative factors dynamically and non-intrusively, making it possible to foresee negative situations, allowing taking actions to mitigate them. The work developed so far resulted in a useful system for the team manager, who can monitor, in real-time, the level of attention of users.

The door is thus open to intelligent platforms that allow to analyze user's profiles, taking into account their individual characteristics, and to propose new strategies and actions. By providing managers with access to this information, we allow them to better manage their interactions with the users, namely by pointing out the most problematic cases of inattention in real-time.

Acknowledgements. This work has been supported by COMPETE: POCI-01-0145-FEDER-007043 and FCT – Fundação para a Ciência e Tecnologia within the Project Scope: UID/CEC/00319/2013.

References

1. Liao, M., Drury, C.: Posture, discomfort and performance in a VDT task. Ergonomics **43**(3), 345–359 (2000)
2. Carneiro, D., Novais, P., Andrade, F., Zeleznikow, J., Neves, J.: Using Case-based reasoning and principled negotiation to provide decision support for dispute resolution. Knowl. Inf. Syst. **36**(3), 789–826 (2013)
3. Augusto, J., Callaghan, V., Cook, D., Kameas, A., Satoh, I.: Intelligent environments: a Manifesto. Human-Centric Comput. Inf. Sci. **3**(1), 1–18 (2013)
4. Durães, D., Jiménez, A., Bajo, J., Novais, P.: Monitoring level attention approach in learning activities. In: Caporuscio, M., De la Prieta, F., Di Mascio, T., Vittorini, P. Advances in Intelligent System and Computing, vol. 478, pp. 33–40 (2016)
5. Durães, D., Gonçalves, S., Carneiro, D., Bajo, J., Novais, P.: Detection of behavioral patterns for increasing attentiveness level. In: International Conference on Intelligent Systems Design and Applications ISDA 2016: Intelligent Systems Design and Applications, pp. 592–601 (2016)
6. Carneiro, D., Durães, D., Bajo, J., Novais, P.: Non-intrusive monitoring of attentional behavior in teams. In: International Symposium on Intelligent and Distributed Computing IDC 2016: Intelligent Distributed Computing X, pp. 153–162 (2016)

Platform for Indexing Music Albums Based on Augmented Reality Techniques and Multi-agent Systems

Ma Ángeles Muñoz, Daniel H. de la Iglesia$^{(\boxtimes)}$,
Gabriel Villarrubia González, Juan F. de Paz, Álvaro Lozano,
and Alberto L. Barriuso

BISITE Digital Innovation Hub, University of Salamanca.
Edificio Multiusos I+D+I, 37007 Salamanca, Spain
{mmg94,danihiglesias,gvg,fcofds,
loza,albarriuso}@usal.es

Abstract. Image recognition systems combined with augmented reality systems provide new tools to relate to the physical world. Obtaining fast and accurate information about analog elements such as a photo or a picture, allows users to be more and better informed. One of these physical elements, which is in the process of disappearing, is the traditional music album. In this work we propose the use of a mobile application of augmented reality for the recognition and indexing of music albums through its cover. The goal is to index discs through image recognition in order to access them digitally through services such as Spotify or YouTube. The system has a music recommendation module based on the albums recognized by the application.

Keywords: Augmented reality · Image recognition · Multi-agent system
Recommender system

1 Introduction

The audiovisual industry has undergone an important change in its business model in recent years, largely due to the way in which users currently consume these products. There is a transition from traditional physical products to digital products [1]. Currently, there are numerous digital platforms such as Spotify or Netflix that offer content that users previously bought in physical format and now consume on-demand. Among these industries are cinema, videogames, books and music mainly. The case of the music industry is especially dramatic if you look at its figures. It is estimated that only in Spain, from the year 2001 until 2013 there was a decrease in sales of physical discs of 78% [2]. To give new impulse to sales of audiovisual content in physical format, as well as a second life to current products, is to seek an interaction between the physical world and the digital world. In this sense, the techniques of augmented reality and image recognition are a useful tool to interact with these physical elements. Thanks to these systems, it is possible to identify hundreds of thousands of different images in just a few seconds.

© Springer Nature Switzerland AG 2019
P. Novais et al. (Eds.): ISAmI 2018, AISC 806, pp. 397–404, 2019.
https://doi.org/10.1007/978-3-030-01746-0_47

On the other hand, intelligent information management systems based on multi-agent systems are a good alternative when it comes to managing information derived from information indexing systems [3, 4]. These systems allow easy integration with other modules such as recommendation systems or information analysis systems that provide users with greater contextualization of the indexed data. In addition, the latest developments in multi-agent systems allow mobile agents to be embedded in devices with low computational capacity, resulting in more flexible and open architectures.

In the present work, the use of augmented reality technologies and image recognition combined with intelligent information management systems based on MAS systems is combined [5]. Specifically, the use of PANGEA [6–9] as a multi-agent system responsible for the intelligent management of information is proposed. First, an application for mobile devices based on the RA Vuforia [10] framework, capable of recognizing thousands of different images, has been developed. The images to be recognized through the application have been previously pre-processed by the RA framework and consist of 5000 album covers of music obtained from different online sources. When the application recognizes a cover page available in the database, it shows the relevant information of the album (name, artist, popularity and link to play the album in Spotify). On the other hand, a centralized system has been developed where all the information coming from the recognitions of users' albums is processed. For each of the users the discs that have been scanned are stored, a complete list of the available songs of that album is shown and the links of the main streaming platforms are shown for its reproduction. Likewise, the system is able to recommend and suggest new albums and songs to users based on the latest music records registered by the user.

The rest of this work is organized as follows: Sect. 2 reviews similar works, Sect. 3 explains the architecture and its application in the case study. In Sect. 4 we will present the results and conclusions obtained in the research.

2   Background

In current literature it is possible to find examples of research where authors propose the use of augmented reality systems in areas such as industry [11], education [12, 13] or video games [14].

Regarding to the field of education are many studies investigating about the use of augmented reality technology. A recent work is [15] where the authors present an educational application of virtual reality for the exploration of the structure and anatomy of the nasal cavity for medical students. Thanks to this technology it is possible to study in a visual way three-dimensional structures that on paper are not complex to understand. Other works such as [16] analyze different augmented reality techniques for the teaching of different civil engineering subjects, the results reveal how this type of applications contribute to the learning of the students. In the case of the authors of [17] they demonstrate how the use of this technology has a positive impact on students when it comes to learning basic concepts of physics in the classroom.

Regarding the use of these techniques in the industry, we can highlight the recent work of the authors of [18]. In this work we propose the use of AR-based glasses for the training of factory personnel. The authors use the Vuforia framework to create the application with the aim of instructing the operator on how to assemble an object composed of different pieces. Another recent example is the work [19] where the use of a system based on augmented reality for the maintenance of a robot factory is proposed.

3 Proposed System

In this section we present the architecture designed in this work for the management of album covers and music recommendations. The application for mobile devices designed and describes the final system implemented in the development of this work is described.

A MAS (multi-agent system) architecture based on the PANGEA platform [6] is proposed in a scenario of recognition of images associated with the covers of the music discs commercialized throughout the world. Figure 1 shows the virtual organization of agents designed.

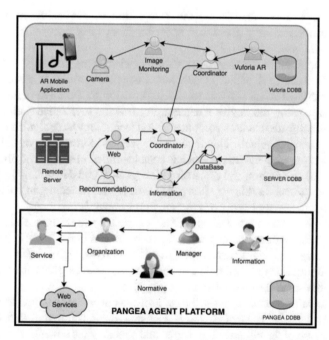

Fig. 1. Multi-agent platform based on PANGEA platform

First of all, we can see the *AR Mobile Application organization* formed by the agents: Camera, image monitoring, coordinator and Vuforia AR. The agent with the camera role will be responsible for obtaining the different frames captured by the camera of the user's mobile device. These images arrive at the agent image monitoring, which is responsible for providing them to the coordination agent. The coordinator is responsible for sending the images to agent Vuforia who in turn makes the request on the AR Vudoria API. In the event that any of the images sent by this agent to the API matches one of the images registered on the platform, the agent Vuforia informs the coordinator, providing the ID of the album found.

The **Remote Server** organization that includes the agents: *Coordinator, web, information, recommendation and database.* The agent with coordinating role is in charge of managing the information that comes from the Vuforia database. The agent whose role is information is responsible for normalizing the data while the DataBase agent is responsible for extracting information related to that identifier and that is stored in the database. For its part, the agent with role recommendation is responsible for, through the data stored in databases, generate new recommendations to users. The agent with Web role will be responsible for displaying the data generated by the recommendation agent, as well as showing through the designed web application, the data relating to all the albums recognized by the user.

3.1   AR Mobile App

The mobile application focuses on the recognition of covers of musical albums, obtaining information from them and showing them later using augmented reality techniques to the user.

Image recognition is a technology capable of identifying an image among several stored in a database in a matter of seconds. Image recognition is closely related to augmented reality as it allows you to add information to the environment by recognizing an image using the device's camera. For this the tool called Vuforia has been used, which is software that uses computer vision technology to recognize and track flat images and simple 3D objects. This functionality allows to introduce virtual 2D and 3D elements overexposed in the image that has been identified in such a way that the object is tracked and its orientation and dimensions are recalculated depending on the position of the terminal camera. In this way it is possible to create the feeling in the user that the virtual element is one more part of the real world scene.

Once the album art in question is focused on the camera of the mobile device, this recognition occurs. After obtaining the information stored in the Vuforia database, it is presented by means of augmented reality. In addition to displaying information such as the name, the artist or the evaluation of the disc, the user is given the possibility of reproducing the same through the Spotify platform.

The information stored in the database grows as users add albums to it, so that a collaborative database is created. Users can add them using the web application, entering the name of the album and the artist and choosing the cover of the disc they are looking for among the possibilities offered by the application. From the same, not only can you add albums, but they can be played and get relevant information about them, their songs and their artists.

3.2   Final System

The infrastructure of the proposed system can be seen in Fig. 2, first, the application for mobile AR devices is observed. This application, as explained above, is in charge of extracting images from the camera to send them in real time to the Vuforia API.

Through the Vuforia framework, matches are obtained based on the images that are registered in the database. If there is a match between the image analyzed by the application and a previously registered image, the identifier of the image is sent back. Through a web service, the remote server deployed for this system is called in which all the data of the registered albums are stored, as well as all the information of the user. At this point the key data of the disk is recovered like the title, the author, the links to YouTube, etc. and they are sent back to the app to be viewed. At the same time, a music recommendation algorithm is deployed on the server. This algorithm is based on collaborative filtering and is responsible for analyzing discs indexed by the user in search of recommendations based on their tastes.

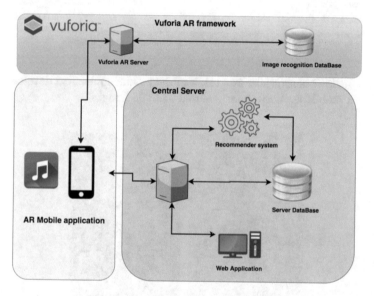

Fig. 2. Overview of the proposal system

Finally, a web application is available to show all the data stored in the system, manage existing data and show the recommendations made by the system to users.

4   Results

This work proposes an image recognition system based on an augmented reality application for indexing music albums. The goal is to recognize the physical disks that users have purchased in order to create a digital index of their physical albums. The use of a multi-agent architecture based on the PANGEA platform is proposed in order to intelligently manage the information of the proposed system.

Figure 3 shows a screenshot of the web application resulting from the system proposed in this work.

Fig. 3. Screenshot of the developed web application

Figure 4 shows screenshots as a result of the mobile application developed in this work. It is a mobile application for Android and iOS badges.

Fig. 4. Captura de la aplicación para dispositivos móviles basada en Vuforia

Acknowledgements. The research of Daniel Hernández de la Iglesia has been co-financed by the European Social Fund and Junta de Castilla y León (Operational Programme 2014–2020 for Castilla y León, EDU/529/2017 BOCYL). Álvaro Lozano is supported by the pre-doctoral fellowship from the University of Salamanca and Banco Santander. This work was supported by the Spanish Ministry, Ministerio de Economía y Competitividad and FEDER funds. The research of Alberto López Barriuso has been co-financed by the European Social Fund and Junta de Castilla y León (Operational Programme 2014–2020 for Castilla y León, EDU/128/2015 BOCYL).

References

1. Fleisch, E., Weinberger, M., Wortmann, F.: Business models and the Internet of Things (Extended Abstract), pp. 6–10. Springer, Cham (2015)
2. Koh, B., Murthi, B.P.S., Raghunathan, S.: Shifting demand: online music piracy, physical music sales, and digital music sales. J. Organ. Comput. Electron. Commer. **24**, 366–387 (2014). https://doi.org/10.1080/10919392.2014.956592
3. De La Iglesia, D., Villarubia, G., De Paz, J., Bajo, J.: Multi-sensor information fusion for optimizing electric bicycle routes using a swarm intelligence algorithm. Sensors **17**, 2501 (2017). https://doi.org/10.3390/s17112501
4. De La Iglesia, D., De Paz, J., Villarrubia González, G., Barriuso, A., Bajo, J., Corchado, J.: Increasing the intensity over time of an electric-assist bike based on the user and route: the bike becomes the gym. Sensors **18**, 220 (2018). https://doi.org/10.3390/s18010220
5. Villarrubia, G., De, P.J., La, I.D., Bajo, J.: Combining multi-agent systems and wireless sensor networks for monitoring crop irrigation. Sensors **17**, 1775 (2017). https://doi.org/10.3390/s17081775
6. Zato, C., Villarrubia, G., Sánchez, A., Bajo, J., Manuel Corchado, J.: PANGEA: a new platform for developing virtual organizations of agents. Int. J. Artif. Intell. **11**, 93–102 (2013)
7. Murciego, Á.L., González, G.V., Barriuso, A.L., de la Iglesia, D.H., Herrero, J.R.: Multi agent gathering waste system. ADCAIJ Adv. Distrib. Comput. Artif. Intell. J. **4**(4), 4:9–4:22 (2015). ISSN-e 2255-2863
8. Hernández De La Iglesia, D., Villarubia González, G., López Barriuso, A., Lozano Murciego, Á., Revuelta Herrero, J.: Monitoring and analysis of vital signs of a patient through a multi-agent application system. ADCAIJ Adv. Distrib. Comput. Artif. Intell. J. **4**, 19 (2016). https://doi.org/10.14201/ADCAIJ2015431930
9. la Iglesia, D.H.D., De, P.J., Villarubia, G., de Luis, A., Omatu, S.: Pollen allergies prediction through historical data in mobile devices. Int. J. Artif. Intell. **13**, 74–80 (2015)
10. Vuforia| Augmented Reality. https://www.vuforia.com/. Accessed 29 Jan 2018
11. Palmarini, R., Erkoyuncu, J.A., Roy, R., Torabmostaedi, H.: A systematic review of augmented reality applications in maintenance. Robot. Comput. Integr. Manuf. **49**, 215–228 (2018). https://doi.org/10.1016/J.RCIM.2017.06.002
12. Bower, M., Howe, C., McCredie, N., Robinson, A., Grover, D.: Augmented reality in education – cases, places and potentials. EMI Educ. Media Int. **51**, 1–15 (2014). https://doi.org/10.1080/09523987.2014.889400
13. Juanes, J.A., Hernández, D., Ruisoto, P., García, E., Villarrubia, G., Prats, A.: Augmented reality techniques, using mobile devices, for learning human anatomy. In: Proceedings of the Second International Conference on Technological Ecosystems for Enhancing Multicultur-ality - TEEM 2014, New York, New York, USA, pp. 7–11. ACM Press (2014)
14. Das, P., Zhu, M., McLaughlin, L., Bilgrami, Z., Milanaik, R.: Augmented reality video games: new possibilities and implications for children and adolescents. Multimodal Technol. Interact. **1**, 8 (2017). https://doi.org/10.3390/mti1020008
15. Marks, S., White, D., Singh, M.: Getting up your nose. In: SIGGRAPH Asia 2017 Symposium on Education - SA 2017, New York, New York, USA, pp. 1–7. ACM Press (2017)
16. Dinis, F.M., Guimaraes, A.S., Carvalho, B.R., Martins, J.P.P.: Virtual and augmented reality game-based applications to civil engineering education. In: 2017 IEEE Global Engineering Education Conference, pp. 1683–1688. IEEE (2017)

17. Ibanez, M.B., de Castro, A.J., Kloos, C.D.: An empirical study of the use of an augmented reality simulator in a face-to-face physics course. In: 2017 IEEE 17th International Conference on Advance Learning Technologies, pp. 469–471. IEEE (2017)
18. Pierdicca, R., Frontoni, E., Pollini, R., Trani, M., Verdini, L.: The use of augmented reality glasses for the application in Industry 4.0, pp. 389–401. Springer, Cham (2017)
19. Mourtzis, D., Zogopoulos, V., Vlachou, E.: Augmented reality application to support remote maintenance as a service in the robotics industry. Procedia CIRP **63**, 46–51 (2017). https://doi.org/10.1016/J.PROCIR.2017.03.154

Prediction System for the Management of Bicycle Sharing Systems

Juan F. De Paz[✉], Gabriel Villarrubia, Ana B. Gil, Ángel L. Sánchez, Vivian F. López, and M. Dolores Muñoz

Departamento de Informática y Automática, Universidad de Salamanca, Plaza de la Merced, s/n, 37008 Salamanca, Spain
{fcofds,gvg,abg,alsl,vivian,mariado}@usal.es

Abstract. Bicycle sharing systems are very common in urban areas; their goal is to improve citizens' mobility around the city. The efficient management of bicycle stations is the greatest challenge for such systems and makes it difficult to satisfy the users' demand for bicycles. To overcome this challenge, it is important to predict their use and to constantly monitor the number of bicycles available at stations, ensuring that they are distributed according to the demand in those places. In this work, we analyse the demand for bicycles per user and predict the routes users may travel to determine the possibility of predicting the behaviour of users. To this end, meteorological information and historical data on the use of the stations were incorporated into the system.

Keywords: Demand forecasting · Classifiers · Route prediction

1 Introduction

Bicycle sharing systems are used to reduce congestion in urban areas around the world [4], not only in big cities like Madrid (Spain), Washington, DC (USA) [9] and Zhongshan (China) [10] but also smaller towns like Salamanca [7]. These services provide a series of stations throughout the city, allowing users to pick up a bicycle at one station and leave it at another. Bicycle sharing faces difficulties in distributing bicycles according to the demand at different stations. This is because the needs for bicycles change depending on the time of day and the area of the city [12]. The services provided by bicycle sharing systems include system and fleet maintenance and management. Normally, bike sharing companies have employees whose role is to distribute the bicycles according to demand at the stations.

In the literature, we can find studies concerning a variety of subjects in the field of bike sharing systems. Some focus on technical aspects, such as the distribution of bicycles at stations while others look at the health benefits of using a bicycle [8, 13]. One study proposed the introduction of incentives for the users that encourage them to maintain a balance of bicycles at the stations of the system, this is done with the aim of minimizing the redistribution tasks performed by the company's lorry [3]. The system is defined by an optimization problem that allows the company to reduce the maintenance costs of the stations. In another study, neural networks were proposed for

P. Novais et al. (Eds.): ISAmI 2018, AISC 806, pp. 405–410, 2019.
https://doi.org/10.1007/978-3-030-01746-0_48

predicting bicycle use, they perform clustering to identify zones and make predictions [4]. In [5], availability at stations is predicted using the Population Continuous Time Markov Chain, the system does not consider the number of bicycles, but determines whether there will be bicycles available when they arrive at the station, this is because it is designated for users and not system administrators. Simulations were performed in [11] to forecast the demand and availability of vehicles. Other works made a visual analysis of the bike sharing systems [6] or classified bike trips into single and multiple trips [10].

This paper analyses the possibility of predicting the use of stations, specifically the possibility of predicting the number of trips users will make and the routes they will travel. Different classifiers and regressors have been used to perform this analysis. The data used in the work come from the data obtained by the Salamanca Salenbici company and weather information from AEMET has also been incorporated.

The paper is organized as follows: Sect. 1 reviews related work, Sect. 2 describes our proposal, and finally Sect. 3 provides the preliminary results and conclusions drawn from the research.

2   Proposal

To analyse the functioning of the bike sharing service, a multi-agent system has been created to recover information from the Salenbici system and AEMET. Within the system, the *salbici* agent is a crawler in charge of recovering the information coming from *selenbici* and the AEMET agent in charge of obtaining current meteorology data. If the station in Salamanca does not have the required information, the data are obtained from the station in Matacán. Once the data are recovered, the *PreprocessedTravel* and *PreprocessedDestination* agents take the information from the agent Salenbici and AEMET and are responsible for generating the data that will be processed further on. These agents add new variables to the existing data in order to make predictions. Finally, we have the agents in charge of making the predictions, the agents in charge of making the predictions take the *PredictTravel* and *PredictDestination* information and apply different classifiers to calculate the kappa index and accuracy. Figure 1 shows the system diagram responsible for predicting the information from the data obtained by the crawler.

Classifiers have been implemented in the *PredictTravel* agent in order to estimate the number of trips made by users. However, the problem could be considered as a regression problem. In the previous work, a regression analysis was performed to determine the number of trips by applying regressions based on classifiers [7], in this case a class-based analysis was performed, where the classes represent the number of trips. Nominal variables are suitable for measuring the success rate. In the case of the number of trips however, the variables are numerical. In spite of this, the success rate can be measured as the number of trips is low and the success rate and the Kappa index are high. As a result, an approximation of the errors in the estimates can be obtained. In the case of the destination, a classifier can be applied directly.

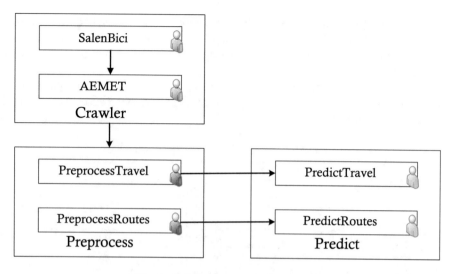

Fig. 1. Information processing system

3 Results and Conclusions

The case study was conducted in Salamanca with data from the company Salenbici [2]. The data used in the case study correspond to year 2016. The data contains information on the movements of bikes, specifically: the station of origin, the destination station, the identifier of the bicycle padlock, the time of departure, the time of arrival, the user id, the bicycle id. In total there are 51,000 records of movements. There are 29 stations in Salamanca, their distribution can be seen in Fig. 2.

Fig. 2. Location of the bicycle stations in Salamanca

In addition, to the data provided by the company, the corresponding data from AEMET were also added, such as maximum temperature, minimum temperature, average temperature, rain, average wind, maximum wind. Also, information on the work schedule was included.

On the first day of the case study, the first step was to predict the number of trips made in the morning and afternoon, considering each user and station. The obtained results can be seen in Table 1.

Table 1. Prediction of the number of trips.

Classifier	Kappa	Accuracy	Execution time
Random Forest	0.707257	96.25021	15.28
Bayes Net	0.7082538	96.22515	1.44
Naïve Bayes	0.7119578	96.25021	0.53
AdaBoostM1	0.725888	96.72624	0.95
Bagging	0.712255	96.40889	3.64
Decision Stump	0.725888	96.72624	0.21
J48	0.7780303	97.12711	1.26
IBk	0.7223367	96.37548	5.88
JRip	0.7323534	96.71789	9.66
LMT	0.7441194	96.72624	1161.87
OneR	0.7136432	96.43394	0.41
SMO	0.7586624	96.86821	416.88

Similarly, the predictive capacity was analysed by analysing the relationship between origin and destination. To make this prediction, different classifiers were analyzed considering the accuracy and the Kappa index. The results can be seen in Table 2.

Table 2. Prediction of destination vs origin.

Classifier	Kappa	Accuracy	Execution time
Random Forest	0.1040868	16.97108	12.97
Bayes Net	0.3171046	37.67123	0.39
Naïve Bayes	0.1519645	22.98326	0.14
AdaBoostM1	0.08543829	19.63470	0.14
Bagging	0.06066634	13.85084	1.46
Decision Stump	0.08418712	19.55860	0.07
J48	0.272095	33.02892	0.45
IBk	0.1061659	16.97108	0.26
JRip	0.4871524	53.80518	3.73
LMT	0.2750774	33.71385	74.35
OneR	0.5205294	56.01218	0.12
SMO	0.0880487	17.96043	13.80

Moreover, we have looked at the possibility of predicting the number of trips made by users per hour, but the obtained results were not good. From the analysis of the data, it was possible to verify that the time of use of the bicycles by the users varied, so it was difficult to estimate this variable with precision. Figure 3 shows the density graph of the trips made by a user, the X-axis shows the hours and the Y-axis shows the density of trips. The different colors represent the days of the week.

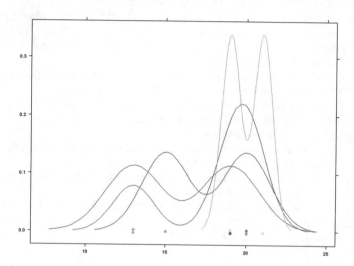

Fig. 3. Density graph with the trips made by a specific user during the first 4 months of 2016.

In conclusion, the proposed system is capable of predicting the use of bicycle sharing stations by users. It has not been possible to provide hourly predictions, but they are predicted for the morning and afternoon period. This information is sufficient to success-fully balance the distribution of bicycles at the stations. In a future work we will examine whether it would be possible for the system to automatically determine the distribution of bicycles at stations. This will allow to minimize the workload of employees and will be of economic benefit for the company.

Acknowledgments. This work was supported by the Spanish Ministry, Ministerio de Economía y Competitividad and FEDER funds. Project. SURF: Intelligent System for integrated and sustainable management of urban fleets TIN2015-65515-C4-3-R.

References

1. http://www.salamancasalenbici.com/
2. Haider, Z., Nikolaev, A., Kang, J.E., Kwon, C.: Inventory rebalancing through pricing in public bike sharing systems. Eur. J. Oper. Res. **270**, 103–117 (2018)
3. Wang, M., Zhou, X.: Bike-sharing systems and congestion: evidence from US cities. J. Transp. Geogr. **65**, 147–154 (2017)

4. Caggiani, L., Camporeale, R., Ottomanelli, M., Szeto, W.Y.: A modeling framework for the dynamic management of free-floating bike-sharing systems. Transp. Res. Part C: Emerg. Technol. **87**, 159–182 (2018)

5. Feng, C., Hillston, J., Reijsbergen, D.: Moment-based availability prediction for bike-sharing systems. Perform. Eval. **117**, 58–74 (2017)

6. Oliveira, G.N., Sotomayor, J.L., Torchelsen, R.P., Silva, C.T., Comba, J.L.D.: Visual analysis of bike-sharing systems. Comput. Graph. **60**, 119–129 (2016)

7. Lozano, Á., De Paz, J.F., Villarrubia, G., De La Iglesia, D.H., Bajo, J.: Multi-agent system for demand prediction and trip visualization in bike sharing systems. Appl. Sci. **8**(1), 67 (2018)

8. Otero, I., Nieuwenhuijsen, M.J., Rojas-Rueda, D.: Health impacts of bike sharing systems in Europe. Environ. Int. **115**, 387–394 (2018)

9. Haider, Z., Nikolaev, A., Kang, J.E., Kwon, C.: Inventory rebalancing through pricing in public bike sharing systems. Eur. J. Oper. Res. **270**, 103–117 (2018)

10. Zhang, Y., Brussel, M.J.G., Thomas, T., van Maarseveen, M.F.A.M.: Mining bike-sharing travel behavior data: an investigation into trip chains and transition activities. Comput. Environ. Urban Syst. **69**, 39–50 (2018)

11. Ji, S., Cherry, C.R., Han, L.D., Jordan, D.A.: Electric bike sharing: simulation of user demand and system availability. J. Clean. Prod. **85**, 250–257 (2014)

12. de Chardon, C.M., Caruso, G., Thomas, I.: Bike-share rebalancing strategies, patterns, and purpose. J. Transp. Geogr. **55**, 22–39 (2016)

13. De La Iglesia, D.H., De Paz, J.F., Villarrubia, G., Barriuso, A.L., Bajo, J., Corchado, J.M.: Increasing the intensity over time of an electric-assist bike based on the user and route: the bike becomes the gym. Sensors **18**(1), 220 (2018)

Author Index

© Springer Nature Switzerland AG 2019
P. Novais et al. (Eds.): ISAmI 2018, AISC 806, pp. 411–413, 2019.
https://doi.org/10.1007/978-3-030-01746-0

Printed in the United States
By Bookmasters